RESEARCH & EDUCATION IN DESIGN: PEOPLE & PROCESSES & PRODUCTS & PHILOSOPHY

*CRC Press/Balkema is an imprint of the Taylor & Francis Group, an informa business*



*Typeset by Integra Software Services Pvt. Ltd., Pondicherry, India*



*Library of Congress Cataloging-in-Publication Data*

Applied for

Published by: CRC Press/Balkema
Schipholweg 107C, 2316XC Leiden, The Netherlands
e-mail: Pub.NL@taylorandfrancis.com
www.routledge.com

ISBN: 978-0-367-49296-0 (Hbk)
ISBN: 978-1-003-04610-3 (eBook)
DOI: 10.1201/9781003046103
https://doi.org/10.1201/9781003046103

*Research & Education in Design: People & Processes & Products & Philosophy – Almendra & Ferreira (eds)*
*© 2020 Taylor & Francis Group, London, ISBN 978-0-367-49296-0*

# Table of contents

*Research & Education in Design: People & Processes & Products & Philosophy – Almendra & Ferreira (eds)*
*© 2020 Taylor & Francis Group, London, ISBN 978-0-367-49296-0*

# Preface

We are delighted to present the book *Research & Education in Design: People & Processes & Products & Philosophy,* which brings together the contributions by the authors who attended the 1st Research & Education in Design International Conference.

The conference was organised by REDES—Research & Education in Design Group and CIAUD—Research Centre in Architecture, Urbanism and Design of the Lisbon School of Architecture, Universidade de Lisboa, and the event was held at the Lisbon School of Architecture on 14 and 15 November 2019.

The event brought together participants from all over the world with a shared interest in design research and education; the authors presented varied perspectives on the topic, always at the intersection of the four areas of design knowledge: people, processes, products, and philosophy.

The diversity of perspectives is illustrated in the book's eight sections. There are studies directly related to the pedagogy of design and the design studio educational setting; in the Visual Design Representations, Research in Design, and Design Thinking sections there are articles exploring specific design research tools and methodologies; and finally, the Beyond the Classroom, Design & Society, and Community-based Research sections look into the impact of design upon society in general.

Design is about connections and relationships: form and function, beauty and utility, people and things, problems and solutions. This book pays particular attention to the connections between research and education and how these create value for society as a whole and contribute to advance human wellbeing.

The editors

*Rita Almendra*
*João Ferreira*

*Research & Education in Design: People & Processes & Products & Philosophy – Almendra & Ferreira (eds)*
*© 2020 Taylor & Francis Group, London, ISBN 978-0-367-49296-0*

# Committee members and keynote speakers

**Conference chair**
Rita Almendra *CIAUD, Lisbon School of Architecture, Universidade de Lisboa, Portugal*

**Organising Committee**
Ana Cristina Dias (FA.ULisboa)
Ana Brígida (CIAUD)
Bruna Ruschel Moreira (FA.ULisboa)
Filipa Nogueira Pires (CIAUD)
Francisco Ramos (FA.ULisboa)
Inês Veiga (FA.ULisboa)
Joana Sousa (CIAUD)
João Ferreira (FA.ULisboa)
Luís Ginja (FA.ULisboa)
Mafalda Casais (CIAUD)
Marco Neves (FA.ULisboa)
Maria Inês Casaca (FA.ULisboa)
Natália Plentz (FA.ULisboa)
Pedro Cortesão Monteiro (FA.ULisboa)

**Keynote Speakers**
Barbara Tversky – *Stanford University*
Gabriela Goldschmidt - *Technion – Israel Institute of Technology*
Ken Friedman – *Tongji University*

**International Scientific Committee**
Aline Souza – *Federal University of Uberlândia*
Ana Cristina Dias – *University of Lisbon*
Ana Moreira da Silva – *University of Lisbon*
Ana Thudichum Vasconcelos – *University of Lisbon*
Bruna Ruschel Moreira – *University of Lisbon*
Carlo Franzato – *Unisinos*
Cees de Bont – *Loughborough University*
Daniel Raposo – *Polytechnic Institute of Castelo Branco*
Eduardo Gonçalves – *European University of Lisbon*
Elisabete Rolo – *University of Lisbon*
Emilio Gil – *European University of Madrid*
Erik Bohemia – *Loughborough University*
Fernando Moreira da Silva – *University of Lisbon*
Gabriel Patrocinio – *State University of Rio de Janeiro*
Henri Christiaans – *University of Lisbon*
Inês Simões – *University of Lisbon*
Inês Veiga – *University of Lisbon*
Janet McDonnell – *University of the Arts London*

João Ferreira – *University of Lisbon*
João Vasco Neves – *Polytechnic Institute of Castelo Branco*
José Silveira Dias – *University of Lisbon*
Kalevi Ekman – *Aalto University*
Letícia Schiehll – *Lutheran University of Brazil*
Livia Rezende – *University of New South Wales*
Marco Neves – *University of Lisbon*
Marieke Sonneveld – *TU Delft*
Mário Matos Ribeiro – *University of Lisbon*
Michele Santos – *University of Lisbon*
Milene Gonçalves – *TU Delft*
Natália Plentz – *University of Lisbon*
Pedro Cortesão Monteiro – *University of Lisbon*
Pekka Korvenmaa – *Aalto University*
Rita Almendra – *University of Lisbon*
Safi Hefetz – *Bezalel Academy of Arts and Design*
Shilppa Das – *National Institute of Design*
Sónia Vieira – *University of Porto*
Thomas Binder – *The Royal Danish Academy of Fine Arts*
Ulrich Lehmann – *Parsons School of Design*
Vasco Branco – *University of Aveiro*
Zoy Anastassakis – *State University of Rio de Janeiro*

*Design & Pedagogy*

*Research & Education in Design: People & Processes & Products & Philosophy – Almendra & Ferreira (eds)*


# Facilitating complex knowledge in design education through design tools

M. Casais
*CIAUD, Lisbon School of Architecture, Universidade de Lisboa, Portugal*

ABSTRACT: Design is a practice-based discipline, which is reflected in its education methodology. In this context, consuming complex knowledge might prove challenging, because designers think and work in a predominantly visual way. However, being able to master precise subjects offers a unique advantage in design, in contrast with a dispersed and superficial approach to knowledge. In this paper, we argue that design education can benefit from using design tools as a didactic way of conveying complex knowledge and making it actionable. To illustrate our proposition, we analysed 14 theory-focused design tools, showing how these convey knowledge and make it usable. In addition, we compared design tools with their respective sources of knowledge to showcase the contrast, proposing a three-part model of knowledge accessibility. Lastly, we describe anecdotal experiences: teaching design classes with and without design tools. The ideas discussed in this paper represent opportunities for further research.

## 1 INTRODUCTION

Design is a practice-based discipline, which is reflected in its education methodology, grounded in project development—notwithstanding the textual components present in design history, theory and critique. As such, making use of dense knowledge might prove challenging, because designers think and work in a predominantly visual way, and use visual representation to organize and communicate their thinking (Cross, 1982; Schön, 1983; Lawson, 2005; Wastiels, Schifferstein, Wouters, & Heylighen, 2013).

The fact that designers are visual thinkers does not mean that they do not benefit from understanding and incorporating knowledge specificity into their practice. On the one hand, multidisciplinarity has become a necessary tenet of our economy, and navigating diverse fields with some fluency and applying that information offers an (apparent) upper hand. On the other hand, it is true that being able to master a precise subject gives a unique advantage in contrast with a dispersed and superficial type of approach (Newport, 2016).

Therefore, we argue, design students can benefit from engaging with specific complex knowledge. This is particularly relevant concerning knowledge that can directly benefit the outcome of design interventions. For example, it is beneficial to deepen knowledge about materials, technology, politics, social issues, users, communities, or society at large, or about predicting or speculating on new and future situations. Its source can be anthropology, engineering, psychology, consumer research, sociology, healthcare, or other fields.

In this paper we propose that design education can benefit from using design tools originating from design research—which articulate design with different fields of knowledge—as an engaging and effective way of conveying complex knowledge and of making it actionable.

Design tools are compact vehicles of data, often with game elements, that deliver methods of working, inspire with ideas or solutions, and summarise complex information in a format that is possible to handle. Such tools have the potential to increase eloquence in intricate matters, by streamlining concepts and theories. They can offer an introduction that demystifies the apparent obscurity of complex knowledge, bridging theory and practice through an action-driven medium.

To illustrate our proposition, we conducted an analysis of information-focused design tools (as opposed to tools that offer only practical methodological support) available in physical format. We demonstrate how these convey information and make it accessible and actionable.

In addition, we provide some comparisons of the analysed tools with their respective sources of knowledge, frequently in the form of scientific articles or books, to showcase the stark contrast. Stemming from this comparison, we present a three-part model, based on the levels of accessibility of knowledge to design students.

Lastly, we share anecdotal experiences: teaching classes on design for emotion and well-being—knowledge based on psychology—with and without design tools. The ideas discussed in this paper represent opportunities for further research.

## 2 DESIGN TOOLS

Creative thinking tools are aids for creative and problem-solving tasks, using visual and textual stimuli to communicate methods, techniques or strategies. These tools often summarise complex information in the form of booklets, card decks, or digital formats, frequently using game elements.

Two widespread examples of creative thinking tools are the Thinkpak card deck (Michalko, 2006), based on the SCAMPER technique (an acronym for *substitute*, *combine*, *adjust*, *modify/magnify*, *put to other use*, *reverse/rearrange*); and 75 Tools for Creative Thinking (Cordoba Rubino, Hazenberg, & Huisman, 2013), a box with five card decks (1. Get Started; 2. Check Around; 3. Break It Down; 4. Break Free; 5. Evaluate & Select) for creative inspiration, also available as a mobile application.

Design tools—a specific form of creative thinking tools—are those developed within design research or practice, articulating diverse fields, such as healthcare, psychology, or consumer research, with design. In general, design tools aim to trigger designers in their process by providing inspiration and information in a flexible way that also allows freedom for them to apply their own understanding in their practice (Sleeswijk Visser, 2009). Design toolkits are a collection of different tools with the same theme/aim.

Several attempts have been made to categorise design tools to understand what these aim to afford in terms of assistance to designers in their processes. Roy and Warren (2018) provided an overview of existing classifications of (card-based) design tools, finding them ultimately focused on small samples and on subjects that arguably fall outside the scope of the design discipline. To tackle this, they proposed their own classification using a sample of 72 card-based design tools, grouping them under the following themes:

Systematic Design Methods and Procedures;
Creative Thinking and Problem Solving;
Human-Centred Design;
Domain-Specific Methods;
Team Building and Collaborative Working;
and Futures Thinking.

Potentially, any of these tools could be what we designate as information/theory-focused tools, that is, vehicles for summarized complex knowledge from diverse fields that can enrich the design practice in many ways.

To have a more focused understanding of how design tools convey such theory-based, dense information and make it accessible and actionable, we conducted an analysis of different specimens, directed at the knowledge-focused potential discussed above.

## 2.1 *Analysis of design tools*

24 design tools were selected through an online search using the terms *design tool[s]*, *design method[s]*, and *design toolkit[s]*. The search results were not overwhelming, and it was easy to discern examples that fitted the criteria of search, namely that the tool or toolkit would have at least partly a physical tangible component, and that it originated from design research or practice.

An analysis on those 24 tools consisted in checking them against a list of elements which could potentially be used to argue for our proposition. The aim of this preliminary analysis was firstly to grasp a range of formats, navigation styles, and presentation of information, and secondly to uncover information/theory-focused tools. Those constituent elements were:

1. *Origin*: Which design/education institution developed and launched the tool;
2. *Format*: Which elements constitute the toolkit (e.g. canvas, posters, cards);
3. *Icons*: Does it use symbols, icons, pictograms to structure navigation;
4. *Images:* Does it use images to provide examples of context or depict other related things;
5. *Infographics*: Does it use graphic representations of information in combination with text;
6. *Linear text*: Does it contain linear text;
7. *Non-linear text*: Does it contain non-linear text, e.g. lists, networked text/words, word-clouds;
8. *Action-focused*: Does it explain a method of working and/or prompt the designer to work in a directive way;
9. *Information/theory-focused*: Does it convey theoretical information or knowledge about a topic;
10. *Colour coding*: Does it use colour to structure navigation;
11. *Graphs/diagrams*: Does it use graphs and/or diagrams to illustrate or explain content;
12. *Game elements*: Does it contain explicit game instructions or elements, or does it tacitly invite the user to get involved in a gamified experience (due to its components, set of instructions, need for partnering, etc.);
13. *Digital components*: Does it contain digital components such as applications, photo albums, etc.;
14. *Other relevant elements*: Does it have other relevant elements, such as instruction manual;
15. *Objectives*: Description of the objectives as provided in the tool itself.

Following the preliminary analysis, we shifted our attention to the focused analysis of information/theory-focused tools, that is, those tools whose primary aim was to convey theoretical knowledge from other fields to enrich the design practice at any level (e.g. understand specific users, or people in a global sense, understand situated contexts, predict outcomes of designs, speculate about future scenarios).

## 2.2 *Results*

The analysis of the 24 design tools (see annex) showed that their origin was not indicative of their content, i.e., a tool developed by an education institution did not aim to, necessarily, convey complex knowledge. However, we did find examples of this type of theory-based tool from design research (see TU Delft examples).

While we aimed for tools with at least one tangible component, the majority had either one or more. Most analysed tools contained a card deck, canvas, posters, or booklets. One third (8) had additional digital components, such as web-based guides. However, all analysed tools were able to be downloaded or accessed digitally, and subsequently printed or ordered in printed format.

Colour frequently played a role in organizing and hierarchizing information. In addition, the tools largely contained both text and image, often both linear and non-linear text.

The tools aimed to provide information and/or methodological instructions, as well as design cases that illustrated these methods or techniques. Specifically, we found that the objectives could be divided into four types: (1) summarizing theoretical knowledge; (2)

providing inspiration/displaying design cases; (3) understanding the user; and (4) providing methodological support.

Lastly, we observed over one third (9) with specific game elements or game-like instructions. We found 14 design tools in our sample that corresponded to the criterion of being information/theory-focused (see annex).

A further analysis of these 14 tools revealed that these have certain traits in common, namely:

Action: accompanying the theory, the tools presented a methodological component which allowed the knowledge to be put into practice.

Format: the tools were presented in a size that is easy to handle (in the design studio or in class), and in a portable format, such as a booklet or card set.

Pictorials: images supplement linear and non-linear text and provide a more vivid illustration of a given phenomenon.

Language: the theory is often accompanied by eliciting conditions (this happens when…) presented in simple terms.

Text: linear text is kept short and turned into graphs or illustrated when appropriate, text is highlighted or enhanced to hierarchize information.

## 2.3 *A model of information accessibility: Three levels of communication and understanding*

When we compare the information/theory-focused design tools with their respective sources, the stark contrast between how the information is presented becomes apparent. Taking the case of two design tools—Positive Emotion Granularity Cards (Yoon, Pohlmeyer, & Desmet, 2015) and SIM toolkit (Casais, Mugge, & Desmet, 2016), we can explore how information about a certain topic is accessible in different ways to design students.

We begin by looking at the original source that led to the design research work. In the two illustrated cases bellow (Figure 1 and 2) the original sources are from the field of psychology and are presented in the form of books and scientific articles. We designate this as level 1 of

Figure 1. On the left side, a scientific article about the determinants of psychological well-being (Ryff, 1989); on the right side the SIM toolkit (Casais et al., 2016), based on Ryff's theory.

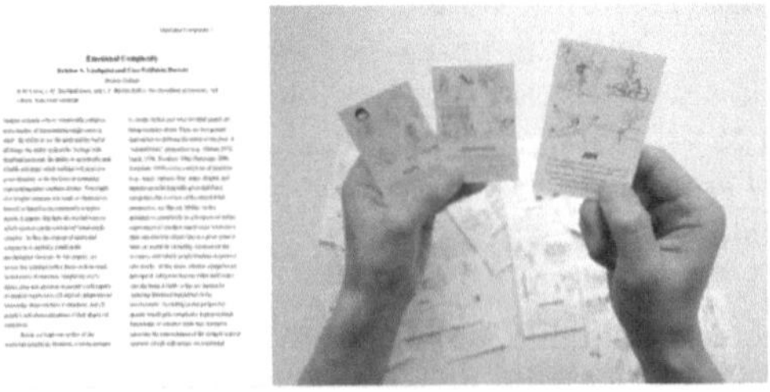

Figure 2. On the left side, a book chapter about emotional complexity (Lindquist & Barrett, 2008); on the right side, the Positive Emotional Granularity Cards (Yoon, et al. 2015), based on Lindquist and Barret's theory.

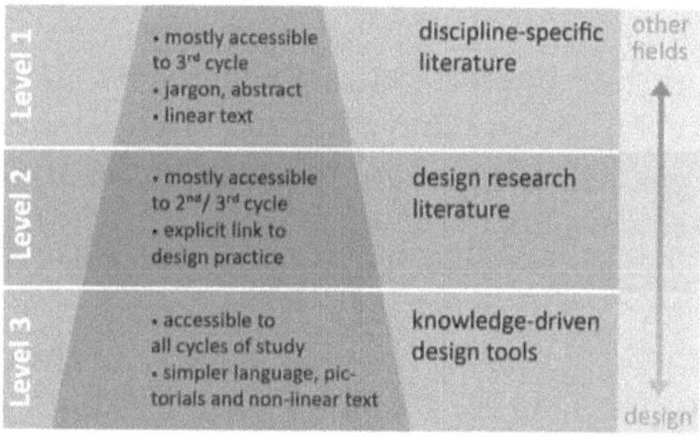


Figure 3. Model of information accessibility.

accessibility, a type of knowledge and presentation of information that is most commonly accessible to 3rd cycle design students (and eventually also 2nd cycle students). This is due to the use of discipline-specific terminology, abstract reasoning, and dense sections of linear text which are not particularly appealing to visual thinkers.

On an intermediate level, we find the design research from which the design tools were developed; we call this level 2 of accessibility, more approachable by 2nd and 3rd cycle design students. In the given cases, the sources are also in the form of scientific articles and books. At this level we find a link from the original source-discipline to the field of design, showing explicitly how there is relevance for design practice, which renders this type of reading more appealing and digestible by 2nd cycle students. However, we often find some dislike for the abundance of linear text, density of information, and abstract reasoning.

At level 3 of accessibility we find the design tools, widely accessible for all cycles of education. At this level, as uncovered in the previous section, knowledge is made available through easy language, pictorials accompanying linear and non-linear text, actionable directions, and portable or manageable formats.

In Figure 3 we find a model of information accessibility summarizing and illustrating the three levels of communication and understanding.

## 3 SOME ANECDOTAL EXPERIENCES

In numerous occasions, the author had the opportunity to use design tools in design education. However, these were not specifically set up to evaluate the effectiveness of this modality in conveying complex knowledge. Rather, design tools were used to complement a traditional type of design class, in which there was a lecture about theoretical knowledge, a moment to apply the theory in project-based exercises, and a discussion moment. The design tools were used to exemplify the application of certain theories into methodological aids, or in discussions and in short exercises.

Nevertheless, these anecdotes reveal the potential impact of information/theory-focused design tools in design education and are encouraging for the pursuit of further investigation on the subject. Two interesting cases of teaching design students about human emotions and wellbeing are described below, one without design tools and one with.

### 3.1 *Case 1: Teaching emotions with traditional didactic instruments*

The first case occurred in a workshop setting, part of a semester-long course, in the 1st semester of the 2nd year of the 2nd cycle of study, with 5 groups of students (each group with 4 to 5 students). Students attended an engineering design master programme from an industrial design engineering faculty.

Immediately prior to the workshop, students attended a lecture on the topic of human emotions and subsequently were asked to read one level 1 publication (a book chapter) and two level 2 publications (a research article and a book chapter) (see Figure 3). Then, they were asked to prepare for the workshop by analysing a product from the user's perspective applying the learned theoretical concepts, and to summarize and visualise their findings.

During the workshop, students had one hour to redesign the product in accordance to their findings, one hour to discuss and present, and 5 minutes to receive focused feedback from the teacher.

During the contact hours of the workshop, the degree of difficulty to communicate about the concept of human emotions and their applicability was high. A suitable analogy to describe the situation is trying to make oneself understood by a foreign person that does not speak the same language or share the same basic gestures or symbols. Furthermore, students found it incredibly difficult to incorporate something as abstract as emotions in their designs.

### 3.2 *Case 2: Teaching wellbeing with theory-focused design tools*

The second case occurred in a workshop setting, part of a two-week elective course, in the 1st semester of the 1st year of the 2nd cycle of study, with 3 groups of students (each group with 4 to 5 students and 1 student facilitator per group). Students attended an engineering design and a strategic design master programme, from an industrial design engineering faculty.

The workshop preparation occurred in class and was focused on setting up a creative session within an innovation team, aiming to expand their knowledge of creative techniques and later experience process consulting with real cases—so no focus was given to the theme of wellbeing or related content.

The workshop was setup and facilitated by the students themselves, and the author provided the *real case*, acting as a client. The case description provided to the students was to think of alternative uses for a theory-focused card set about wellbeing, a topic stemming from the field of psychology. Possible uses could have been, for example, in communication with end-users, clients or design teams, in mental healthcare or counselling, etc. Students had 1:30 hour of creative session, followed by one hour of presentation and discussion.

While this was not a class on human wellbeing, the students' task implied that they understood the theoretical themes being discussed through the theory-focused design tool in order to iterate on the tool's applications in diverse contexts.

To realise whether and how well students understood the topic, the author observed the exploration process of the groups, watched the presentations, and discussed with them. Together, these elements demonstrated the students' remarkable comprehension of the theoretical content in a short time-span. Students also mentioned the simplicity and straightforwardness of the design tool in explaining the topic.

## 4 DESIGN TOOLS AS APPROPRIATE DIDACTIC INSTRUMENTS

Examples of commonly recognised didactic resources can be books, articles, lectures, videos, and workshops. Tools, in the sense that we describe here, are not commonly used as an in-class educational instruments, for the discovery of new topics or deepening of content.

However, we argue that for different reasons this modality presents a great potential particularly in design education. In this section we list some compelling motives in trying to directly answer the question: why should we consider design tools *appropriate* didactic instruments for design education?

### 4.1 *Designers think visually*

Information designer Francis Miller, a practitioner focused on simplifying learning processes (Miller, 2016-2019), proposes a methodology for the treatment of complex information, which he calls *multi-level content*. Based on Christopher Alexander's ideas, multi-level content is

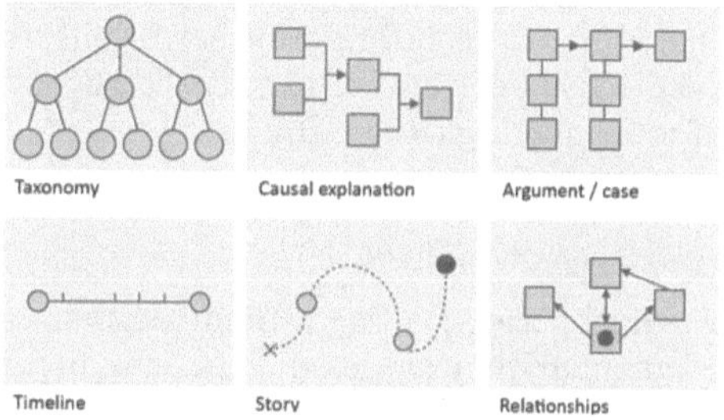


Figure 4. Several types of knowledge structures (based on Miller, 2018, p. 13).

defined as the provision of content over multiple levels of detail ranging from the big picture summary down to a granular detail (Miller, 2018).

Miller argued for his concept of multi-level content by indicating that "making sense of any knowledge requires understanding the structure of the explanation. However, neither spoken or written words on their own are very effective at communicating structure" (Miller, 2018, p. 2).

Clarifying his statement "all knowledge has structure" (p. 13), Miller identified several types of knowledge (Figure 4), some of which are explicit in text, while others are implicit. In large information vessels, such as books, there can be many different types of structures; this means that the challenge is threefold: understanding a (potentially implicit) knowledge structure through a limiting medium like linear text in sequential blocks, uncovering the links between several knowledge structures, and grasping what is being told in an overview. Multi-level content is thus proposed as a solution.

The way Miller presents the knowledge structures in pictorial form is, in itself, a form of multi-level content. Explaining each knowledge structure concept is done succinctly, in a straightforward way, leaving room for other layers of information to be added. The image itself serves to illustrate how striking visualization can work for certain reader profiles, particularly in our case, design students.

Designers, as visual thinkers, can potentially find such simplifications valuable, because they allow the easy and prompt application of ideas into practice. Providing summarized information (linear and non-linear text) with pictorial stimuli provides a more compelling mental image of concepts, concept links, and respective theories, catering especially well to a visual thinking audience.

### 4.2 *Creating mental images supports learning*

Design tools typically provide information through both pictorial means and verbal means, often in combination. This was corroborated by our tool analysis, wherein all design tools had visual cues (images, colours) or a combination of visual and textual cues. Such multidimensional representation of information is easier to grasp and apply compared to text alone (Santos, Pereira Neto, & Neves, 2019).

The dual-coding theory (Paivio, 1986/1990) hypothesized that people code information through two distinct coding channels—a visual channel, also called analogue coding; and a verbal channel, also called symbolic coding. These two separate systems processes input differently, images are processed by the visual coding channel, whilst words have a particular path according to their nature: "concrete words are processed by both systems, whereas abstract words are processed primarily by the verbal system" (Kounios & Holcomb, 1994, p. 804).

This corroborates our proposal that design tools are situated at a third level of accessibility (Figure 3), a wider and more approachable level of language, format and presentation, that can serve the purpose of conveying complex knowledge in simplified and concrete terms aided by pictorials, which in turn support the creation of mental images for theoretical concepts.

The creation of mental images improves information retention, storage, and retrieval, therefore supporting learning, since a given concept can be recalled by either code, verbal or visual, providing a better chance of remembering (Reed, 2007).

### 4.3 *Synthesis makes knowledge (more) actionable*

Some design tools were developed because, while certain "academically oriented design guidelines and frameworks" have potential, they are long, dense, use discipline-specific language, and thus become difficult to apply in practice (Hornecker, 2010; Deng, Antle & Neustaedter, 2014).

Design tools may provide, through the way they synthesize information, a potential solution for this. They summarize dense information and deconstruct it in different elements, such as definitions, application examples and anecdotes, eliciting conditions, behavioural manifestations, illustrating theories with the aid of graphs, diagrams and pictorials, etc. This synthesis was observed on all analysed tools, and while some were more information dense than others, overall, they summarised somewhat complex knowledge—for example, from psychology—into straightforward and succinct content.

This synthetic display of information potentially stimulates connections that lead to the structured or serendipitous finding of certain routes for design, or to novel solutions to design problems, or even to the uncovering of surprising opportunities and links within the context of work or study.

### 4.4 *The physical format makes people interact differently*

In her studies on design practitioners, Rogers (2004) found that "a frequently cited complaint was that designers wanted more guidance and ways of communicating about theories and techniques" (p. 39), suggesting that language and format were also an issue.

We encountered several design tools available in digital formats (webpages, video, etc.), making them updatable, easily accessible, and offering a dynamic type of interaction. However, the printed format offers certain advantages that might be diluted with intangibility.

Specifically, printed information makes people interact differently, and "demonstrate[s] a propensity for manipulation by people, modifying communication effect" (Neves, 2019). Nevertheless, the printed format alone does not guarantee engagement.

The comparison of the original source of knowledge with its summarized, visual, and actionable modality—the tool—showcases potential it can have in motivating design students, as discussed above (see Figure 1 and 2).

The advantages of physical elements of design tools, such as card sets, have to do with navigation, hierarchization, flexibility—for evaluation, pairing or comparison, with the ability to provide an overview, and freedom to apply or interpret their use in different ways (e.g. Deng et al., 2014; Casais et al., 2016; Yoon et al., 2016).

Another important aspect associated with tangibility is the possibility of having game elements. This can be used as a strategy to entice design students to engage and learn or apply the tool's contents.

### 4.5 *Figure captions*

Always use the Figure caption style tag (10 points size on 11-points line space). Place the caption underneath the figure (see Section 5). Type as follows: 'Figure 1. Caption.' Leave about two lines of space between the figure caption and the text of the paper.

## 5 CONCLUSIONS

While conclusions about the effectiveness of theory-focused design tools in a design education setting might be premature at this point, some preliminary thoughts might be considered.

From the analysis of design tools, we learned that design tools make dense knowledge streamlined, actionable and accessible; and that they present a lot of information in a small and portable size that can be used in multiple ways (particularly card sets). Furthermore, this modality tends to communicate with images and concrete text which helps create strong mental images and aids with better learning. Tools illustrate dense knowledge with various modes of communication: symbols, icons, graphs and diagrams, eliciting conditions, behavioural manifestations, anecdotes, pictures, strategies, solutions, etc.

From the comparison of the design tools and their sources of knowledge, we developed a three-part model of information accessibility. This model summarizes three levels of communication and understanding that design students use, namely: level 1 – the knowledge from other fields other than design, often resorting to discipline-specific complex language, linear text and abstract reasoning; level 2 – the knowledge from design research that articulates design with other fields, makes evident the relevance of such knowledge to design practice but often remains obscure and abstract, communicated through linear text; level 3 – the streamlined, simplified, and actionable version of the knowledge, the design tool, more widely accessible to students.

Further research on the topic could focus on interviewing design teachers that already use such aids in education setting, or those that face the challenge of teaching design students' knowledge from other fields, and together with them set up studies to verify whether the potential effectiveness and engagement that is proposed in the current paper holds true.

## REFERENCES


Casais, M., Mugge, R., & Desmet, P.M.A. (2016). *Design with symbolic meaning for user happiness card set (SIM)*. Delft: Delft University of Technology. ISBN/EAN: 9789461866387.

Cordoba Rubino, S., Hazenberg, W., & Huisman, M. (2013). *75 Tools for Creative Thinking*. Booreiland. ISBN: 978-90-6369-275-9.

Cross, N. (1982). Designerly ways of knowing. *Design Studies*, 3(4), 221–227.

Deng, Y., Antle, A.N., & Neustaedter. C. (2014). Tango Cards: A card-based design tool for informing the design of tangible learning games. In: *Proceedings of the 2014 conference on Designing interactive systems DIS'14: Crafting Design*, (695–704), Vancouver, June 21-15, 2014. New York, NY: ACM Press.

Hornecker, E. (2010). Creative idea exploration within the structure of a guiding framework: The Card Brainstorming Game. In: *Proceedings of the fourth international conference on Tangible, embedded, and embodied interaction TEI'10*, (101–108), Cambridge, MA, January 24-27, 2010. New York, NY: ACM.

Kounios, J. & Holcomb, P.J. (1994). Concreteness effects in semantic processing: ERP evidence supporting Dual-Coding Theory. *Journal of Experimental Psychology: Learning, Memory, and Cognition*. 20 (4), 804–823.

Lawson, B. (2005). *How designers think: The design process demystified* (4th ed.). Oxford, UK: Architectural Press.

Lindquist, K. A., & Barrett, L. F. (2008). Emotional complexity. In L. F. Barrett, M. Lewis, & J. M. Haviland-Jones (Eds.), *Handbook of emotions* (3rd ed., 513–530). New York, NY: Guilford Press.

Michalko, M. (2006). *Thinkpak: A brainstorming card deck*. Ten Speed Press. ISBN: 978-1-58008-772-8.

Miller, F. (2016-2019). *Making knowledge easier to understand, remember and implement*. Retrieved from http://www.francismiller.com/

Miller, F. (2018). Organising knowledge with multi-level content: Making knowledge easier to understand, remember and communicate. Retrieved from http://www.francismiller.com/organising_knowledge_paper.pdf

Neves M. (2019). Design for (Inter)Actions: A teaching experience in graphic design. In: F. Rebelo & M.M. Soares (Eds). *Advances in Ergonomics in Design* AHFE *2018: Advances in Intelligent* Systems *and* Computing, (Vol. 777, 422–431). Springer.

Newport, C. (2016). *Deep Work. Rules for focused success in a distracted world*. New York, NY: Grand Central Publishing.

Paivio, A. (1990). Mental representations. A dual coding approach. New York, NY: Oxford University Press. (Original work published 1986).

Reed, S.K. (2007). Cognition. Theory and applications (7th ed.). Belmont, CA: Thomson Wadsworth.
Rogers, Y. (2004), New theoretical approaches for human-computer interaction. *Annual Review of Information Science and Technology*, 38(1), 87–143.
Roy, R., & Warren, J. (2018). Card-based tools for creative and systematic design. In: C. Storni, K. Leahy, M. McMahon, P. Lloyd, & E. Bohemia (Eds.). *Proceedings of the Design Research Society conference DRS2018: Design as a Catalyst for Change*, (Vol. 3, Sec.10, Tools of Design, 1075-1087). Limerick, June 25-28, 2018. London: Design Research Society.
Ryff, C.D. (1989). Happiness is everything, or is it? Explorations on the meaning of psychological well-being. *Journal of Personality and Social Psychology*, 57(6), 1069–1081.
Santos C., Neto M.J.P., Neves M. (2019) The Influence of Infographics in Accessing Information: Multidimensionality in Visual Representation and Configuration of Different Media. In: F. Rebelo & M.M. Soares (Eds). *Advances in Ergonomics in Design AHFE 2018: Advances in Intelligent Systems and Computing*, (Vol 777, 497-508). Springer.
Schön, D. (1983). *The reflective practitioner: How professionals think in action*. New York, NY: Basic Books.
Sleeswijk Visser, F. (2009). *Bringing the everyday life of people into design*. (Unpublished doctoral dissertation). Delft University of Technology. Delft, The Netherlands.
*Three-cycle system*. (2016, September) Retrieved from http://www.ehea.info/page-three-cycle-system
Wastiels, L., Schifferstein, H. N. J., Wouters, I., & Heylighen, A. (2013). Touching materials visually: About the dominance of vision in building material assessment. *International Journal of Design*, 7(2), 31–41.
Yoon, J., Pohlmeyer, A.E., & Desmet, P.M.A. (2015). *Positive Emotional Granularity Cards*. Delft, Delft University of Technology. ISBN: 978-94-6186-440-6.

*Research & Education in Design: People & Processes & Products & Philosophy – Almendra & Ferreira (eds)*
 *ISBN 978-0-367-49296-0*

# Teaching interaction design: A theoretical framework

M. Neves
*CIAUD, Lisbon School of Architecture, Universidade de Lisboa, Portugal*

ABSTRACT: Education in interaction design is growing and gaining importance. Given the diversified origins of the area, courses which have emerged in higher education have different approaches. In this article, we present a theoretical framework for a curricular unit of introduction to Interaction Design Project, at master's level, at the Lisbon School of Architecture, University of Lisbon. For this purpose and this specific context, we will assume the nature of interaction is multidisciplinary and that training at this level should refer to designers' abilities to think and act in different circumstances. In establishing this framework, we isolated four items that such teaching-learning system should assure: fundamentals, process, media and tools. Together they form a context upon which a syllabus can be developed for teaching interaction design practice.

## 1 INTRODUCTION

Interaction design, as a comprehensive area, has had a diversified professional presence and a permanent need for updating. Faced with an emerging demand for professionals, in which very specific skills are valued, several educational institutions in several countries have organized courses on interaction design.

The area has grown in the curricula of higher education institutions, all over the world. With a greater presence in institutions where design teaching takes place several decades ago, but also now in new courses, in institutions where this teaching is more recent. Here there has been an attempt to implement a few possibilities of contact with this area, usually isolated and typically dragged by urgency to demonstrate technology-based training.

This emergence of interaction design related courses, is a justifiable procedure, since the area is going through a stimulating moment, of knowledge intersection, which allow creation of new languages and the arrival of new media and applications. Interaction design professionals now work in diverse settings and assume different involvements, sometimes in tasks for which they did not always had specific training.

The moment is appropriate for an organization of various practices and inherent contents so they may be transmitted to those who seek to dedicate to interaction design.

The present framework is intended for matters comprised in a master's degree course, which corresponds to a second cycle of training in higher education.

According to the current higher education system, most courses at this level allow application of students from different areas without specific initial training in course subjects. Although some candidates may already have higher training in design, there is no set of contents taught at country level, which approach interaction design concerns.

In this way, the present framework, is intended for the first curricular unit of theoretical-practical scope, at the beginning of the course. Regarding the collection and comparison of various references, four essential topics stand out: fundamentals, process, media and tools.

## 2 INTERACTION DESIGN MASTER DEGREE

One of the intentions of interaction design is to create material and functional conditions for behaviour change. Still, this master degree sets out to train future professionals who can use necessary means in a creative way, and to combine with scientific methods and design procedures to obtain systems and artefacts, able to produce several levels of interaction with users, which can translate into pleasant and effective experiences.

## 3 INTERACTION PROJECT CURRICULAR UNIT: OBJECTIVES

The first curricular unit dedicated to project development, for which the present theoretical framework was made, allows contextualization, where structural and fundamental concepts can be presented. This was established considering a suitable engagement of students in the course subjects, outlining guidelines for work to be carried out in following semesters, but also thinking about students whose basic training may have been different, with respect to interaction design.

Therefore, the approach of this curricular unit aims to identify, define and inform individual practices of interaction design. In this way, our proposal intends to instigate a kind of research, which drives students to express personal interests.

For this to be accomplished, it is our purpose that every student carry out and examine a design process, developing and executing an interaction work. With the aim of helping future designers to structure and organize several necessary stages, for presentation and distribution of a qualitative and distinct solution, which transposes correctly and effectively his ideas.

On the other hand, experiencing and mastering a design process becomes useful to carry out a successful job. To include it in the first semester can be decisive to an ambition of professional training, but also for each students' autonomy.

Expected learning outcomes

Know how to observe and analyse interaction artefacts, generating a critical approach in project development.

Dealing, developing and mastering a work process and its different stages.

Perform several stages that meet rigor expectations.

Expand visual thinking and behavioural strategies, using schematic and expressive elements and its transposition for effective contact.

## 4 FUNDAMENTALS

### 4.1 *Integrating knowledge for user experience*

An interaction project can be mainly defined by our understanding of what interaction is. Beyond any technological constraint, we may accept interaction as a succession of actions and physical reactions between a human being and that which involves him, objects or other human beings (Heeter, 2000). An action triggers a reaction and therefore, a reciprocal exchange of messages between the two parties occurs. Also, interpretation comes into play, since interaction is always directed towards a purpose, even though its context may always differ (Müller-Prove, 2002).

When interaction was considered for association with the work some designers were developing, the expression 'interaction design' emerged (Moggridge, 2007; Cooper, Reimann, and Cronin 2007). In such impetus, it would quickly designate "designing for everyday life" (Moggridge 2007, p. xi). From its origin to present day, the activity to which this expression refers to, has accompanied other design areas in their nature, that of integrating outside knowledge. Design is an expansive field in contemporary culture, which attempts to cover various activities (Buchanan 1992), but its breadth implies inclusion of various knowledges for its practice:

"Design is "integrative" in that, by its lack of specific subject matter, it has the potential to connect many disciplines" (Swanson 2010, p.3).

Interaction design has gathered knowledge from psychology, sociology or education (Preece, Rogers and Sharp 2002) and has simultaneously involved tasks and elements from more established areas in the design field, such as information architecture or graphic design (Saffer, 2010) and information design (Cooper, Reimann, and Cronin 2007).

This integration has given rise to a more stable core for interaction design as a discipline. This also allows it to be frequently associated with other expressions, which broaden its study concerns: attention with formal treatment given to machines and their interface, beyond its material, with aesthetics of interaction (Overbeeke, et al., 2002); to consider people, as end users, not only in a physical and functional way, but also their emotional presence (Norman 2004); or the comprehensive purpose of understanding user experience (Hassenzahl et al. 2013).

Trying to come up with a specific explanation, Siang (2017) synthesizes five inherent dimensions of interaction design: words, visual representations, space, time, and behaviour. All of them present themselves as necessary to understand and produce interaction, although the differentiating element is behaviour (Cooper, Reimann, and Cronin 2007). For Arvola and Artman (2007), both experience and behaviour are far more important notions to consider than materials or media. To anticipate behaviour is to arrange conditions for interaction with and through products and services (Arvola and Artman 2007). By defining expected behaviours, interaction design paves the way for a sustained relationship between individual and artefact (Siang 2017).

For Saffer (2010) interaction design is an applied art, which simply makes things useful and fun. This more relaxed view is not, however, naive. Fun is a condition for behaviour and should be present within an interactive system in the same way as usefulness or usability. An interaction project is moreover based on design of systems that change state over time (Goodwin 2009). People are a considerable part of these systems, individually or in groups (Forlizzi and Ford 2000; Battarbee and Koskinen 2005), since they are a driving force in such change. They are one of the fundamental requirements for the distinction and adequacy of the system to be created, assuming their role as users (Preece, Rogers and Sharp 2002).

As such, there is a need to address methods of collecting and understanding target groups so that any carried out solution can be based on reliable information about human factors implied in the system (Saariluoma 2005). People become users through products and systems and through design (Redström 2005) and in a complementary way, users provide awareness of their needs, whether they are expert users, general users or just simply available to experiment (Colborne 2017). This becomes a fundamental element in the creation of useful and effective interactive systems (Norman 1983).

Focusing on people and alternate behaviours, between object and its user becomes a common vision and involves stages of understanding the context of use, specifying requirements for a given situation and design solutions that must be tested and evaluated, in accordance with requirements. Such importance given to people and on them becoming users leads us to user experience and to the relevance of its holistic interpretation (Anderson 2011; Krishna 2015). A matter which raises professional concerns, since to estimate experience we must rely on knowledge from very different areas and because the designer is often a user himself (Fischer 2002; Kolko 2011).

User experience is understood as a set of user perceptions and responses to the use of a product, system or service (Bevan et al 2016). Norman (2013) adds that is also constituted by design of such products, systems or services. It balances subjectivity and hedonism, concerned simultaneously with functional and emotional features that change over time (Battarbee and Koskinen 2005).

Experience is also a social phenomenon, about which it is not enough to understand emotions or empathy. The first factor to highlight for this is interaction: "When people act together, they come to create unpredictable situations where they must respond to each other's actions creatively." (Battarbee e Koskinen 2005, p. 9).

Objects, products and systems are no longer simple creations, obtained through a linear process. They are embedded in business and institutional contexts, sometimes of high complexity. To understand user experience, and therefore to anticipate it through interaction, designers may need to become thinkers of great flexibility.

For this purpose, interaction design has appropriated some fundamentals, especially from industrial and graphic design (Saffer 2010), which serve today as guidelines for establishing a process. Constant research, generation of several alternative proposals, prototyping as an initial method for testing and the importance of ideation.

## 5 PROCESS

### 5.1 *Working between education and profession*

Some authors approach interaction design in its entirety, establishing a framework (Saffer 2010), a set of methods and elements (Cooper, Reimann, and Cronin 2007; Shneiderman & Plaisant, 2004) or several stages for an iterative procedure (Preece, Rogers and Sharp 2002). Both Norman (2013) and Nielsen (2000) do not direct their attention to the activity itself, but they rather try to regulate it somehow. Moggridge (2007) and Buxton (2007) approach its richness in another direction. Moggridge (2007) draws a general blueprint through many leading figures, while Buxton (2007) chooses a vision where design of products and systems is related to a business side or a broader domain of design.

There is a common concern to explain interaction design as responsible for functional systems and artefacts, but above all, how they respond and relate to people.

Still, the greater agreement among these positions seems to be implicit in details and in how they refer to ways of acting. It allows us to verify a certain consensus regarding required stages in an interaction design process: research on work situations and research on people involved; the need for information on context of use; information organization, its accesses and its dynamics; combination of visual form and content to favour expected behaviours; evaluation of solutions and their iteration.

Verplank (2009) proposes four key stages for an interaction design process, each with two variables: motivation, meaning, models and mapping. Some features are obvious: the demand for research, concern with user observation, use of predictive methods, definition of requirements and prototyping. But Verplank's scheme summarizes, it defines an operational side, but does not detail an understanding of the design project where evidence is needed or where advances and setbacks are often required.

For Preece, Rogers and Sharp (2002) it is important to identify the problem and to elaborate interactive prototypes that will be evaluated. But they are too brief when mentioning design. However, certain procedures have become necessary and complementary. Information architecture organizes and categorizes content (Saffer 2010), while visual media gives an interaction project the ability to realize potential solutions (Goodwin 2009).

Still, interaction and the consequent provided experience can be difficult to explain. Objects or artefacts are prepared to be requested and respond to people's actions. But this relationship, of continuity and variable behaviour, does not lend itself to a simple and unidirectional description.

The use of any interactive system assumes a time of learning, although uncertain, due to the complexity of its information and possible tasks to realize. As such, it is necessary to verify if the system is adequate to the expected performance, if it includes operating errors, speed in the flow of information, retaining knowledge and satisfaction by users. This usually implies making several prototypes, at different moments of the process, either low or high fidelity (Shneiderman & Plaisant, 2004). It is therefore noteworthy if the designer develops an ability to generate prototypes (Coughlan, Fulton Suri and Canales 2007), of diverse complexity, and that they are done together with evaluation methods.

Stabilizing in a process to educate students leads to choices. This is mainly because, every designer tends to define his profession by what he does. Therefore, design practices vary from company to company or even from one professional segment to another. It becomes difficult to teach professional practice only through applied problems. Kelly (n. d.) refers to his distrust of proposing professional problems to students, because they originate already established ideas and consequently a "stereotyped thinking".

It remains appropriate, students should acquire technical skills and formulate a notion of professional practices. Lewis and Bonollo (2002) describe to this respect, the most valued skills by clients and employers: ability to execute the design process (which includes interpretation of the problem, generating ideas, refinements, communication of results), total compliance with the process, achieved results and perceived quality, project management, professional responsibility. This attention to the process should, however, mark a separation of responsibilities, in which in school the greatest purpose is to learn and acquire a set of skills. Being professional will happen when students start their work path. In this sense, Kelly (n. d.) argues a greater presence of students' individual concerns, since they are more likely to get involved and consequently raise classroom productivity (Kelly n.d., 91).

Learning an interaction design process is valuable and increasingly justified, given the incidence on behaviour. In a way, it opposes education in other design areas, where modern practice in design influences design education (Davis 2008; Swanson 2010) and it mainly understood design as giving form to an object and not by the way it is obtained (Frascara 2004). Indeed, for Yagou: "The modernist tradition in particular has created a series of misunderstandings which encompasses two aspects: the functional and the stylistic." (Yagou 2005, p. 4). The vision of form as a dominant element, reduced some production to aesthetic appreciations devoid of context and still today, it is by the final shape, we measure designs' success (Stolterman 2008). What may be reasonable for the public, but for designers, this is just one of several stages.

It is important to focus on the process and its procedures (Lewis and Bonollo, 2002) or to describe them (Kelly n. d.). In this way, design can be studied and interpreted not only in the sense in which it produces objects, but also in that they derive from a process which allows good production. This is even more important because, unlike the generic context of interaction design, where all deadlines seem to be insufficient and all skills quickly required, in a teaching context it is important to evaluate students' final results, as well as the process fulfilled to reach such results, along with accumulated knowledge.

## 6 MEDIA

### 6.1 *Technology meets users*

Interaction design has an intimate relationship with technology, which is regularly updated and imposes consequences on the way we handle artefacts and on communication, causing a (disguised) discomfort. Teaching interaction design forces, in a way, a 'paralysis' to determine lexicon, grammar, elements and even syntax, which can be sustained as a basis of disciplinary work. How can we teach interaction design, if its assumptions change so quickly due to this considered or unfairly attributed dependence on technology?

It is uncertain whether interaction design process can be equally reproduced in any technology and there are always specific skills to consider, which may be required in a circumscribed professional context related to the adoption of defined technologies. So, to base a design educational plan in this possible divergence, seems of little help (Bonsiepe 2012).

At the same time, it is well-known several areas such as engineering, ergonomics or computer science have also addressed the scope of information and communication technologies, studying them with distinction.

On the other hand, the presence of technology in interaction design is not completely clarifying. Both Moggridge (2007) and Cooper, Reimann, and Cronin (2007) understand interaction design as the creation of digital artefacts. Manovich (2001) considers this a redundancy, since in his study of new media, there is an interaction inherent in them. Grudin (1990) claims the history of interaction is the way the computer communicates. It is interaction that allows a comprehension away from pure machinery and directed towards human contact. Although we tend to confuse computers with digital technology, the first models were analogue (Dourish 2001).

However, with historical development and emergence of several devices, there has been a tendency to associate digital technologies, to areas such as interaction design, human-computer interaction, web design, interfaces and user experience. However, if web design and human-computer interaction seem to be terms easily understood as dependent on digital technology, interaction, interface or user experience, do not detain features that completely binds them.

So, Norman (2013), guided by peoples' use of objects, detaches himself from this apparent commitment to digital technologies. He then describes the possibility of interaction occurring in bottom-up (something in the world triggers the action of a person) or top-down (a person has a goal that leads to action) as presented by Dubberly, Haque and Pangaro (2009) in their review. Here they suggest the relationship between man and artifact is considered in an evaluation of the design project, as they try to relate interaction with a static object and with a dynamic system.

The connection of technology with interaction design has helped understand which media a design project in such area uses. Although interaction does not exclusively belong to digital technologies (Hallnas 2004), they have enabled some of the most obvious features. One good example for the study and development of interaction has been the graphical interface of computerized systems, especially Graphical User Interfaces (GUI) for personal computers operating systems.

Computer interfaces in the 1950s and 1960s were designed for exclusive use of experts. Interfaces for these systems were intended to accompany computational problem-solving (Saffer, 2010). As the use of these systems gradually ceases to be carried out by specialized technicians, a concern with the relationship between people and machine begins.

This association with computer media marked interaction design development right from the start. Because these interfaces shaped behaviours and because conversely, they were shaped by people's use. The use of GUI is still currently one of the main expressions of digital interaction, supported by two essential elements, the desktop and the mouse (Moggridge 2007).

In fact, the most familiar transition in interface model's development is perhaps that of textual interaction for graphic interaction. This passage meant changing from a one-dimensional posture of text insertion to a two-dimensional representation of information in the interface and consequent relation with images and icons. In the textual interface, there was only one sequence of lines formed by input and output. By contrast, graphical interface makes use of screen space and focus of attention may be scattered on the screen or in several ones.

These interfaces have become conventional (Dourish 2001) and their novelty and update is deceiving, as several elements remain very much unchanged in the last decades. Still, their elements and intensive, continuous work have rated them as an important stand of user-centred design (Muller-Prove 2002).

## 7 TOOLS

### 7.1 *Shaping possibilities*

Design has been mentioned as a tool or instrument for other areas (Bonsiepe 2000). However, a design project it is always dependent on the means and materials employed in its development (Holmlid & Arvola 2007). To able to manage them within the scope of their project, designers need to know and use tools. Not only as instruments to shape ideas, but also of perceiving limitations and possibilities.

In an interaction design project, there are several tools and they can arise at different moments of the process. An interaction designer will mainly need tools to aid design, visual development, prototyping, interaction production, evaluation and communication.

Benyon (2010) addresses early stages requirements and recommends the use of storyboards, several sketches, mood boards, navigation schemes and scenarios. Coughlan, Fulton Suri and Canales (2007) associate these stages with further developing ones and mention observing users, brainstorming and rapid prototyping.

It is interesting to note most designers still sketch on paper for early stages of their process (King & Magoulas 2016) and use prototypes for three-dimensional models (Nam, Park & Verlinden 2009). However, these initial steps may not encompass an organization for information, intended for continuous and alternating behaviour. In cases where navigation is required, a visual scheme for structure detail is used to effectively transpose an articulation between screens or elements of a system.

After obtaining this structure, sketches should be progressive, generating different working hypotheses, including various prototypes. The making of prototypes is understood as a frequent procedure in design to develop products and services. They are a decisive tool, with advantages in organization of artefacts and in the capacity to generate qualitative changes throughout the project. Prototypes should enable quick modifications, allow direct handling and not condition styles (Benyon 2010); create tangible versions, allow to acknowledge inadequacies and address new behaviours (Coughlan, Fulton Suri & Canales 2007).

Ultimately, making several prototypes in early stages of the process opens the way to making versions with greater fidelity. It is not surprising the amount of software that allows today a convenient simulation of final intended interaction (Benyon 2010), given the spectrum of uses and the various moments in which they are needed.

These high-fidelity prototypes should be submitted to evaluation procedures. Martin and Hanington (2012) consider usability testing as an evaluation assessment that allows observing individual experience performing a task or set of tasks. So, it focuses on people and their actions.

Finally, designers will produce outcome which needs to be communicated (Benyon 2010). Presenting ideas is the ultimate skill in need of a good collection of techniques interaction designers must know.

## 8 CONCLUSION

To ensure acquisition of knowledge and skills in interaction design, we must ensure a proper syllabus planning, understanding the extension of the area, but also its most common features. In order to establish initial training in an interaction design master's degree, we developed a theoretical framework which sought to isolate the pressing topics, reflecting on their presence and approach in such training context.

Interaction design is a study of shared and reciprocal behaviours between users and systems or artefacts. In this sense, it is important to define and apply a work process that considers fundamental elements and strategies to interact and all the involved stakeholders. Such process must include several stages, allowing students to detail them, to implement them and to adequate methods, media and tools.

It is beneficial to know the media by which interaction projects can occur. This knowledge becomes important because of what each medium can bring in defining behaviours and what interferences can lead to accomplish tasks and operations.

Likewise, carrying out any part of the process is dependent on a set of tools. These allow production of interaction, and numerous work evidences throughout the process.

The main highlights of this framework complement and allow the necessary completion of a design project in interaction, giving knowledge about the area and providing the necessary skills to develop future work.

## ACKNOWLEDGEMENTS

The author would like to thank the funding support by the Foundation for Science and Technology of the Ministry of Science, Technology and Higher Education of Portugal under the project UID/EAT/04008/2013.

## REFERENCES

Anderson, S. P. (2011). *Seductive Interaction Design: Creating Playful, Fun, and Effective User Experiences.* New Riders.

Arvola, M. & Artman, H. (2007) Enactments in Interaction Design: How Designers Make Sketches Behave. *Artifact*, 1(2), 106–119.

Battarbee, K. & Koskinen, I. (2005). Co-experience: user experience as interaction. *CoDesign: International Journal of CoCreation in Design and the Arts*, 1:1, 5–18.

Benyon, D. (2010). *Designing Interactive Systems: A Comprehensive Guide to HCI and Interaction Design.* Pearson, Edinburgh.

Bevan, N., Carter, J., Earthy, J., Geis, T., Harker, S. (2016). New ISO standards for usability, usability reports and usability measures. in: Kurosu, M. (ed.) *HCI 2016*. LNCS, vol. 9731, Springer, Cham, 268–278.

Bonsiepe, G. (2000). Design as Tool for Cognitive Metabolism: From Knowledge Production to Knowledge Presentation. In S. Pizzocaro, A. Arruda, & D. De Moraes (Eds.), *Proceedings of the Politecnico di Milano Conference.* Presented at the Design Plus Research, Milano.

Bonsiepe, G. (2012) A Step Towards the Reinvention of Graphic Design. in Downs, S. (2012) *The Graphic Communication Handbook*, New York: Routledge, 8–13.

Buchanan, R. (1992). Wicked Problems in Design Thinking. *Design Issues* 8(2), 5–22.

Buxton, B. (2007). *Sketching user experiences: getting the design right and the right design*, San Francisco: Morgan Kaufman.

Colborne, G. (2017). *Simple and Usable Web, Mobile, and Interaction Design.* 2ed. Berkeley, CA: New Riders.

Cooper, A., Reimann, R. & Cronin, D. (2007). *About face 3: the essentials of interaction design*, Indianapolis: Wiley Publishing.

Coughlan, P., Fulton Suri, J. & Canales, K. (2007). Prototypes as (Design) Tools for Behavioral and Organizational Change: A Design-Based Approach to Help Organizations Change Work Behaviors. *The Journal Of Applied Behavioral Science*, 43 (1), March 2007, 1–13.

Davis, M. (2008). Why do we need doctoral study in design? *International Journal of Design* 2(3), 71–79.

Dourish, P. (2001). *Where the action is: the foundations of embodied interaction.* Cambridge, Massachussets, London: The MIT Press.

Dubberly, H., Haque, U. & Pangaro, P. (2009). *What is interaction? Are there different types?* http://www.dubberly.com/articles/what-is-interaction.html (access July 18, 2012).

Fischer, G. (2002). Beyond 'couch potatoes': from consumers to designers and active contributors. *First Monday* 7(12).

Forlizzi, J. & Ford, S. (2000). The Building Blocks of Experience: An Early Framework for Interaction Designers. *DIS '00*, New York: Brooklyn.

Frascara, J. (2004). *Communication Design: Principles, Methods and Practice.* New York: Allworth Press.

Goodwin, K. (2009). *Designing for The Digital Age: How to Create Human-Centered Products and Services.* Indianapolis: Wiley Publishing, Inc.

Grudin, J. (2007). A Moving target: the evolution of human-computer interaction. in: A. Sears and J. A. Jacko (Eds.). *Human-computer interaction handbook: Fundamentals, evolving technologies, and emerging applications* (3rd edition). Taylor & Francis.

Hallnas, L. (2004). Interaction design aesthetics: a position paper. in O. W. Bertelsen, M. G. Petersen & S. Pold (eds.) *Aesthetic approaches to human-computer interaction: proceedings of the NordiCHI 2004 workshop.* Aarhus: University of Aarhus.

Hassenzahl, M., Eckoldt, K., Diefenbach, S., Laschke, M., Lenz, E., & Kim, J. (2013). Designing moments of meaning and pleasure. Experience design and happiness. *International Journal of Design*, 7(3), 21–31.

Heeter, C. (2000). Interactivity in the context of designed experiences. *Journal of Interactive Advertising*, 1(1). American Academy of Advertising, 4–15.

Holmlid, S., & Arvola, M. (2007). Developing a thematic design curriculum as a Bologna master. in *International Conference on Engineering and Product Design Education*, 13-14 September 2007, Northumbria University, Newcastle upon Tyne, UK.
Kelly, R. R. (n. d.)Collected Writings of Rob Roy Kelly. *In Memory of Rob Roy Kelly*. RIT Library. In http://www.rit.edu/library/archives/rkelly/html/index2.html
King J. & Magoulas R. (2016). *2016 Design Salary and Tools Survey*. O'Reilly Media.
Kolko, J. (2011). *Thoughts on Interaction Design*. Burlington: Morgan Kaufmann.
Krishna, G. (2015). *The best Interface is no Interface: The Simple Path to Brilliant Technology*. USA: New Riders.
Lewis, W.P. & Bonollo, W. (2002) An Analysis of Professional Skills in Design: Implications for Education and Research. *Design Studies*, Vol. 23: 385–406.
Manovich, L. (2001). *The language of new media*. Cambridge and London: MIT Press.
Martin, B. & Hanington, B. (2012). *Universal Methods of Design*. Beverly: Rockport Publishers.
Moggridge, B. (2007). *Designing Interactions*, The MIT Press.
Müller-Prove, M. (2002). *Vision and reality of hypertext and graphical user interfaces*. Hamburg: Universität Hamburg.
Nam, T.-J., Park, S. & Verlinden, J. (2009). A model to conceptualize interactivity. *International Journal on Interactive Design and Manufacturing (IJIDeM)* 3, 147–156.
Nielsen, J. (2000). *Electronic books: a bad idea*, http://www.useit.com/alertbox/980726.html.
Norman, D. A. (1983). Design principles for human-computer interfaces. *CHI '83 Human Factors in Computing Systems* (Boston, December 12-15, 1983), New York: ACM, 1–10.
Norman, D. (2004). *Emotional design: why we love (or hate) everyday things*. New York: Basic Books.
Norman, D. (2013). *The design of everyday things*. New York: Basic Books.
Overbeeke, K.; Djajadiningrat, T.; Hummels, C., & Wensveen, S. (2002). Beauty in Usability: Forget about Ease of Use! In: Green, W. and Jordan, P. (Eds.), *Pleasure with Products, Beyond Usability*, New York: Taylor & Francis.
Redstrom J. (2005). Towards user design? On the shift from object to user as the subject of design. *Design Studies*, 27 (2), 123–139.
Preece, J., Rogers, Y., & Sharp, H. (2002). *Interaction design: beyond human-computer interaction*. New York: John Wiley & Sons, Inc.
Saffer, D. (2010). *Designing for interaction*. Berkley: New Riders.
Saariluoma, P. (2005). Explanatory frameworks for interaction design. in. A. Pirhonen, P. Saariluoma, H Isomaki, & C. Roast (Eds.) *Future interaction design*. London: Springer, 67–83.
Shneiderman, B. e Plaisant, C., (2004). *Designing the User Interface: Strategies for Effective Human-computer Interaction*. 4a Ed., Boston: Pearson/Addison Wesley.
Siang, T. (2017). *What is Interaction Design*. Interaction Design Foundation. < https://www.interaction-design.org/literature/article/what-is-interaction-design> (access December 13, 2017).
Stolterman, E. (2008). The nature of design practice and implications for interaction design research. *International Journal of Design* 2(1), 55–65.
Swanson, G. (1994) Graphic Design Education as a Liberal Art: Design and Knowledge in the University and the 'Real World'. *Design Issues* 10(1), 53–63.
Verplank, B. (2009). Interaction design sketchbook. in <http://ccrma.stanford.edu/courses/250a-fall-2004/IDSketchbok.pdf> (access October 8, 2015).
Yagou, A. (2005) Rethinking design history from an evolutionary perspective. *AKTO Art and Design*.

*Research & Education in Design: People & Processes & Products & Philosophy – Almendra & Ferreira (eds)*
*© 2020 Taylor & Francis Group, London, ISBN 978-0-367-49296-0*

# Pastiche plus; styles, philosophies and advanced design skills

N. Middelham & W. Eggink
*University of Twente, The Netherlands*

ABSTRACT: This paper describes the Pastiche+ learning method for design skills on a master education level. By using philosophies and stylistic elements of art and design movements the method tries to extend one's skill and knowledge about design tools, principles and processes in an elaborate way. The paper describes the purpose and context of the method in the introduction and method section. Pastiche+ is then put to the test in a case study around graphic design and communication. This resulted in a profound set of posters, advocating using the train. After an extensive reflection and evaluation of the process and results we conclude that the method can be considered effective as a high-level individual skills developer. The research results about communication are inherently subjective, however the insights from the personal reflection can be valuable in informing the design process of every designer who has a message to tell.

## 1 INTRODUCTION

In this paper we introduce a learning method for designers that uses art and design movements throughout the ages as a source of new knowledge, philosophies, visual language and methods. The learning method is based on the pastiche. A work can be called pastiche if it imitates a style from another work, artist or era, and unlike parody it does not criticize. The proposed method lets the design practitioner do pastiche, however also includes the movements defining philosophy and circumstances as additional elements in the design practice, hence the name Pastiche+.

Pastiche+ is developed as a learning technique for design skills in a higher education environment. Especially in the context of master education, merely skills training is not preferred (Q.A.N.U., 2014). We therefore introduced several elements of reflection and research in the different steps of the technique to make it suitable as a Capita Selecta for the Master Industrial Design Engineering. The result can be seen as an approach that is a combination of the reflective practitioner by Schön (1983) and the Research Through Design methodology (Findeli, 2010; Frayling, 1993).

In the approach, through the analysis of various styles, philosophies, circumstances and possible interpretations, an artistic playing field is created. The main goal of Pastiche+ is to let the practitioner explore this playing field, which is done by investigating the nature of multiple movements, using the movement related knowledge to design with, and evaluating the different processes and results.

The Pastiche+ method is relevant for designers that seek to explore and expand their knowledge and skills. Someone who undergoes this method is assumed to be overloaded with new ways of thinking and working rather than focusing on what one already knows. Additionally, the designer will increasingly be able to put stylistics elements in their historic context.

This paper explains the individual steps of our Pastiche+ approach in the Method section and provides a summary of the Case Study along two of the chosen movements. The case study will be reflected upon in the Discussion section, followed by the Conclusion.

## 2 METHOD

One of the teaching principles the Pastiche+ is based upon is repetition. We think that doing a Pastiche+ cycle with, for example, just one movement would not train a practitioner enough to use and reflect on the effects of movements. The Pastiche+ should be seen as a skill of using art and design movements for one's own practice, it is about being inspired and doing something with that inspiration. In such way, it fits in a methodical approach of using insights from design history in contemporary design practice (Eggink, 2009, 2016). The actual process of the Pastiche+ learning method consists of seven steps:

Selection of movements and message
Movement research
Ideation
Designing and iteration
Personal evaluation
Exhibiting and survey
Survey evaluation and conclusion

During the entire process, two variables are controlled for: [1] The designed artefact (e.g. poster, video, website) has to be consistent across all movements. [2] The designed artefacts need to have the same general purpose or goal (e.g. to entertain or persuade), but each movement can approach this shared purpose in its own way (e.g. persuasion through entertainment or persuasion by informing).

### 2.1 *Step 1: Selection of movements and message*

The selection of the movements is an exploration in itself, the Pastiche+ practitioner is invited to select his or her own preferred topics. However, there are guidelines that are assumed to aid the explorative aim of the method:

Choose many movements, repetition is more important than finesse;

Choose movements with a variance in time in history, country of origin, and stylistic elements;

Choose movements with a variance in scope of the definition (e.g. Modernism versus Bauhaus).

Also note that one could play with the variance of variances, it could be evenly interesting to compare two largely different movements as it would with two which are much alike.

The purpose should be achieved in the manner the movement dictates; thus it can happen that the purpose could be entirely inconceivable when it is one of the defining traits of that art movement. It is recommended to select a topic that could relate to multiple art movements, a topic that for example only relates to social networks would not work well with movements that mainly focus on nature. However, it could prove to be an interesting angle to work with, as the method is about exploring the playing field.

### 2.2 *Step 2: Movement research*

For every movement, one or multiple relevant books or other sources should be consumed and summarized, later to be used as inspiration. Movement associated pieces of work should also be collected and potentially be used as reference.

### 2.3 *Step 3: Ideation + step 4: Designing and iteration*

Ideation starts by doing short experiments with the movements process, motives and styles, in order to understand the implicit implications. Then these experiments are set aside, and compositions are made. These compositions will be inspired by the philosophy of the movement, they dictate which motives are selected, how they are placed and how they are communicated. After compositions are made, they will serve as inspiration of the creation process, this process is influenced by the art movement and selected motives. Artists can select their preferred tools, for the case study in this paper the tools photography, Adobe Photoshop, -Illustrator, -InDesign and Blender were used.

The next iteration is applying the style of the movement. Some movements have strict defining stylistic elements (e.g. Swiss School) while others do not (e.g. Romanticism). When it is the first of the two, the stylistic characteristics of the style will be used, when it is the latter, one or more artists that are associated with the style will be used as reference. If possible, the results should be discussed with a design teacher or peer and iterated upon.

### 2.4 *Step 5: Personal evaluation*

The personal reflection looks at the effects of the different movements on the act of designing. As reflection is an important aspect of learning (Bishop-Clark & Dietz-Uhler, 2012), it is also advised to do this in a structured manner (Procee & Visscher-Voerman, 2004). In the Pastiche + method, the design process and results are therefore reflected upon alongside a predefined set of qualities that are related to the chosen artefact and the purpose.

### 2.5 *Step 6: Exhibition and survey*

The exhibition is an invitation for others to view and give their opinion on your designs. This is done for motivation and for having an additional source of insights.

To gain these insights a survey has to be filled-in by guests of the exhibition. The survey can have both qualitative and quantitative statements to have a variety in kinds of insights. The set of statements should of course also be related to the chosen artefact and purpose, and can be comparable to the set of qualities from the personal evaluation.

### 2.6 *Step 7: Survey evaluation and conclusion*

The answers of the survey of step 6 are going to be reflected upon. Results will be compared, and new insights will be written down. This should lead to a better understanding of the acquired design skills in relation to the movements, philosophies, designs and the purpose.

## 3 CASE STUDY

In the case study, the design skill to develop was graphic design. The artefact that was chosen as subject to be designed was posters, and the chosen purpose was communication. The message to be communicated through the posters was chosen to be: “Take the train because it is good for nature.” Where the message implies that compared to other modes of transport the train damages the earth less.

This message was selected because it could broadly relate to society, politics, nature and technology, which are subjects that are often used in art and design movements. In the design process the philosophy aspect of the movements dictates which part of the message needs to be highlighted. Some art movements will focus on the train part more, others on the nature part.

Chosen movements were: Romanticism, Ukiyo-E, Futurism, Constructivism, Dadaism, Swiss School, Pulp (paperbacks), Post-Modernism, and Pop-Art. Romanticism was chosen because of its relative old age. Ukiyo-E was included to incorporate a movement that did not inherit its characteristics from western culture. Futurism, constructivism and Dadaism were chosen because of their visual qualities and closeness in time. Swiss School (or International Style) was chosen because of its lasting influence on graphic design. Pulp, and specifically Pulp paperbacks, was chosen because of its unique stylistic elements and focus on advertisement. Post-Modernism was chosen because it is a large umbrella term that incorporates many styles, unlike very specific styles like Swiss School and Pulp paperbacks. Finally, Pop-Art was chosen because its visual style is still widely used.

### 3.1 *Case study goals*

With a case study the teaching method is tested. The goals are to explore and reflect on the various qualities of the method, finding its up- and downsides, and to identify points of improvement.

## 4 MOVEMENT RESEARCH AND DESIGN

For every chosen movement style research was performed, based on literature. Both visual characteristics and philosophy of the movements were incorporated in the investigation. Because of the size limits of the paper we describe only two examples of the findings of such research, followed by the designs that were based on these findings.

### 4.1 *Romanticism*

Romanticism has multiple defining characteristics. Some define it as a Gothic novel genre while others would define it by their perception of Catholicism. There even exist decision models that place a piece of work on the Romantic spectrum (Ferber, 2010).

Although there are so many differences, most Romanticists tend to focus on its new way of viewing and living with God. At the time, the dominant interpretation was that one God ruled all, this was portrayed by the older Classic movement. A growing number of philosophers philosophized that god was not only somewhere up in the sky, but was everywhere, and more importantly, everything. He was the sky above and the people the ants below. Some scientists used this philosophy to research nature without the obstruction of the church (Ferber, 2010). This version of Romanticism was used in this project.

Like these philosopher and scientists, the Romantic artists focused on nature, discarding the holy compositions of the Classicists. Romanticism often saw emotion as something pure and holy, promoting introspection and letting emotions flow, choosing emotion over ratio.

#### 4.1.1 *Style*

Romanticism is more about their philosophical interpretation of God, or about letting their emotion flow, it does not have defining rules about style. Stylistics elements differentiate from medium to medium and artist to artist (Wolf et al., 2010). The painter Francisco Goya and Henry Fuseli focused more on the inner emotional aspect (depicted respectively with the *Black paintings series* and *The Nightmare*), but other known painters like William Turner and Caspar David Friedrich focused more on large terrific environments (depicted respectively in *Snow Storm* and *Wanderlust*). Their paintings are often described as sublime, which the poet William Wordsworth defines as "Arouses terror at the vastness and wild, ungovernable nature" (Busch, 2017).

#### 4.1.2 *Design*

William Turner was used as most important influence for the poster design (Figure 1). His paintings often include a sun that acts as the biggest source of energy. He often paints terrifying storms and plays with the focus of the eye by adding blurry brushwork.

This poster tries to merge the train with nature, saying in a visual manner, that the train and the nature are not different from each other and can live side by side. Therefore, a train passenger should not feel guilty, instead rather pleased. The merger of nature and train is done by blending their colours, blurring their edges and decreasing the contrast between them.

The text pushes the message a little further, "the train neither vandalizes nor destroys" implies that other ways of transport do vandalize and are therefore worse. A dramatic tone was chosen, instead of a rational argument, because of the emotional side of Romanticism.

Figure 1. Poster design "Romanticism".

## 4.2 *Constructivism*

Constructivists chose order over chaos, ratio over emotion and collective over individual worth. They saw the new machinery as a way to enter new domains and go across the limits of the human body. Constructivism often celebrated Communistic principles but it would be a mistake that all works are driven by them (Hötte et al., 2013).

### 4.2.1 *Style*

New Typography liberated typography from classical rules, adopting Modernistic principles. It was adopted by multiple movements (Futurism, Dadaism, Constructivism). Constructivists choose to use thick geometric sans-serif type and often placed them diagonally, both in 2 and 3 dimensions.

Like New Typography, New Objectivity liberated photography from classical rules, also adopting Modernistic principles. It was also adopted by multiple movements, Constructivism being one of them. This new look on photography was made possible by the emergence of technology (Hötte et al., 2013). Now artists, like Paul Schuitema, could change things up like neglecting the usage of a clear horizon, not having a breast or eye perspective and experiment with different camera distances (Maan, 2017). Other factors that were used in Constructivism were Euclidian geometry, differentiating brightness, precise and streamlined reality.

### 4.2.2 *Design*

Because most work is a celebrates the machine, in most work traces of machinery can be found. This can be either done by showing marks that reveal the usage of the machine but it was mainly done by having machine-like geometry. The machine parts of the poster (portrayed by the train, track and typography) have nature at their centre (portrayed by the tree), saying in its subtext that this machinery has nature at its hart. The text "Mother Earth Machine" then declares that this is a machine of the world, implying that it would not harm it (Figure 2).

Figure 2. Poster design "Constructivism".

Figure 3. Poster designs for respectively "Swiss School", "Post-Modernism" (Memphis), "Pulp", and "Pop-Art".

Figure 4. Poster designs for "Futurism" (left) and "Dadaism" (right).

Figure 5. Impressions of the poster exhibition, organised for the survey evaluation.

The poster uses New Objectivity by placing images without the classical rules of positioning, scaling and colouring. It also uses a little bit of New Typography, by adding perspective to the text and using an unconventional manner of placing. The composition uses a horizon which is off and a pointed perspective, imitating Nina Krogan's *Composition 2*, which is a Supremacist painting, however that style does share visual qualities with Constructivism (but not its fundamental philosophy). The composition is also heavily influenced by the usage of photography in Dick Schuitema's work.

## 5 DESIGN RESULTS

In line with the methods basic rule of repetition, a total of nine posters was designed. The poster designs resulting from six of the other movements are pictured in Figures 3 and 4. The poster associated with Ukiyo-E is visible in the impression of the exhibition in Figure 5 (left).

## 6 PERSONAL EVALUVATION

The personal reflection looked at the effects of the art movement on the act of designing a poster. To structure the reflection process, six qualities were thoroughly compared for each movement.

### 6.1 *Ease of creating meaning*

This quality is about how easy it was to combine the philosophy and style to create and communicate the idea that would communicate the message. A movement could be either very direct in communicating the message, very indirect or somewhere in-between.

Swiss School, Pulp-paperback and Pop-Art were easy to create meaning with. What they have in common is a clear usage of elements to portray meaning. Swiss School and Pulp-paperbacks have this inherit quality because they were made for advertisement (Ruder, 1959; Schreuders, 1981). Pop-Art uses pop-culture (thus also advertisement) which means the designer has a wide variety of already created images and ideas with meaning, which makes it easy.

Dadaism was hard to create meaning with. The first reason is that I wanted to add randomness like Tristan Tirza with his poems (Tirza, undated). The second reason is the anti-commerce and anti-art side of Dadaism, making it hard to let the poster have a persuasive commercial meaning. In the same manner, many Post-Modernists went with the 'function follows form' mantra which made it hard to create meaning with. However, the art movement did have tools to work with, like the combination of references from 'high' and 'low' culture (Joris et al., 1984) and the theory of Dual Coding (Jencks, 1984).

### 6.2 *Ease of creating composition*

This quality is about transforming the message into stylistic elements that have to be placed. Some art movements have strict composition rules, while others do not. Both could actually either aid or hinder the crating process.

Numerous styles (Romanticism, Ukiyo-E, Dadaism, Swiss School and Pop-Art) were easy to create composition with. Romanticism and Ukiyo-E were easy because they were mostly filled with landscape (Ferber, 2010; Schlombs, 2007). Dadaism was easy because meaning was largely non-existent which made the composition more about visual play. Swiss School was the easiest to design for because it followed the grid structure method (Müller-Brockmann, 1999). Pop-art was easy because the object of popular culture needs the most attention, all the other parts just need a way of being there without taking away the attention.

Because Post-Modernism is mainly about style (Adamson & Pavitt, 2011) it should have been an easy movement to create a composition for, however it dealt with many stylistically different geometric forms, making it hard to tell a story while maintaining balance in the composition. Both Futurism and Constructivism are a lot about machinery (Martin, 2005; Hötte et al., 2013), making it hard to create a composition which included nature.

### 6.3 *Ease of replicating the style*

After the philosophy is used and the composition is created the design process goes through a stylistic phase that tries to replicate the style of an artist of the movement. Some of them require a lot of artistry, with for example a lot of painted detail, while others use techniques that are quicker to apply, like assemblage.

Ukiyo-E, Swiss School and Pop-Art were easy to design with. Ukiyo-E's style is easy to replicate because the stylistic elements are simple, however, it does need a lot of detailing. Swiss School made it easy, their stylistic elements are limited and the grid system makes it simple to put together. Pop-Art was easy because it has no inherit style, but copies styles from popular culture. In that sense, Pop-Art is often already pastiche. Post-Modernism was hard because Memphis mainly designed physical products, and the poster would only be in the two dimensions. Also the 'Universal Dynamism' of Futurism was a difficult style to master.

Dadaism was contradictory, Dadaists used a lot of randomness but their work did seem to have a kind of order to it. Making the piece was easy because of mixing randomness and order, but questions about if the philosophy is applied right do remain. Assuming there even is a way, because Dada is also anti-Dada (Elger, 2004).

### 6.4 *Ease of creating impact*

This quality is one of the more subjective ones. It is about how doable it is to create images that captivate people in a consumeristic context. In this context it is assumed, for the sake of simplicity, that a design should be readable from afar and have a high amount of contrast to do so.

Dadaism, Swiss School, Pulp-paperbacks and Pop-Art had a style that was helpful in creating impactful images. Dadaists and their events (e.g. the ones in Circus Voltaire and Berlin) and other works (e.g. Marcel Duchamp's Fountain), tried to shock the perceiver and therefore used big contrasting imagery. Swiss School has a style that is crafted to create impactful imagery with the least amount of effort and time. Pulp-paperbacks thrived on shocking the perceiver, sometimes even misrepresenting the actual content, so that the consumer bought one of their novels. Pop-Art was easy because one only had to find the impactful imagery in popular culture and give it the main stage.

Romanticism was terrible to design with concerning impact, while the reference paintings often seemed too smooth.

### 6.5 *The pleasure in creating*

This quality is about if the act of creation was something that was enjoying and worth doing.

Ukiyo-E, Dadaism and Post-Modernism were fun to create with. Ukiyo-E is a method that is relaxing, it is forgiving and the repetition of drawing small details makes one forget the time. Dadaism was fun because it did not have a wrong or right way of doing it, making the process freeing and effortless. Post-Modernism had a difficult part which was finding the right forms; however, it was made highly pleasurable by denying any Modernistic process.

Swiss School's method was too methodological to have the pleasurable creative freedom. Pop-Art was fun to find the right images with, and rewarded you quickly with results but proved itself too simple to be a fulfilling process.

### 6.6 *The perceived duration of the production*

This quality is about how productive a style was perceived in terms of design per hour ratio.

Swiss School and Pop-Art lend themselves for quick production. You might say the first was made for it, with methods like the grid system that helped making the right decisions fast. Pop artists re-use a lot of work from other people, sometimes not even taking the time to remix it. This made for fast production, where the most time was spend finding the right imagery.

The other styles were less 'productive'. Post-Modernism took a long time to get the story told in the complex 3-dimensional geometry while the artwork of Pulp-paperbacks demanded a lot of artistry. Ukiyo-E took a very long time for drawing all the details and Romanticism to execute William Turner's style, which arguably still is not really achieved. With Dadaism it took time because of the multiple phases, but because there was no 'bad' way of doing there was no stress, which made it perceived as short.

## 7 SURVEY EVALUATION

To gain more insight in the effects of the different movements on the design and communication of the message, the posters were presented to an audience in a small one-day exhibition (Figure 5). Visitors were asked to fill in a survey for one or more posters. In this case, a total of 38 surveys were filled with quantitative and qualitative feedback. The results were then summarized and used to either confirm, enhance, or reject the conclusions of the personal reflection. These results added to the research through design aspect of the project, it turned out for instance that the Pulp poster was much disliked because of the sexist connotations of the 'philosophy' of the movement. The message of the Swiss School poster was intentionally the clearest and the Romanticism poster was surprisingly liked the most.

## 8 DISCUSSION

In the case study, Pastiche+ proved to be excellent for exploring different design methods and styles. The circumstances and philosophies of the movements also helped explain why some style elements are the way they are, increasing the applicability when a practitioner decides to use them. In that sense it helped improve the designers 'design sense', which could be applied in other projects. The reflection and evaluation also helped to gain insight in the 'design opportunities' of the different movements' styles and philosophies. However, the method's result will be always highly depended on the practitioner. Thus, to gain academic merit, multiple case studies need to be done, evaluation and compared, preferably with more controlled values.

Movements had drastically diverging views on the design process, this helped in evaluation each style because they could be compared. This variance of chosen movements felt contributing to the explorative aspect of Pastiche+. However, the amount of movement often felt overwhelming, processing every movement proved to be time-intensive, which could hinder the time for repetition which is one of the goals of Pastiche+. A solution could be to lower the amount of movements and increase the number of designs per movement, in that case a practitioner needs less time to research and would thus have more time to spend on making designs and reflecting.

It is hard to capture all facets of an art or design movement, there are diverging historic views plus a variety of possible perspectives. However, the research of the movements as philosophies, rather than as mere styles added much to the abstraction level of designing and the quality of the results. This is best visible in the different ways in which the intended message is communicated and integrated in the visuals, from very direct in the Swiss School poster to rather 'hidden' in Romanticism.

## 9 CONCLUSION

In this paper, the Pastiche+ method was described, then examined by a case study and discussed in the end. The Pastiche+ method showed potential as a learning tool for extending one's knowledge and skill in graphic design and visual communication. In the design case, the method did execute its goal of letting a participant explore the playing field created by the various elements of the art and design movements. The research results are inherently subjective, however the insights from the personal reflection can be valuable in informing the design process of every designer who has a message to tell.

## REFERENCES


Adamson, G. & J. Pavitt (2011), *Postmodernism: Style and Subversion*, London, V&A Publishing.

Bishop-Clark, C., & B. Dietz-Uhler (2012). *Engaging in the Scholarship of Teaching and Learning*, Sterling, Virginia (US): Stylus.

Busch, W. (2017). *De Romantiek In het Noorden: van Friedrich tot Turner [Romanticism in the North: from Friedrich til Turner]*, Groningen, Uitgeverij Wbooks & Groninger Museum.

Eggink, W. (2009). A Chair to Look to the Moon: What We Can Learn from Irrational Design History for Contemporary Design Practice. *Design Principles & Practices: an International Journal*, 3(1), 103–114.

Eggink, W. (2016). Design History Education in the Netherlands: the Historic, the Cultural, and the Methodical., *Proceedings of the Nordic Forum For Design History Studies Conference 2016: Design Education In The Nordic Countries*, Aalto University School of Arts, Helsinki, Finland, 23–24 Sept.

Elger, D. (2004), *Dadaïsme [Dadaism]*, Berlin, Taschen.
Ferber, M. (2010). *Romanticism: A Very Short Introduction*, Oxford University Press.
Findeli, A. (2010). Searching for Design Research Questions: Some Conceptual Clarifications. In: R. Chow, W. Jonas & G. Joost (Eds.), *Questions, Hypotheses & Conjectures; Discussions on projects by early stage and senior design researchers* (pp. 278–292): iUniverse.
Frayling, C. (1993). Research in Art and Design. *Royal College of Art Research Papers*, 1(1), 1–5.
Hötte, D. W. et al. (2013), *Utopia 1900-1940: Visies op een Nieuwe Wereld [Utopia 1900-1940: Visions of a New World]*, Rotterdam, nai010.
Jencks, C.A. (1984 (1977)). *The language of Post-Modern architecture*, London: Academy Editions.
Joris, Y., et al. (1984). *Memphis-Design*, 's Hertogenbosch: Dienst Beeldende Kunst.
Maan, D. (2017). *Paul Schuitema: Visual Organizer*, Rotterdam, nai010.
Martin, S. (2005), *Futurisme [Futursim]*, Berlin, Taschen.
Müller-Brockmann, J. (1999), *Grid Systems*, Salenstein (Switzerland), Niggli Verlag.
Procee, H., & I. Visscher-Voerman (2004). Reflecteren in het onderwijs: een kleine systematiek. [Reflection in education: a small systematics]. *VELON magazine for teacher training*, 25(3).
Q.A.N.U. (2014). *Industrial Design, Faculty of Engineering Technology, University of Twente*. Q0441. Utrecht, Quality Assurance Netherlands Universities.
Ruder, E. (1959), *The Typography of Order*, New York, Allworth Press.
Schlombs, A. (2007), *Hiroshige 1797-1858*, Berlin, Taschen.
Schön, D. (1983). *The Reflective Practitioner: How professionals think in action.*, New York: Basic books.
Schreuders, P. (1981), *Paperback, U.S.A.: Een Grafische Geschiedenis 1939-1959 [Paperback USA: A Graphic History 1939-1959]*, Amsterdam, Loeb Uitgevers.
Tirza, T. (undated). How to Make a Dadaist Poem, original work. via: https://fleursdumal.nl/mag/tristan-tzara-to-make-a-dadist-poem, accessed 12 May 2019.
Wolf et al., N. (2010). *Romantiek [Romanticism]*, Berlin, Taschen.

*Research & Education in Design: People & Processes & Products & Philosophy – Almendra & Ferreira (eds)*
*© 2020 Taylor & Francis Group, London, ISBN 978-0-367-49296-0*

# A process which captures insight and nature's silent design lessons

A. Stephen
*Texas State University*

ABSTRACT: Design education, design research and design practice are vitalized by embracing tools through which people take lessons from nature. We use a case study to illustrate this point. Two phenomena have come together to compel case for expanded scope of design research as part of design problem solving and education: (1) multi-facetedness and complexity of the world we inhabit and (2) ardent new role design has acquired recently as a way of cross-disciplinary/cross-professional problem solving. What all this means is (1) designers should be trained about and (2) learn to use tools which enable probing into how phenomena are constituted or constructed. This is particularly useful for phenomena that occur silently, organically, but if we have interrogated them, give informed perspectives about how independent and interdependent phenomena occur or might be predicted. Proficient designing includes recognizing (through probing and discovery) interdependencies.

## 1 INTRODUCTION

Nature is *the great* designer.' We say a thing is *by design* when we wish to express that it is the way nature intended or planned for it to be. If knowledge about design is a human goal, then people must possess capacity of discovering lessons of nature. Perhaps an area of growth in design education is cultivating sensitivity towards nature's quiet lessons through expansion of the range of tools we bring to engagement of nature. The same critique as above can be levelled at mainstream design research, where mainstream traditional researchers are battling to preserve the old status quo and fighting an upstream battle against the fact that design research must continue to expand its boundaries if it is to continue to be successful in its significant role as a way of cross-disciplinary and cross-professional problem solving, as a border-crossing discipline. We position this as provocation for design education as well as design research.

In this paper, we present a case which illustrates the point that there are silent phenomena of nature that are discoverable and can provide instruction for us. We are able to employ lessons of this sort for purposes of understanding and interpreting our environment, its phenomena and how people behave in it or coexist with it (and make decisions–scientific or more prosaic–about it). We close with cross-disciplinary implications of our case. We should strive to understand design of the world we occupy operationally through interactions with physical, structural or other formational contextual phenomena.

The scope of design is sometimes underestimated, its function sometimes undeservedly constrained. That, for instance, is the case when design is considered in its didactic function. Design's instructive role plays out in, among other things, (1) acts of representation and (2) acts of description. In representation, the designer constructs a form of mediation by which a complex, elusive or intangible phenomenon is rendered concrete–or, at least, legible. In the current paper, we reference plots–in particular, plots which have been designed to be used to make intelligible a silent phenomenon [*IC*, as discussed later]. In description, the designer, if skilled in analyzing, is compelled to take apart or decompose a phenomenon into its constitutive parts, thereby creating an intellectual space for elaboration and commentary through careful, systematic or least-confounding investigation.

### 1.1 *Recurrent event*

In course of investigation of dynamics of human encounter and choice in social space and strategic diffusion of information, using an agent-based model, we observed a recurring event–a design of nature. Within a modulating environment of emergent formations, when there is a principal entity (a breed, an idea, etc.)–dominant by a combination of design and chance–which, by expectation or logic, ought to either stand out uniquely or distinctly equally eviscerate all other phenomena of its type, there would often emerge, in concurrence, a second, less-dominant one from among the others, which would often not be as prominent as the dominant one, but would be *notably* detached from the rest of the lower order entities. With observed proportions of occurrence of this autonomically-forged event (see Figure 1), we concluded that the phenomenon was not to be merely trivialized, that it merited further exploration. We termed the phenomenon, *intersectional capitalization (IC)*: it occurs in the theoretical and functional space of interaction between phenomena and it is outcome (it appears) of capitalization on the same.

Dynamics of diffusion bear import for a wide range of subjects and problems as diverse as diffusion of innovation, disease spread, spread of rumour, information spread on the Internet, culture diffusion, spread of a riot, study of invasive species, spread of a fad, decision making and more. Insights aid researchers, planners, strategists and others in interpretation, control and study of events which involve diffusion or spread. Above all, such insights are invaluable within the paradigm of design–since design is seen currently, today, as a discipline of extensive, cross-disciplinary problem solving in the world.

## 2 NETWORK TOPOLOGY AND SOCIAL ELEMENTS OF DIFFUSION

In social space, at some point, what" other people do or say matters with regard to choice-making. For instance, sometimes when a decision has to be made between two alternatives, particularly when the decision maker has limited information about the matter at hand or limited ability to process available information, that decision maker would observe actions of other people or listen to other people in order to make her own decision (Watts, 2002). There is an influence of others. Social influence is a core element of diffusion–of information and other things. Structure variables related to social influence which have been examined include such things as roles played by key individuals (e.g. innovators), homogeneity of groups (homophily), characteristics of message source and characteristics of recipient/adopter (Kaigler-Evans, Leavitt, & Dickey, 1978).

### 2.1 *Roles in adoption*

One of the most prominent models of diffusion is the one presented by Rogers in *Diffusion of innovations* (see Sahin, 2006; see Figure 2). The model presents sequence of adoption (left to right), named adoption cohorts and proportions of population adopting at different stages.

In our delimited models below, innovators are of interest. Model-wise, they represent the inaugural 2.5% who are keen to adopt an innovation (or message). In terms of cross-community

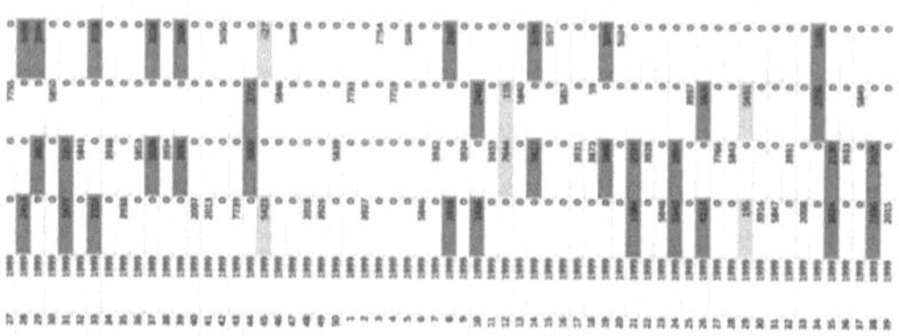

Figure 1. Recurrence of IC (dark gray pairs, by column).

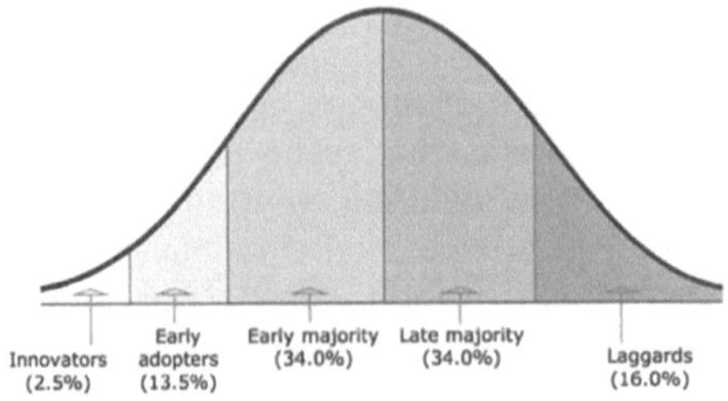


Figure 2. Adopter categories (based on innovativeness).
Source: Rogers, in Sahin, 2006. (Recreated by author of current paper.)

diffusion, innovators are the ones who are instrumental in bringing an innovation from outside into the system (Sahin, 2006). In our models, we will use them as cross-breed transmitters.

### 2.2 *Message source and message adoption*

An important dimension of diffusion is *source* of message to be diffused. Alpert and Anderson (1973) noted that effectiveness of communication could be indexed by degree to which "source exercises influence over the actions and reactions of message recipients" (p. 328) and asked strategically if perceived source-receiver dissimilarities might have instrumental value in enhancing communication effectiveness.

Ideas of similarity and dissimilarity are often expressed in terms of homophily and heterophily, respectively. According to Rogers and Bhowmik (1970/71), homophily "refers to the degree to which pairs of individuals who interact are similar with respect to certain attributes, such as beliefs, values, education, social status, etc." The relational contrast is heterophily: "the degree to which pairs of individuals who interact are different with respect to certain attributes" (p. 526).

In a study, Alpert and Anderson (1973) defined similarity in terms of degree of homophily-heterophily. Their findings indicated that "maximally effective communication occurred when the source was perceived as neither highly homophilous nor highly heterophilous, but somewhere between" (pp. 338-339). Accordingly, they marked that point as optimal heterophily (also see Gatignon & Robertson, 1985). We will employ lessons of their operationalization of source in our exploration below.

### 2.3 *Diffusion dynamics: Simple contagion, complex contagion, weak ties, strong ties*

In an information age, when designers are locked in an entanglement with structural matters surrounding information distribution, questions about diffusion are of interest. Phenomena such as information or disease spread belong to the topological class of "simple contagion," where a single contact is usually sufficient to effect transmission or adoption (Centola, 2010). By contrast, "complex contagions" require contact with multiple sources (Centola, 2010, p. 1194; Zhou, Zhao, & Lu, 2015) before adoption takes place. Social behaviour which requires *social reinforcement* for adoption is considered to be a complex contagion (Centola, 2010; Weng, Menczer, & Ahn, 2013). Diffusion in this latter case (i.e. in complex contagion) is often hypothesized to be facilitated by the clustered network, within which there are more redundant ties between agents (nodes/vertices) and "neighbourhoods" are more tightly knit (Centola, 2010). These types of ties are called "strong ties"–in the *relational* sense (see Centola & Macy, 2007). Strong ties of this relational understanding "connect close friends or kin whose interactions are frequent, affectively charged, and highly salient to each other." Strong ties increase mutual influence among those who are thus close to one another (Centola & Macy, 2007, p. 703)

There is another understanding of strength within the topology, a different type designated as a *structural* one. This form of strong tie, according to Centola and Macy (2007), is the one with capacity to "facilitate diffusion, cohesion, and integration of a social network by linking otherwise distant nodes" (p. 704). The authors elaborated:

Although casual friendships are relationally weak, they are more likely to be formed between socially distant actors with few network "neighbours" in common. These "long ties" between otherwise distant nodes provide access to new information and greatly increase the rate at which information propagates, despite the relational weakness of the tie as a conduit. (p. 704)

A critical insight is that ties which are weak in the relational sense can be strong in the structural sense because they "provide shortcuts across the social topology" (Centola & Macy, 2007, p. 704) and "bridge otherwise distant subgroups" (Larson, 2017, p. 1).

An influential conceptual point of view stems from the foregoing: *weak relational* ties (but *structurally* strong) are more effective for quicker and farther diffusion in a wide network than are *strong relational* ties as might be found in clustered networks, where close relationships result in redundancy through sharing contact with same neighbours. Redundancy in information sharing might inhibit intersection with new ideas from distant settings. In short, this view suggests that strong relational ties are seen here as effective for diffusion within a cluster, but might inhibit diffusion beyond the cluster, across groups.

A cautionary finding called for modification of that point of view. As described by Centola and Macy (2007), Watts and Strogatz discovered that a small fraction of long ties present in a clustered network can enable the latter to function as a network with strong *structural* ties, in which information is diffused far and quickly. They wrote: "Information and disease can spread very rapidly even in a 'small world' composed mostly of tightly clustered provincial communities with strong in-group ties, so long as a few of the ties are long" (p. 705).

Centola (2010) carried out a study bearing influence of that insight and found the clustered network to be actually more effective for spreading behaviour. "Not only is individual adoption improved by reinforcing signals that come from clustered social ties," Centola wrote, "but this individual-level effect also translates into a system-level phenomenon where by large-scale diffusion can reach more people and spread more quickly in clustered networks than in random networks" (p. 1197). He added a cautionary note that it is possible, though, that we need to understand effects of "other topological features... [such as] density" (p. 1196). In one of the models below, we controlled presence of long ties through proportion of each population that was allowed to be homogeneous before diffusion operations were enabled.

## 3 METHODS AND FINDINGS: IC IN DIFFERENT ENVIRONMENTS OF DIFFUSION

We present two models operating under different structural conditions. We are able to show that IC has presence in different operational environments. During explorations described in this paper, we completed more than 1700 model runs in a bid to further describe the phenomenon of interest.

### 3.1 *"Simple" decision making: "Strict-majority"*

Given that source is one established influential variable of message diffusion, we took it as granted and strategically manipulated it to determine if IC would persist under changes in this core variable. As described above, Alpert and Anderson (1973) described–hence, operationalized–source as a function of homophily. Since homophily and heterophily are grounded upon attribute similarity, we adhered to the point of view that the definition and design described by Alpert and Anderson are admissible in other environments involving attribute description–and not strictly communication environment alone.

We performed a stress test to determine persistence of IC under different degrees of homophily. Our range spanned the continuum from total homophily to total heterophily. We introduced grades of constraint on breed homophily by controlling proportion of each population that was allowed to be homogeneous (i.e. we *designed* to systematically carve out or save some space for long ties to be able to operate) before cross-breed diffusion operations were enabled. We borrowed the criterion of "strict majority" from Chang and Lyuu (2009), setting the rule that a node or agent would adopt a message, role or less when the latter is owned or transmitted by more than half of the former's immediate neighbours.

We created four breeds to represent four homophilic groups, but also as representations of four messages or any other phenomenon to be transmitted. Using all breeds, we populated a model world of 1 sq. mile with a total of 7776 plus one agent. (For a city population density, we conveniently selected density of Helsinki, at 3,002.44 persons per $km^2$ or 7776.33 persons per $ml^2$ in 2017 [Statistics Finland, 2018]. Each of the four breeds had 1,944 agents). The one extra agent was our seed.

We introduced a fundamental seed of one agent. All agents moved randomly. When an agent encountered the seed, it became an *innovator*. Once our seed generated *innovators* that constituted 2.5% of our population, it ended its inducements. The *innovators* then became drivers of diffusion, working with "long ties" capacity (not contained by own breed alone). In this model, we did not allocate a 2.5%-specific-generated number of original seed-induced innovators per breed; we simply allowed all originally-created innovators (all breeds together) to be 2.5% of the entire model population. We left it to the multiple number of independent runs performed (100 per condition) to neutralize favouring a particular breed. That design was chosen to reflect the relatively random nature of encounter within a city, where members of multiple groups/breeds move freely and concurrences (corporeal, ideational, material) are based on *accidents*. We monitored imprint of diffusion on every breed. In particular, we looked for presence of *IC*: a dominant breed (or phenomenon) and legible presence of a subordinate dominant breed, sufficiently distinguished from the lowest two.

We initiated several levels of this model. In the first, basic level, we programmed a condition such that all agents within a breed were proselytized (representing complete homophily) before cross-diffusion (across breeds) began. Subsequent model levels were built on a homophily-then-heterophily gradient: A specified proportion of the breed was allowed to adopt first (representing homophilous diffusion) and after that, the idea, message or phenomenon represented by the breed was allowed to percolate across the population (across other breeds). At end of each model run, we consistently checked for IC. Each model run lasted 2000 time-steps. We determined that time period by measuring the number of time-steps it took to "infect" a strict-majority of the entire population (i.e. take [7776 * 0.5) + 1]), starting with a single seed and allowing for effect of "long ties" as in the models we would be running. We took a mean of twenty runs, which yielded just over 1000 time-steps. We doubled that time length for final model runs so as to comfortably account for outlying situations. Degree of diffusion of each phenomenon/message was indicated by end count of the breed which embodied it.

After completion of 100 runs for each level of the model that we explored, we examined our data for IC. Table 1 image (above) presents rate of occurrence of IC.

Table 1. Presence of IC in diffusion model applying "strict majority".

| Run ID[1] | Number of runs | Proportion of runs presenting IC |
|---|---|---|
| 1944/1944 | 100 | 12% |
| 1500/1944 | 100 | 18% |
| 1000/1944 | 100 | 13% |
| 500/1944 | 100 | 14% |
| 100/1944 | 100 | 18% |
| 50/1944 | 100 | 22% |
| 10/1944 | 100 | 20% |
| 5/1944 | 100 | 24% |
| 0/1944 | 100 | 26% |

1 *Run ID is grade of homophily before "long ties" enabled: proportion = count of homogeneous agents/breed population*

Results for multiple levels of the model are also presented in graphs in Figures 3 and 4 above. In order to delineate presence of IC, we generated the graphs by uncoupling count from breed so as to facilitate ordering hierarchy of count independently.

Figure 3 shows a typical plot. Ordinate data represent run-end count (for each breed, count at the 2000th time-step). Run identities are displayed on the x-axis, from 1 to 100. Each complete plot (each panel in Figure 4) presents the result of 100 runs. Counts per run have been ordered hierarchically in order to express spatio-graphic relationship between levels of diffusion represented by counts. Counts are indicators of diffusion and plot lines articulate the relationship between diffusion levels across 100 runs.

In Figure 3 and the first seven panels of Figure 4, a dominant form ("principal") is evident in the emerald/jade green line. Consistently attendant is also a sub-dominant ("minor")–visible as a light gray line. These plots mark presence of IC. It can also be seen that the homophily-heterophily gradient has an effect on the sub-dominant. The greater the level of homophily before cross-breed diffusion, the less "aggressive" the sub-dominant.

Panel 8 of Figure 4 presents differently. It is the condition of complete diffusion within cluster before extra-cluster diffusion. Under this condition, the general expectation ought to be that, in balance (i.e. all variabilities "neutralized" by multiple runs), every cluster would settle at its cluster population at end count. For easier legibility, we have modified the color palette on this plot: the dominant is a hollow red line and the sub-dominant is a black, dotted line. It

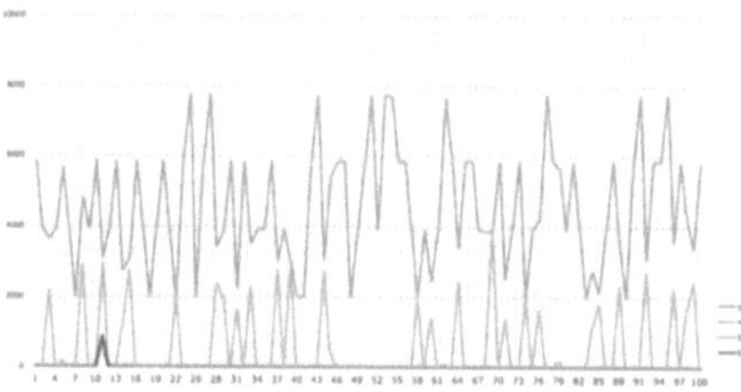

Figure 3. Manifestation of IC in cross-cluster diffusion (long ties) in a model which employs "strict majority" specification.

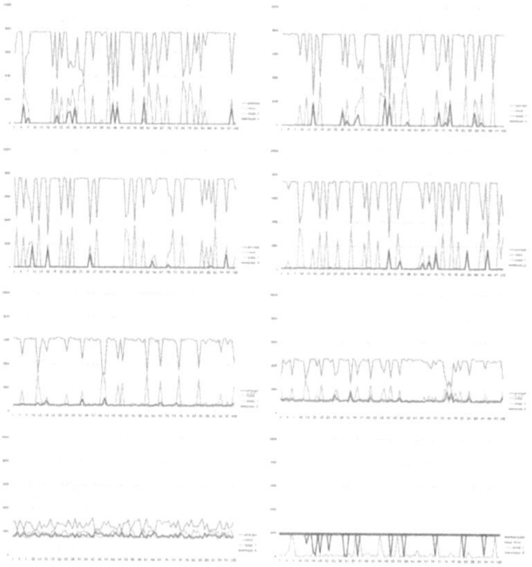

Figure 4. Manifestation of IC in cross-cluster diffusion (long ties) under different initial conditions of cluster homophily in a model which employs "strict majority" specification.

can be seen that the dominant generally flat-lines at the cluster population of 1944 (because we have capped it that way). Even in this condition, the sub-dominant distinguishes itself from the other two breeds/clusters. Instead of demonstrating the variations evident in the latter two, the sub-dominant actually adheres to behaviour of the dominant.

## 3.2 *Choice making in a more complex environment: Multiple messages, memory*

The graphs we present in this section are from an earlier investigation (Stephen, 2019). That investigation raised the original interest in occurrence of IC. We present its relevant results here as essential argument because the model represents a different set of structural conditions (from the above) under which, despite differences, IC is found to be present (Figure 5).

In the current model, four breeds were also initiated. (Each breed is construable as a message or phenomenon to be diffused.) One dominant breed was assigned and weighted for efficacy across different model conditions. Agents moved along designated paths and encountered one another. With every encounter, an agent recorded the message of the other and coded it as a weighted index in own memory function. (So, here, we created a more complex performance by introducing a new "vitality": memory.) That internal index itself was modulated (an added layer of intricateness) by an owned personal characteristic–based on a real-life survey and which we referred to as a personal "urge" –that regulated how much the individual would be personally impelled to adopt an external proposition. This encounter-inform social dialogue continued until an agent made decision to adopt a message. The adoption decision would be based on a comparison of information in its memory. Each run was terminated at 1200-time steps–based on a gentle, 20-minute traverse of one mile. Each condition was run 100 times.

Each point of ordinate datum in these plots represents mean of 1200 data points, one mean for each breed (or cluster). Each of the 1200 data points mentioned represents count-at-a-point-in-time of number of agents of each breed that is consigned to a location. So, the mean represents an average count over 1200-time steps. (In the model, that "location" just mentioned above is called an "attraction." The number of agents consigned to, or grounded at, each of four attractions is, given the conceptual framework of the current paper, an indicator of [level of] diffusion–a constructed equivalent of the number of agents who have adopted a message.) All of the foregoing belongs to one run (of 1200-time steps). Each complete plot (each panel in Figure 5 below) presents the result of 100 runs. Run identities are arrayed along the x-axis. Means per run have been ordered hierarchically in order to express spatial-graphic relationship between levels of diffusion. In summary, means are indicators of diffusion and plot lines express a comparative gauge of those levels dynamically across 100 runs.

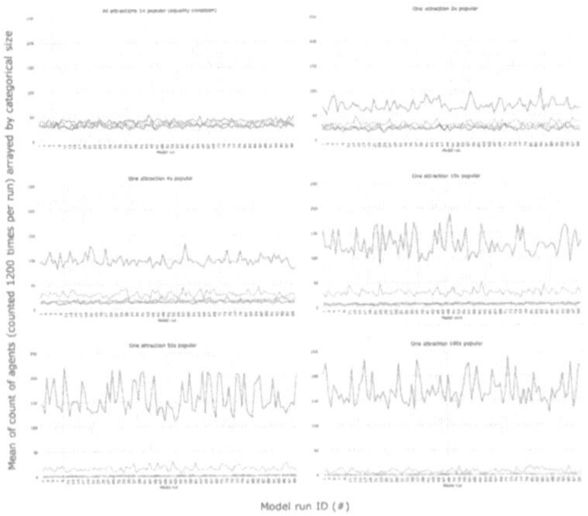


Figure 5. IC in adoption model based on popularity of message.

Attractions in this model were invested with different levels of drawing power to lure agents. (The idea is construable here as effectiveness of a message.) As can be seen, IC, while not clearly defined in the first panel, becomes quickly manifest when rest of these panels are regarded.

## 4 DISCUSSION

### 4.1 *Implications*

As indicated at beginning of this paper, study of the environment of diffusion and its processes crosses multiple disciplines–a demonstration of its importance. Efforts to understand conditions, circumstances, features, facets, attributes and dynamics of that environment are contributions to that greater endeavour.

Awareness of the possibility of IC raises some thought-provoking prospects. Under assumption that, in IC, the dominant and *sub*-dominant are structurally linked, we introduce a few.

If a commercial entity was engaged in efforts to establish a dominant presence, it is possible that, unwittingly, a competitor who benefited from exertions or toil of the former to be dominant might be forged from those very efforts. That competitor might be relatively strong, closely or more distantly, but its own presence is enhanced, nonetheless.

In monitoring spread of a disease, it might be the case that an attendant malady has been facilitated–unbeknown to observers–based simply on the combination of present conditions which have generated the large disorder in focus.

Under a condition of political domination, might some conditions which have maintained that domination also be connected to conditions which might foster a second (supportive or resistant) movement–structural conditions, perhaps, which do not merely originate from cognition-based resistant or supportive political views?

A prospect we find to be stirring is surrounded by the idea of species survival. If we knew the right conditions to look for under circumstances of declared potential or real species extinction, perhaps we might be able to find a "hidden," resilient, related community. Perhaps in the case where an invasive species has apparently decimated an existing population, an undisclosed beneficiary might be found.

Finally, in terms of the urban environment, it is to be seen as possible that a major event might "spring" another smaller, but noticeable event. An accumulating large gathering (a celebration, a riot, etc.) might spur a significant other gathering. With a sudden large outbreak of an illness across communities, a possible other ailment might be anticipated. It might be a critical issue for those managing those types of situations (and for researchers) that, while their energy is devoted to solving a large extant problem (e.g. disease, riot), there might be a potential other in the offing.

The diffusion environment is an important one in view of its potential for broad application. It is, not unexpectedly, an area which can command multiple research questions and design can offer insight.

### 4.2 *Design education, design research: Expanding boundaries of interrogation and knowledge: People embracing processes, processes sharpening design insights, interpretation and problem solving*

People are movers of design through their knowledge and abilities. The *nature* of things, of itself, is design because the world operates in a way that we say things happen *by design*. In these insights, we bring together people, processes and philosophy (occurrence in nature as *by design*). Nature, however, is greater than people and so people benefit by taking lessons from it. People have to seek and construct tools that enable them to see the lessons or insights of nature which can help them interpret the environment and how things are happening within it. In that way, people can learn how to design and manage human behaviour in the environment and other environmental phenomena. Design education and design research must recognize such a need in a world where design has suddenly assumed a central role.

## REFERENCES


Alpert, M. I., & Anderson, T. A. (1973). Optimal heterophily and communication effectiveness: Some empirical findings. *The Journal of Communication, 23*, 328–343. https://doi.org/10.1111/j.1460-2466.1973.tb00952.x

Centola, D. (2010). The spread of behavior in an online social network experiment. *Science, 329*, 1194–1197. DOI: 10.1126/science.1185231

Centola, D, & Macy, M. (2007). Complex contagions and the weakness of long ties. *American Journal of Sociology, 113*, 702–734. Retrieved from https://www.jstor.org/stable/10.1086/521848

Chang, C., & Lyuu, Y. (2009). Spreading messages. *Theoretical Computer Science, 410*, 2714–2724. doi:10.1016/j.tcs.2009.03.032

Gatignon, H., & Robertson, T. S. (1985). A propositional inventory for diffusion research. *Journal of Consumer Research, 11*, 849–867. Retrieved from https://www.jstor.org/stable/2489212

Kaigler-Evans, K., Leavitt, C., & Dickey, L. (1978). Source similarity and fashion newness as determinants of consumer innovation. *Advances in Consumer Research, 5*, 738–742. Retrieved from http://acrwebsite.org/volumes/5695/volumes/v05/NA-05

Larson, J. M. (2017). The weakness of weak ties for novel information diffusion. *Applied Network Science, 2*: 14. DOI 10.1007/s41109-017-0034-3

Rogers, E. M., & Bhowmik, D. K. (1970/71). Homophily-heterophily: Relational concepts for communication research. The Public Opinion Quarterly, 34, 523–538. Retrieved from https://www.jstor.org/stable/2747708

Sahin, I. (2006). Detailed review of Rogers' diffusion of innovations theory and educational technology-related studies based on Rogers' theory. *The Turkish Online Journal of Educational Technology, 5*, 14–23. Retrieved from https://files.eric.ed.gov/fulltext/EJ1102473.pdf

Statistics Finland (2018). *Statistical databases*. Retrieved from http://stat.fi/tup/tilastotietokannat/index_en.html

Stephen, A. (2019). Steering gently: Crowd management with a non-confrontational philosophy. In *The Design Journal, 22 (Supp. 1): Running with scissors: Proceedings of 13th International Conference of the European Academy of Design*. Available: https://doi.org/10.1080/14606925.2019.1594938

Watts, D. J. (2002). A simple model of global cascades on random networks. *Proceedings of the National Academy of Sciences of the United States of America, 99*, 5766–5771. Retrieved from http://www.jstor.org/stable/3058573

Weng, L., Menczer, F., & Ahn, Y. (2013). Virality prediction and community structure in social networks. *Scientific Reports, 3*, Article # 2522. DOI: 10.1038/srep02522

Zhou, C., Zhao, Q., & Lu, W. (2015). Impact of repeated exposures on information spreading in social networks. *PLoS ONE, 10*, e0140556. doi:10.1371/journal.pone.0140556

*Research & Education in Design: People & Processes & Products & Philosophy – Almendra & Ferreira (eds)*
*© 2020 Taylor & Francis Group, London, ISBN 978-0-367-49296-0*

# A framework for introducing emerging technologies in design studio classes

M. Lewis
*The Ohio State University*

ABSTRACT: This paper presents a framework for introducing emerging technologies into design studio classes. After an overview of motivations, three projects are described including their structures, typical outcomes, successes, and challenges. The first project exposes students to different conceptions of emerging technologies as well as a generative design process. Visualizations of possibility spaces are created that communicate students' overlapping research interests. The second project challenges students to engage with unfamiliar technologies of their choosing. Student attention is directed toward their learning process and hurdles encountered, more than toward what they are able to design. A final project presents both strategies for prototyping emerging technology systems as well as the use of emerging technologies for prototyping. Permission to take risks and document failures is strongly emphasized. The paper concludes with general observations and plans for improvement.

## 1 INTRODUCTION

The field of design education, like many others, wrestles with how best to adapt to the flood of emerging technologies. Adjacent disciplines in the arts, humanities, and engineering similarly are engaging with technology and education in myriad ways, both experimental and applied. The AIGA Designer 2025 report (Armstrong et al., 2017) observes that, "technology plays an outsized role in the future of design" and that, "the field risks losing influence to other disciplines if colleges and universities do not reset their expectations of design curricula." (pp. 2-3). This paper presents an emerging framework for giving design students at both graduate and undergraduate levels positive studio experiences integrating unfamiliar software and hardware into their design practices, regardless of their prior experience with technology.

The strategies discussed have been evolving in two different studio contexts over the past two years. At the graduate level, students with a wide range of technical backgrounds begin their design research MFA degree and start to explore how their research interests intersect with graduate-level expectations and possibilities. A second group of students are undergraduate design majors in their third year in a collaborative studio. The class is one of their few opportunities to work with students from design areas different from their own. Small collaborative groups are assigned such that a mix of students from different design backgrounds (i.e., visual, interior, and industrial) are on each team. Most of these students have limited exposure to the emerging technologies discussed.

The primary learning goals involve awareness of the capabilities of emerging technologies, confidence learning about them, and gaining a better intuition for what is easy or difficult. Students learn mainly through *making*, are "active in developing their knowledge" (Millis, 2012, p. 1), and "make uncertainty safe" and "embrace complexity" (Ambrose et al., 2010, pp. 180-181), moving forward despite the number of unknowns. Most design students are proficient at exploring solutions to interesting new problems within their domains of growing expertise, but they are often less comfortable in areas demanding problem *discovery* combined with unfamiliar tools.

Through the semester students are introduced to the existence of different currently emerging technologies. While they are provided with many examples, they are encouraged to discover their own usage instances rather than sticking to instructor provided material. The undergraduate teams are given an additional constraint: a specific facility on which to focus their design efforts. Their design ideas are generated with the goal of assisting a new campus institute that was created to facilitate collaboration between data analytics researchers. Design studio students visit technology spaces for tours and demonstrations. The undergraduates are in turn visited by guest stakeholders, including the individuals in charge of the data institute's operations, outreach and student engagement, space coordination, business development, project and resource management, strategic directions, and external partnerships. The design studio courses are strategically divided into three stages, described below, with roughly one month allocated for each project.

## 2 VISUALLY REPRESENTING POSSIBILITY SPACES

Students begin by exploring the intersections of their design interests and potentially relevant emerging technologies. The graduate students work individually and have (nearly) complete freedom for choosing research topics, subject to the categorical constraints below. The undergraduates working in their multidisciplinary teams additionally must negotiate compromises and combinations of their groups' members' potential interests.

### 2.1 *Structure*

Students are asked to start by creating a ***collection*** of three unrelated research topics. The first should be an emerging technology that they are curious about. The second topic is either a current social issue (for the graduate students) or an issue relevant to the needs of the research institute (for the undergraduates.) The third topic selected can be any additional design or research interest.

Once these three research topics are identified, each student or team is given a set of ***organization*** tasks: (1) identify a set of primary concepts for each topic; (2) research potential connections between all of the possible *pairings* of their three topics, discovering examples of properties, problems, and/or projects existing within these overlaps; (3) find existing problems and projects that connect all three topics.

Finally, students iteratively develop a visual ***representation*** of these overlapping possibility spaces. The students explore, consider, and communicate design research interests through iterative making, using whatever methods they are most comfortable with. Instead of writing a report, students use familiar tools (*not* unfamiliar technologies) to generate visual representations of their findings. Though the structure of the assignment lends itself minimally to a simple Venn diagram, students are instead encouraged to explore alternative visualizations of their data that reveal the complex relationships relevant to their emerging research interests.

### 2.2 *Making*

Many students select technologies currently in the news such as virtual and augmented reality (VR/AR), virtual assistants, or machine learning (ML). Some students are more curious about perpetually emerging technologies like 3D printing, computer vision, crowd sourcing, interactive displays, locative media, connected devices, projection mapping, or robotics. A few are particularly interested in broader technology-adjacent topics relevant to their interests, like data visualization or app development. Examples of social issues selected include climate change, inequality, mental health, and sustainability. Research institution issues include environment personalization, collaboration, and the availability, accessibility, efficiency, flexibility, and scheduling of shared spaces. Finally, a few individual research topic examples include entrepreneurship, journalism, narrative, curation, and community.

Students then explore ways of combining and connecting their unrelated chosen topics. If a student is interested in VR, but also healthcare, their research will quickly reveal a large space of projects using VR simulation for medical training and device prototyping. Other examples of topic overlaps include AR and wayfinding, personalization and connected devices, community and VR, and AI and public services. As a final step, the students are then able to investigate the intersection of all three of their topics. By first researching the intersecting spaces formed by each pair of topics, it is significantly easier for students to find connections between these already narrower spaces, than when they attempt to directly find links between all three high-level topics.

Students with architectural or industrial design backgrounds might create visual representations of these abstract possibility spaces using wood, paper, and string. Graphic designers make images and posters, and product designers might create 3D models (see Figure 1.) There are laser cut maps, twine chord diagrams, and musical data-to-sound mappings. Standard information visualization approaches are also common: word clouds/clusters, network diagrams, circular tree maps, arc diagrams, bubble maps, 3d printed heightfields, and concept heat maps. Images may show usage scenarios, for example, AR data visualizations in first or third person perspectives, created by interior designers.

### 2.3 *Evaluation*

Two related challenges are initially encountered by many students. The first is understanding that the target outcome is a *possibility space* visualizing many solutions, topics, and projects, instead of the identification of a single design problem. If this first issue is thought of as *zooming in too far* from the research topics to a single problem, the second challenge can be considered as the inverse: "What if I already know exactly what I want to do?" A student may start with a single complex design problem, then struggle to *zoom out*. Understanding that a single design problem can be represented as one point in the center of an abstract space of possibilities, this project can then be viewed as first choosing three orthogonal topic axes, and then considering other relevant projects, concepts, and connections existing in the surrounding spaces that are implicitly defined by these three dimensions. Students can begin to consider innovation in terms of the "adjacent possible" (Johnson, 2011).

When one of their interests is too broad (e.g., "healthcare") students quickly discover there are too many possible properties, problems, and projects to attempt to connect to their other topics. Instead of going narrower, a few students stick with the complexity and tackle the project of manually creating visualizations of very complex possibility spaces, then document their challenges with doing so.

Some students get stuck selecting a third topic because they cannot think of one that "fits well" with their first two. Students need to be encouraged to start with *independent* research topics and take a leap of faith that for *any* topics that interest them, they will be pleasantly surprised to discover myriad ways of fitting them together. In this context, the concept of generative design is introduced: developing processes that can be "run" to generate possibilities. "Generative" here is aligned more with the broader sense found in the term "Generative Art" as defined by Galanter (2003), rather than the narrower sense more recently employed by Autodesk ("What is Generative Design," n.d.).

## 3 UNFAMILIAR TECHNOLOGY

In a second project, students end up somewhere along a continuum between actually using an unfamiliar technology, to learning about a technology's capabilities by creating non-functional prototypes. While a few students use specialized university resources (e.g., VR headsets, 360-degree spherical cameras, or Bluetooth beacons) most rely primarily on their own hardware, downloadable software, and sometimes purchase relatively inexpensive kits or devices, e.g., a microcontroller and sensor kit, or a second-hand gesture tracker.

### 3.1 *Structure*

Most graduate students apply the technology they select to their previously identified research topics. Undergraduates apply the technology they choose towards the needs of the research institute facility. The undergraduate students are introduced in a hands-on workshop to several technologies unfamiliar to them that can also optionally be used for prototyping. WebXR ("WebXR", n.d.) enables browser-based distribution of VR/AR experiences. Students are specifically introduced to Sketchfab ("Sketchfab", n.d.), A-Frame ("A-Frame", n.d.), and Glitch ("Glitch", n.d.), as well as 360-degree spherical photography. A few additional constraints are given to the undergraduates to provide additional guidance. Their prototyping must include video showing classmates interacting with their simulated systems. They also need to design both a physical component and an online accessible component. It is emphasized that it is completely acceptable to *fail* to get a technology working successfully, as long as their investigations and learning are thoroughly documented.

The Designer 2025 report's (Armstrong et al., 2017) *complexity* section recommends that design students "visually map the interdependent relationships among people, places, things, and activities in a complex system." (p. 3). To be able to create and discuss their projects, students learn about creating *system diagrams* to visualize the components interacting in the complex systems they design. This helps students become self-sufficient by being able to break complex problems into manageable tasks. After identifying the relationships and parts within their systems, students choose which to explicitly build.

### 3.2 *Making*

Graduate students individually explore technologies they think might help with their thesis research. Many dive into learning the 3D interactive authoring software Unity ("Unity", n.d.) specifically for working with VR or AR (see Figure 1) Individuals may learn how to get their 3D abstract possibility space visualization into VR to walk through at architectural scale. Others are interested in screen-based AR (marker-based or markerless, sometimes interactive.) Graduate students with related technology interests (e.g., VR) optionally form a group with different individuals focusing on different aspects. VR may involve 360-degree spherical photo backgrounds, hand tracking, interactive physics, and spatialized sound. Students interested in machine learning vary in their technical backgrounds and thus pursue different paths. They may investigate programming libraries, demo applications, or interactive web tools. Applications of machine learning include text generation, object recognition and tracking, and style transfer. Other technologies explored include microcontroller and sensor circuits, Bluetooth beacons and iOS app programming, data mapping, and web development.

Technologies undergraduate students explore or simulate include augmented reality, indoor tracking, voice interfaces, face recognition, identity tracking, and smart networked objects. The WebXR technologies introduced are applied in several ways. Some project documentation is shared via linked, modified, and annotated 360-degree spherical photo environments served online. Screen recordings are captured of students pretending to interact with web-based AR 3D objects as if the objects were either physically present or virtually visible to them. Basic Photoshop or After Effects skills are also used to present first person or third person user journeys with simulated AR glasses showing menus or virtual wayfinding guides, people and space virtual annotations (potentially interactive:

Figure 1. Research possibility space virtual sculpture by Sana Behnam (left). Greenhouse gas awareness AR poster by Rhys Gruebel (middle). Mental health smartphone app prototype by Maya Jenkins (right).

e.g., contact this person or reserve this room) providing information, or revealing animated virtual assistants. Stop motion or concealed students are used to animate physical objects to show large scale intelligent object movement concepts (e.g., reconfigurable rooms and intelligent objects.)

### 3.3 *Evaluation*

Students require encouragement (or deadlines) to embolden them to shift from tutorials and videos to making. This transition may begin (but hopefully does not end) with tutorial content substitutions personalized to their topics. Discovering what they can actually create using a technology like VR, AR, or ML in only a couple weeks is a revelation to many. Many students select the same technologies but investigate them at different levels. They may be more interested in understanding *how* it works, or what it can be used for. For machine learning (ML), students might be more interested in the statistics involved, or how to work with ML programming libraries, or using software providing ML techniques.

Undergraduate teams more commonly succeed in *simulating* technologies of interest using prototypes but, along the way, learn additional technologies in creating their simulations. Their emphasis on prototyping over functionality is likely the result of instructor guidance and their more constrained project requirements. This can increase the tension between the self-imposed pressure to create finished design solutions, versus taking the opportunity to learn new skills. Only a couple students have been completely frustrated and unsuccessful getting a specific technology to work. They do however learn a great deal about what is involved with that technology, including terminology, processes, and concepts involved. They have usually learned more than students who merely followed tutorials substituting their own content. The "unsuccessful" students are inclined to feel like they failed, but it has been important to reward their risks, acknowledge the difficulties, and foreground their successes.

## 4 EMERGING TECHNOLOGY PROTOTYPING

The third project gives students the opportunity to focus on prototyping with emerging technologies of their choice. Graduate students work primarily in the space of their evolving design research topics, while undergraduates designed ways the research institute could move forward, in addition to forwarding their own design education goals.

### 4.1 *Structure*

The subject of the third project offers multiple alternatives, with each individual or team choosing which would be most productive for them. They may choose to return to visualizing spaces of design opportunities, or continue to engage with unfamiliar technologies, or they can focus on building prototypes in service of their selected research topics. More importantly, they can explicitly combine these options. The prior two projects likely ended with students feeling that they had just gotten started but ran out of time. Projects end with discussions of what students would do next *if* they had more time. The third project provides that opportunity, if they choose. Having seen what classmates are able to accomplish, they likely also have ideas for technologies or applications that they too would like to try. This final project gives them an open opportunity to do so. It is strongly suggested that students take the chance to learn something that interests them *and* will move their design skills forward.

To lessen the anxiety that often accompanies *complete* freedom, the undergraduate teams are again directed to prototype emerging technology systems that could be built and deployed within the needs of the data research institute. They are asked to consider possible issues and benefits for different populations within the community. There remains a clear emphasis on *making*, with a balance between prototyping, functionality, speculative concepts, and research.

They are told that their design processes, decisions, potential benefits, and concerns should be communicated clearly.

### 4.2 *Making*

For the selected ***technologies***, a few students choose to learn about current mobile and web app development tools, while the majority investigate forms of virtual, augmented, or mixed reality (XR) technologies, including sight, sound, touch, and even scent. Additional technology interests include interactive physical/digital objects, wearables, smart connected furniture, machines, and spaces with interactive displays and surfaces. These involve varying combinations of machine learning, AI, and computer vision.

The ***applications*** selected for these technologies are diverse. Social challenges selected include rural poverty, sustainability, education, and public policy. A significant number of projects intersect with health care topics such as stress, pain, aging, loneliness, and depression. Designing for the data research center generated workplace ideas for collaboration and resource allocation including reserving, optimizing, and finding available rooms or seating. Design processes also serve as application areas including interior design, co-design processes, information and data visualization, and UI/UX design.

For ***prototyping***, students create first or third person images, video, and animations of scenarios using standard commercial image, video, and 3D software. App interfaces are developed using familiar UX software (e.g., XD, Axure, Principle, or Sketch). VR/AR experiences are primarily built with Unity ("Unity", n.d.) though other open source or online solutions are used, e.g., Sketchfab ("Sketchfab", n.d.). Physical elements are created with cardboard or 3D printing. Larger physical objects (e.g., reconfigurable walls and whiteboards) are pushed by students as if motorized and intelligent. Transparent sheets simulate AR. Slideshows and animation coordinated with choreographed gestures appear as functional app interfaces. Screen recordings of augmented reality apps simulate AR glasses. Inexpensive "Makey Makey" ("Makey Makey", n.d.) keyboard simulation devices make objects touch sensitive, while projection mapped surfaces act as displays. Finally, actors, app mockups, and virtual content combine to give the impression of interactive user experiences (see Figure 1.)

### 4.3 *Evaluation*

While emphasis is on *learning by making*, it remains difficult for some students to apply newly acquired skills to their own topics without specific requirements. When they inevitably encounter problems, additional class structures to encourages cross-group collaboration would be helpful. Sometimes similar problems have been encountered and solved by individuals on other teams, but the instructor is sometimes the only bridge making connections between groups. Requiring regular journal updates and participation in a shared online discussion forum can help with cross-group problem awareness. Through such sharing, students also discover new tools that their peers from adjacent design areas use routinely. A few students then may choose to work with technology that is unfamiliar to them but well established within other disciplines. "Emerging" is, in this respect, quite subjective. Regardless, documenting their journey is considered more important in the class than "selling" their final product. This seems to be an atypical request for many students, and often requires some convincing. Students seeing their peers' learning processes remains more valuable than critiques of final creations.

## 5 CONCLUSION

Traversing the above structure, students acquire basic skills needed to continue learning on their own, as both their interests and technology shift. Students are given opportunities to discover what is easy or difficult, and more importantly, that this frequently misaligns with their intuitions. They are excited to learn how quickly they can go from discovering new technologies to, in most cases, actually using them. They become confident that they could do so again, exhibiting a *growth mindset* (McGuire, Angelo, and McGuire, 2015) that many may not have had at the start of the class.

To adapt to accelerating technological change, it is less important to reward students for producing exceptional designs using skills they already have than it is to make them comfortable with the risks of creating the unpolished work of beginners. As Kevin Kelly (2017) points out, this will become the new norm as we all "remain in the newbie state...because the cycle of obsolescence is accelerating...you won't have time to master anything before it is displaced." (p. 11). Giving permission to fail has been paramount in encouraging experimentation. Students are willing to jump into the unknown once convinced that they are *allowed* to fake anything they cannot actually make, *reveal* their challenges, and *adapt* as needed.

Students express relief at the freedom to attempt things that they cannot know in advance whether they will be able to get working. Prioritizing breadth of investigation over perfectly crafted functional solutions has been critical. Warnings that students will only value assignments with portfolio-ready outcomes fortunately have been unfounded. Students currently have many opportunities for portfolio building and welcome chances for personalized learning. Just as important, students are encouraged to allow their *initial* work with new technologies to *emerge* when possible from the capabilities they discover, instead of first choosing what to make and then struggling to implement that goal. It is striking how significant this shift can be for gaining confidence with new techniques and technologies. I have avoided emphasizing student final outcomes in both this paper *and* the classes, instead focusing on attention to processes and decisions. Formally evaluating learning outcomes will require studying the ease with which students engage unfamiliar emerging technologies in the future.

Many of the technologies mentioned so far involve collecting or accessing data about people, places, and things in one way or another. They may track or control resource access, usage, and availability. Conversation, presence, emotion, and identity all can be treated as data. The origins of nearly all such technologies can be considered in the context of what Zuboff (2019) calls "surveillance capitalism." Design has an increasingly unique responsibility for the impact of practitioners' efforts as user experience and interfaces are shaped. The Designer 2025 report (Armstrong et al., 2017) stresses, "data-aware devices open new avenues of design research into patterns of human activity, while at the same time raising questions of privacy, transparency, and trust that designers must address." (p. 2). There is an emerging flood of thought-provoking writing imploring designers who work with technology to give greater consideration to the potential ethical impacts of their efforts. But where to place technology ethics in the design curriculum? In a specific course, as in engineering? Should it be covered first before design work begins? In every class? Even threaded through each topic? Addressing these questions needs to shift from being a supplementary topic (as in this paper) to the front of the queue, and is the next priority for this project.

The strategies above are entering their third year of development. Prototyping this approach is still resulting in significant changes each semester, with new groups of students and the accessibility of technologies continuing to change as well. From one perspective, software that required lab workstations a few years ago now runs on most student laptops, and is making its way to mobile devices, browsers, and even microcontrollers. On the other hand, there will *always* be emerging technology requiring arbitrary resources, slowly shifting through such hierarchies. For academic research and development efforts, there are always many options when selecting and implementing technologies. Should a given technology be able to run in a web browser? On a mobile device? From the command line? Require programming? Which students will be able to use these tools on their own? After they leave the university? Answering questions like these impacts choices about software and licensing, equipment expense thresholds, and interface accessibility. The concepts and considerations that have been discussed will hopefully prove to be useful to those introducing emerging technologies into design studio classes. It is further hoped that this paper will encourage ideas and dialog about the remaining challenges mentioned, as we work to develop design curricula that engage with technologies for the future.

## REFERENCES


Ambrose, S. A., Bridges, M. W., DiPietro, M., Lovett, M. C., & Norman, M. K. (2010). *How Learning Works: Seven Research-Based Principles for Smart Teaching*. John Wiley & Sons.

A-Frame. (n.d.). Retrieved July 21, 2019, from A-Frame website: https://aframe.io

Armstrong, H., Blume, M., Chochinov, A., Davis, M., Dubberly, H., Kincaid, K., … Yap, M. (2017, August 22). AIGA Designer 2025: Why Design Education Should Pay Attention to Trends. Retrieved from https://educators.aiga.org/wp-content/uploads/2017/08/DESIGNER-2025-SUMMARY.pdf

Galanter, P. (2003). What is generative art? Complexity theory as a context for art theory. In GA2003–6th Generative Art Conference.

Glitch. (n.d.). Retrieved July 21, 2019, from https://glitch.com/

Johnson, S. (2011). Where Good Ideas Come From: The Natural History of Innovation (Reprint edition). New York: Riverhead Books.

Kelly, K. (2017). The Inevitable: Understanding the 12 Technological Forces That Will Shape Our Future (Reprint edition). New York, New York: Penguin Books.

Makey Makey. (n.d.). Retrieved July 21, 2019, from Makey Shop website: https://makeymakey.com/

McGuire, S. Y., Angelo, T., & McGuire, S. (2015). Teach Students How to Learn: Strategies You Can Incorporate into Any Course to Improve Student Metacognition, Study Skills, and Motivation. Sterling, Virginia: Stylus Publishing.

Millis, B. J. & IDEA Center. (2012). Active Learning Strategies in Face-to-Face Courses. Retrieved from https://eric.ed.gov/?id=ED565290

Sketchfab. (n.d.). Retrieved July 21, 2019, from Sketchfab website: https://sketchfab.com

Unity. (n.d.). Retrieved July 21, 2019, from Unity website: https://unity.com/frontpage

WebXR Device API. (n.d.). Retrieved July 21, 2019, from https://www.w3.org/TR/webxr/

What is Generative Design (n.d.). Retrieved July 19, 2019, from https://www.autodesk.com/solutions/generative-design

Zuboff, S. (2019). The Age of Surveillance Capitalism: The Fight for a Human Future at the New Frontier of Power (1st edition). New York: PublicAffairs.

*Research & Education in Design: People & Processes & Products & Philosophy – Almendra & Ferreira (eds)*
*© 2020 Taylor & Francis Group, London, ISBN 978-0-367-49296-0*

# Research as a link between the 'what' and the 'how' in fashion design education

A. Neto
*CIAUD, Lisbon School of Architecture, Universidade de Lisboa, Portugal*

A.C. Broega
*University of Minho, Portugal*

ABSTRACT: While it has been widely recognized that the fashion industry has many negative impacts on our planet and communities, fashion design practice and education hasn't changed enough to reverse the situation. Sustainable fashion practices are more important than ever and we need fashion graduates to be prepared for that task. This paper compares the barriers for the integration of sustainability in the textile and apparel education identified in a study conducted in 2014 with the ones of the present day, recognizing that there isn't much difference in this 5-year period. We conclude by suggesting that Research for Design can mitigate the difficulties faced by educators when teaching the skills and competencies required for students to design fashion in a challenging world.

## 1 INTRODUCTION

The world population is facing the growing consequences of the contamination and depletion of natural resources, for which the fashion industry contributes significantly. Many fashion brands have started including sustainability on their agendas, which reflects the increasing consumer demand for ecologically and ethically cleaner clothes. Fashion sustainability has been approached by assessing the social and environmental impacts in all phases of the product lifecycle, after which action is taken in order to mitigate them. This is often aimed at a specific phase of the product lifecycle, material or process, as it can be overwhelming to tackle all the issues at once.

It is as difficult for businesses to manage all sustainability concerns of their activity as it is for universities to effectively ingrain sustainability in Fashion Design higher education courses, despite the urgent need for better equipped designers in the apparel industry. Common approaches are related to a switch to organic cotton or other natural fibres to avoid plastics, or even opting for recycled materials, a technological advance that is still trying to gain scale. Cosette Armstrong and Melody LeHew conducted a study in 2014 that pointed out several barriers to implementing a sustainable mindset across the whole curriculum in fashion higher education courses. In this paper we will note how five years after their study, the discussion remains very much the same (according to several 2019 related conferences), and we will suggest that research for design within the university can help feed the content regarding Fashion Sustainability, as well as enabling educators to foster in students the skills they may need to thrive in a world of uncertainty.

## 2 THE PERCEIVED CONFLICT BETWEEN SUSTAINABILITY AND FASHION

One of the main perceived barriers to make the shift in the fashion design discourse is the deep connection of economic growth and consumerism to the fashion business activity, the "central impediment being the conflict between sustainability and an emphasis on profit maximisation or sales increase" (Armstrong & LeHew, 2014). Indeed, as Professor Klaus Bosselmann explains, "at its core, sustainability relates to the basic human need to maintain and to nurture the conditions on which life depends", and yet, for most, the term is perceived as "the antithesis of development, a threat to progress and well-being" (Grober, 2012, Foreword, p.8), which makes it difficult to implement both in business and education contexts. Some may argue that we need to shift "from an orientation on profit to an orientation on people and planet" (Pasricha & Kadolph, 2009, p.125) but, in fact, this mindset perpetuates the perceived conflict between all the pillars that must be considered for a balanced human life. The economic viability must be there, and cannot be left out. If students do not address the economic side, the projects they develop in university context will lack a crucial link to reality. At the Beyond Next: Circularity Festival, held in Amsterdam in February 2019, one of the proposed challenges was how to equip fashion students for a new reality — a circular, sustainable one. This challenge was discussed among students, teachers and fashion professionals, and students still argued there's a lack of connection between university and industry.

This also leads to the discussion whether sustainability should be learned across the whole curriculum or as an add-on subject, the latter being the most common approach among schools (Pasricha & Kadolph, 2009). "A well-developed sustainability module added to a course that becomes a lens through which the other course content may be viewed, reframing the entire course" (Armstrong & LeHew, 2014, p.78). On the other hand, having an add-on subject may imply that what is being learned has no practicality, as students can miss the connections to other subjects, and more conscious practices may be perceived as ideal but unrealistic.

After graduating, young designers get into the industry and, instead of being capable to implement, whenever possible, the strategies and mindset they learned, they feel they need to start over, a new blank canvas where they learn the economic-led approach to their activity. This is why school experiences of sustainable fashion product development must be as close as possible to the corporate reality. This may mean, for instance, to create the full garment specs of what students create to assess the mass-production feasibility, which implies an understanding of production processes, costing and planning. Without a strong "reality check", a sustainable fashion project at school can raise awareness among students, but a) that awareness may not follow through the professional practice due to the lack of compatibility with the economic reality of the industry and b) it may lead to frustration and abandonment of the sustainable mindset learned in university, as students are not capable of finding opportunities for disruption that may appear within their professional context.

Thus, pursuing social, cultural and environmental balance in the design project cannot be made at the cost of economic viability. However, the economic growth and profit-led discourse can become more realistic in respect to our planetary boundaries, an example being Herman Daly's view on economic goals, which are (sustainable) scale, (just) distribution and (efficient) resource allocation (Daly, 1992). Furthermore, economic perspective in the design field can align with the notion that competitive advantage "rests not on static efficiency nor on optimizing within fixed constraints, but on the capacity for innovation and improvement that shift the constraints" (Porter & Linde, 1995, p.98), regardless of environmental regulations that may be imposed. Therefore, the idea of environmental and social issues as drivers to innovation within a viable economic context can be fostered among design students. A better framing of economic concerns alongside environmental and social issues has been promoted with the concept of circular economy, which opposes to the current linear model of make-use-dispose by cutting out waste and making a wiser use of materials while envisioning their future lifecycles (McDonough & Braungart, 2009). The Circular Economy as a concept has been mainly boosted into the fashion industry by the Ellen McArthur Foundation, and has been embraced by several globally-known fashion brands, making it even more relevant to introduce this mindset in the fashion design education discourse.

## 3 THE COMPLEXITY OF THE TASK AHEAD

Use t While there is urgency to transform the entire approach instead of simply "greening" the discourse, there are several challenges ahead, such as the time and effort needed to develop and implement alternatives (Armstrong & LeHew, 2014). During the last University Business Forum held in Lisbon in February 2019, participants promptly acknowledged that changing the curriculum in a disruptive way implies new accreditation, which is a barrier for deeper change. A participant noted that we must be careful not to change the curriculum specifically for the industry as it can become too narrow, pointing out that for that purpose "in-company learning" would be more suitable. Each part (business and education) have their own role and responsibility. While students need to learn on the basis of what will be expected of them as professionals, the education path must allow more time for reflection as part of the design process, "to help the student understand what has, and what has not, been a successful design outcome and why" (Gully, 2010, p.44). At the Beyond Next: Circularity Festival, speaker Gwen Cunningham, Lead for the Textiles Program at Circle Economy, used Aesop's fable "The Moon & Her Mother"[1] to question how universities can create an education model for an industry that is in constant change. In both conferences, it was recognized that this is too fast of a pace for Universities to be able to 'tailor' their courses to the needs of the industry, concluding that Higher Education actually will do best by developing courses that promote resilience, adaptability to changes and a high tolerance for ambiguity — the capacity "to deal with novel, complex, or seemingly unsolvable tasks" (Robinson et al., 2018, p.1) —, also stressing the need for a flexible life-long learning system, also suggested by Deborah Andrews (2015).

## 4 THE BLURRY BOUNDARIES OF FASHION DESIGN

This leads us to what many may consider the most manageable option of reframing contents within existing curriculum. But fashion design is related to an array of specialised disciplines, and it can be difficult to articulate sustainability within all these areas. Furthermore, when sustainability is approached in the different subjects, different aspects of it are covered and others are left out, as some subjects are more prone to discussion than others (Armstrong & LeHew, 2014). Indeed, fashion design knowledge is built upon a huge amount of different areas, related to: understanding human behaviour on garments consumption and use through history, psychology and sociology; knowledge of textiles and garment production through sewing, fabric cutting, pattern cutting, manufacturing efficiency, costing, textile engineering, fibres and fabrics; and research and drawing as the primary inputs and outputs of the design process. Furthermore, students generally complain they lack tools related to all aspects of business such as operations management, sourcing, buying, marketing, communication and sales (Armstrong & LeHew, 2014; Amed & Mellery-Pratt, 2017). Still, learning all these matters doesn't mean the designer must be a specialist in all the other subjects, as design has its own territory. But, indeed, when making a design decision upon a tip of the iceberg, the designer can't ignore the submerged part of that iceberg. In other words, during decision-making, designers can't be oblivious to the context in which that decision will take place, nor the consequences it will bring (Andrews, 2015). On generic terms, it means a lifecycle

[1] The Moon once begged her mother to make her a gown. 'How can I?' replied she; 'there's no fitting your figure. At one time you're a new moon, and at another you're a full moon; and between whiles you're neither one nor the other.' – retrieved from a 1994 edition of Aesop's Fables, published by Wordsworth Editions Limited.

assessment is needed, not for every designed garment, but for every garment that is being designed. This involves all the disciplines mentioned above. The designer, as a problem-solver and a developer of products and systems will definitely benefit from a broader knowledge — the alphabet — of the social and scientific areas in which these products or systems will exist and other areas which they will connect with and impact. "Designers have a role in exploring new territories to design and make the futures we dream of, rather than making what we have less bad." (Williams, 2013, n.p.)

On the present day, both professionals, educators and students must keep in mind the biggest challenges that lie ahead: The UN has set 17 Sustainable Development Goals to be met by 2030. How can this translate into fashion design decisions? The European Commission is calling out for climate neutrality in Europe by 2050. How can fashion design follow suit? The Paris Agreement central aim is to keep "a global temperature rise this century well below 2 degrees Celsius above pre-industrial levels and to pursue efforts to limit the temperature increase even further to 1.5 degrees Celsius" (United Nations, 2015, p.3). How does this impact on fashion design? While at first, these concerns may look far from the fashion design discipline, it is widely understood that the design phase of a garment determines most of its social, economic and environmental impact, which highlights the importance of design when tackling the challenges stated above, as addressed in the Fashion Industry Charter for Climate Action (2018).

## 5 BEYOND THE KNOW-HOW IN FASHION DESIGN

Current fashion courses have been receiving a mix of Millennials and Gen-Z students, a youth (and young adults) with some key traits in common: they're hyperconnected and have a strong sense of self boosted by social media; they are more aware of social and environmental issues and they value activism, transparency and inclusivity, far more than the previous generations; they have been facing difficulties with traditional systems of education and workplace that aren't flexible to their fluid thinking, which is seen from outside as an ever-decreasing attention span. (O'Connor, 2018; Salfino, 2018)

In their study, Armstrong and LeHew (2014) identified some barriers to fashion sustainability education that were related to students, such as the unwillingness to accept content perceived as being outside of their field boundaries; the burden that comes with going through complex and often scientific issues, often leading to a depressing inertia, doesn't match the perceived idea of fun, creativity and glamour related to fashion; they also lack critical thinking skills (to think in a nonlinear way and solve complex problems), foundational understanding and awareness about global issues, specifically social responsibility and ecological literacy.

Fast forward to 2019, and teachers participating in the Beyond Next: Circularity Festival noted that there is a lack of resilience and focus among students, arguing that they are not involved in the real-world impact of their decisions and are not connected to the real world innovation. Furthermore, when faced with the immensity of the issues, students were said to feel powerless and hoping to overcome this feeling with the expectation that the uncomfortable learning experience would be soon followed by a 'how-to guide' with immediate solutions, in form of readily applicable practical knowledge brought by educators. During her presentation at the festival, Gwen Cunningham (2019) argued that that would be trading possibility for certainty, while Dilys Williams and Katelyn Toth-Fejel (2017) deem it to be insufficient to deal with the complexity and unpredictability the future holds. It makes it even more challenging for students to realize there are no quick fixes, nor certainty on how to navigate through the fog. Cunningham (2019) reframed 'change' as being part of the system, rather than a bug or anomaly we wish to extinguish, suggesting that we need courage, curiosity and humility to work with it. Williams goes further in predicting that, for students to be prepared to face the instability and uncertainty of a world affected by climate change, "imagination, ingenuity, improvisation, empathy, the ability to contribute to and shape convention, will be at least as important as technical know-how in design, materials and processes." (Williams, 2013, n.p.)

At the University Business Forum, participants recognized the urgent need for students to learn skills such as active listening; communication; problem understanding, analysis and solving; critical thinking and resilience, which were all considered entrepreneurial skills. But how? A participant suggested this must be embedded in the curriculum, as extra-curricular activities may be missed opportunities for many students. Another participant argued that when trying to embed too many things on their short-timed program, other contents are left out, for instance theory, which mustn't happen, as educators already feel students lack theory grounding to evolve in their learning process. In the end, despite the general acknowledgement of the need of all these skills and aptitudes to be learned in higher education, questions were kept unanswered on how to teach, assess and recognize them.

Partnerships between universities and the industry were frequently pointed out by participants on both Conferences as a means to improve the practice of students, professionals and educators alike, which had already been suggested in previous studies (Armstrong & LeHew, 2014; Williams & Toth-Fejel, 2017). While field trips are not an uncommon way to link both worlds, it is barely enough for each part to understand the development opportunities that come with cooperation in creating a shared vision and "explorative space to realise present and future necessities and possibilities" (Williams & Toth-Fejel, 2017, p.80). One of the techniques suggested was the use of challenges created by brands to be solved by students. At the Beyond Next: Circularity Festival, a teacher explained how this happens in The Amsterdam Fashion Institute, while a speaker at the University Business Forum, Mr. Ruben Habraken, explained how it is done at the Erasmus Centre for Entrepreneurship in Rotterdam, Netherlands. The two participants clearly pointed out that in this problem-based learning scenario there's a need for the Educational Institution to manage the expectations of both industry and students: as they suggested, this can be achieved, on the one hand, by clarifying the companies on what kind (and level) of goals are intended for the students, within a specific timeframe, while conversely specifying to students the importance of the process and clarifying the rewards that they may expect (for instance, credit points rather than money, entrepreneurial guidance to develop a business rather than a job placement at the company involved). Williams & Toth-Fejel (2017) identify the tension between the short term problem solving often desired by the industry and the need for time to question and reflect in the learning process, and recognizes that "for universities and businesses to investigate deep change requires us as academics and businesses to undertake the same transformative processes that we encourage in our students" (Williams & Toth-Fejel, 2017, p. 82)

## 6 AN OVERWHELMING TASK FOR EDUCATORS? BRIDGING RESEARCH AND EDUCATION

One of the most important barriers to effective ingraining of sustainability in fashion design higher education was identified by educators regarding their own practice as they claim lack of time, money, resources, support and incentive to engage in a more structural change on the content they teach. Lack of interest in the subject and resistance to change were also mentioned by some educators; research activity as the main path to academic career development, which is time-consuming, was also pointed out as a barrier, besides the lack of a common ground around the concepts among all educators on the course (Armstrong & LeHew, 2014).

In the midst of their own uncertainty on an ever-evolving fashion industry, educators are confronted with the need to:

1. Make the most up-to-date knowledge available to their students, to foster knowledge-based decisions. An example is the common notion that not enough different fibres are being used in fashion: universities could make the bridge between their students and industry innovation, by supplying (or facilitating access to) innovative materials for students to work with, as dealing with physical materials increases knowledge of their real performance and suitability for each design project, more than theoretically knowing about their existence. This practical knowledge can drive further demand for these diverse and potentially

better materials, promoting production scale and cost decrease. In a wider picture, educators must promote in students the social responsibility and ecological literacy much needed to our present and future well-being, while helping them to frame these inputs together with economic viability to aim for sustainable development.

2. Map out together (among educators of the same course) who covers what, to ensure a systematic education on the matter. Educators need to improve the connections between disciplines and help students map the relations across different subjects, to foster problem solving as 'How can this translate into fashion design decisions?'. A simple example may be the subject of fair trade and what it means to fashion designing, when supplier selection and order negotiations are often made by the commercial and sourcing departments, rather than the designer. Students might feel it is an issue out of their reach, but it is important to understand the implications of designing an overly complicated style to match a very low price point (that adds to the supplier pressure for high labour intensity at a low cost) or the implications of last-minute changes to a given style, which may mean replanning production and costs for the supplier with the added pressure of meeting established deadlines, possibly leading to subcontracting unregulated manufacturing services.
3. Nurture in their students the "skills and competencies that may not yet be marked on job descriptions or interview questions" (Williams, 2016, p.2) but are already recognized as needed to successfully navigate through an uncertain world. If coaching soft skills is to be provided during the different projects along the course, it implies understanding how to coach and assess them. This may also include managing students' expectations and emotions, such as the powerlessness and depression when dealing with the issues and lack of confidence to make and own their decisions.
4. Provide students with the tools that can help them connect theory and practice, bridging sustainable values and feasible actions, desirably translated into a 'navigational system' (Williams & Toth-Fejel, 2017). This is closely connected to the skills required to better translate theoretical and technical knowledge into decision-making during the fashion design project. It is also important for educators to help students understand the richness of their professional field and the influence they can have within their role in a corporation.

No wonder educators may feel overwhelmed with the task at hand. This is where academic research groups are crucially needed. Academic research for design can keep educators up to date and feed curriculum content directed to different subjects, while aiming for cohesion across all disciplines of fashion education and providing guidance for students. Furthermore, researchers within universities can help educators to find new techniques to foster soft skills and assess them alongside their students' knowledge-led achievements.

## 7 CONCLUSIONS AND RESEARCH OPPORTUNITIES

The future is never a blank space ahead of us; it is littered with what the past has thrown in. (Willis, cited by Williams & Toth-Fejel, 2017, p. 85)

Current barriers to ingrain sustainability in fashion design education curriculum aren't much different from those identified five years ago. And yet, responsible fashion design is more important than ever to promote sustainable production and consumption systems. Students need strong technical knowledge together with the capacity to make the connections between all the fields that converge to fashion design territory. Moreover, they need soft skills that can help them navigate through the uncertainty of their professional future. Higher education institutions, and mainly fashion and apparel educators, need to provide the knowledge, skills and tools for students to become socially responsible and ecologically literate and be able to translate this knowledge and values into their design projects. There are several barriers for educators to be able to systemically implement sustainability in fashion design courses, many of which amount to lack of time to research everything about content and teaching methods. Researchers, therefore, have an important role to play: their research can provide new guidance for teaching and learning. Further research areas that are closely connected with fashion design education include:

1. Knowledge update, directly related to the fashion field, to feed content that educators provide to students. This means being aware of innovation that is taking place regarding materials and processes. But it also means being aware of the fast shifts that are happening now in consumer behaviour and how the industry is responding to them. It means researching into latest opportunities and challenges among the industry, and how that can influence design decisions, yearly, monthly, weekly. This way, not only will students learn theory and history, but also understand the fast pace of the industry they will get into, make constant connections and become innovation-savvy.
2. Knowledge update, related to new ways of coaching and assessing soft skills needed to design for a changing world. Critical thinking, active listening, communication, problem analysis and solving, adaptability to change, high tolerance for ambiguity, courage, curiosity, humility, imagination, ingenuity, improvisation, empathy, the ability to contribute and shape convention, resilience. What do each of these mean within the fashion designing process? How can each of these be purposefully incorporated in the fashion design assignment? How can they be assessed? How can knowledge on climate change be passed on to students as an opportunity for better design instead of as a burden with no manageable clear solution?
3. Development of a navigational system that students can 'unlock' along their path through the three cycles of design education, "a means to start from different places and undertake a range of journeys according to participant skills, time and scope for risk or ambition. Such a navigation system might bridge current contexts with future requirements" (Williams & Toth-Fejel, 2017, p. 85) and promote a confident and strong knowledge-based and values-led decision making in their design projects. Again, this shouldn't be seen as a 'how-to-guide', but the set of practical knowledge and skills that will become foundational for the development of change-makers.

Moving towards a sustainable fashion system requires not only openness but also willingness to engage in a fast-paced, lifelong learning journey. This is true for fashion students and practitioners as much as it is for educators. Academic research can and must fill in the current gaps, which will allow fashion educators to confidently improve not only the contents they teach, but the methods to do so.

## REFERENCES


Amed, I. & Mellery-Pratt, R. (2017, September 28). Is Fashion Education Selling a False Dream? *Business of Fashion. Retrieved* April 10, 2019, from https://www.businessoffashion.com/articles/education/global-fashion-school-rankings-2015

Andrews, D. (2015). The Circular economy, design thinking and education for sustainability. *Local Economy*, *30*(3), 305–315. DOI: 10.1177/0269094215578226

Armstrong, C. & LeHew, M. (2014). Barriers and Mechanisms for the Integration of Sustainability in Textile and Apparel Education: Stories from the Front Line. *Fashion Practice*, *6*(1), 59–85. DOI: 10.2752/175693814X13916967094830

Cunningham, G. (2019). Speaker presentation. Beyond Next: *The Circularity Festival*. 08 February 2019.

Daly, H.E. (1992). Allocation, distribution and scale: towards an economics that is efficient, just and sustainable. *Ecological Economics*, *6*, 185–193.

*Fashion Industry Charter for Climate Action*. (2018). Available from https://unfccc.int/sites/default/files/resource/Industry%20Charter%20%20Fashion%20and%20Climate%20Action%20-%2022102018.pdf

Gully, R. (2010). Cognition and Process vs. Design Artifact in Fashion Design Pedagogy. *Melbourne Cumulus Working Papers*, Publication Series G. Aalto University.

Grober, U. (2012). *Sustainability: A Cultural History*. Totnes, UK: Green Books.

McDonough, W. & Braungart, M. (2009). *Cradle to Cradle: Re-making The Way We Make Things*. London, UK: Vintage.

O'Connor, T. (2018, November 30). Navigating Gen-Z: The ABCs of Fluidity. *Business of Fashion*. Retrieved April 16, 2019, from https://www.businessoffashion.com/articles/video/gal-dem-recens-wallet-navigating-gen-z-the-abcs-of-fluidity

Pasricha, A. & Kadolph, S. (2009). Millennial generation and fashion education: a discussion on agents of change, *International Journal of Fashion Design, Technology and Education*, 2:2-3, 119–126, DOI: 10.1080/17543260903390183

Porter, M. & Linde, C. (1995). Toward a New Conception of the Environment-Competitiveness Relationship. *Journal of Economic Perspectives*, *9*(4), 97–118.

Robinson, J., Workman, J. & Freeburg, B. (2018). Creativity and tolerance of ambiguity in fashion design students. *International Journal of Fashion Design, Technology and Education*. DOI: 10.1080/17543266.2018.1516807

Salfino, C. (2018, July 31). Will Millennials Force Apparel Industry into Transparency? *Sourcing Journal*. Retrieved April 16, 2019, from https://sourcingjournal.com/topics/lifestyle-monitor/millennials-apparel-transparency-113786/

United Nations (2015). *Paris Agreement*, available from https://unfccc.int/sites/default/files/english_paris_agreement.pdf

Williams, D. (2013). Fashion Education in Sustainability: Change Through Experiential Crossings. In: *TEESNET Global Citizenship Sixth Annual Conference*, 11 July 2013, London South Bank University. Retrieved from http://ualresearchonline.arts.ac.uk/6305/1/TESSNET_Dilys_Williams_-_Fashion_Education_in_Sustainability.pdf

Williams, D. (2016). Transition and Transformation in Fashion Education for Sustainability. In: Filho, W. & Brandli, L. (eds.). *Engaging Stakeholders in Education for Sustainable Development at University Level*. World Sustainability Series. Springer. 217–232. DOI: 10.1007/978-3-319-26734-0_14

Williams, D. & Toth-Fejel, K. (2017). The Will and the Skill in Education for Sustainability. In: *Handbook of Theory and Practice of Sustainable Development in Higher Education*. Springer, 79–94.

*Research & Education in Design: People & Processes & Products & Philosophy – Almendra & Ferreira (eds)*
 *ISBN 978-0-367-49296-0*

# Ecoliteracy: Shaping the design process from a systems-based perspective

R. Lopez-Leon
*Autonomous University of Aguascalientes*

A.G. Encino-Muñoz
*University of Leeds*

ABSTRACT: Multiple literacies have emerged in the 21st Century broadening the meaning of literacy. These literacies describe the professional skills required to meet the challenges of contemporary and future society. This paper focuses on Ecoliteracy as a skill to understand the organizational principles of ecological communities in order to be able to apply these principles to the improvement of human communities. Design as a discipline plays a fundamental role in the development of such communities, and so, this paper presents a systems-based design process developed from Ecoliteracy perspective, reflecting upon how ecoliteracy could nurture the design process and aid in developing educational strategies to train eco-aware professionals.

## 1 INTRODUCTION: UNDERSTANDING ECOLITERACY

The term Ecoliteracy was coined by Fritjof Capra to define the comprehension of "the principles of organization that ecosystems have developed to sustain the web of life" (1999, p. 1). He describes that being ecoliterate means "understanding the basic principles of ecology and being able to embody them in the daily life of human communities (1999, p. 2). Therefore, being ecoliterate implies, not only knowing ecological systems but a deep understanding of how they work in order to build communities aligned with the same characteristics.

The term emerges as an answer to environmental problems, but within its core, it embraces not only ecological concerns but also human and social ones. This focus implies a difference between environmental literacy –which is often linked to environmental awareness and concern-; Ecological literacy –which is focused on ecological knowledge to make the best decisions based on sciences and systems thinking- and Ecoliteracy (McBride et al. 2013). For this paper, it is not essential to deeply explore conceptual differences of these three terms, focusing only on the elements that Ecoliteracy comprises and how they can nurture the design process.

Capra (1999) emphasized the importance of understanding ecological principles to build sustainable communities and transforming educational systems. According to Capra's perspective: "an ecoliterate person is prepared to be an effective member of a sustainable society, with well-rounded abilities of head, heart, hands, and spirit, comprising an organic understanding of the world and participatory action within and with the environment" (McBride et al. 2013, p.14).

Systems thinking is one of the cornerstones of ecoliteracy, setting principles of connection and interdependence as main values for this philosophy. According to its perspective, linear thinking has to be left behind and move, from studying separate parts of a problem to the whole, from objects to relationships, from quantity to quality, from structure to process (Stone & Barlow 2005, p. 21). From this perspective, ecoliteracy "means to apply the principles of relationships, connectedness, and context" (Semetsky 2008, p. 1).

Becoming ecoliterate means developing a set of competencies: cognitive, emotional, active, and connectional (McBridge 2014, p. 15), which can be described as follows:

Cognitive: It allows approaching issues from a systems perspective. This focus means taking into account invisible links and cycles. Also, it is essential to think critically to assess impacts and ethical effects of human actions.

Emotional: It is linked to empathy: encourages respecting living things, and committing to equity and justice.

Active: It is related to materiality and creativity. When it comes to materializing a project, it is essential to optimize resources by assessing and adjusting the use of energy and resources.

Connectional: On one hand, the connectional ability is related to personal experience by feeling a kinship with the natural world and trying to invoke this into others. On the other hand, it is also about feeling reverence for living things and connect strongly with places and organisms.

This set of competencies highlights two main challenges for design practice. On the one hand, it shows the urgent necessity to develop design methods and process that promote activating these competencies instead of blocking them or constraining their development. On the other hand, it emphasizes the gap between what design education is about nowadays and how design needs to be taught. A systems-based education promoting these competencies should impact in the curricula and in what kind of assignments and goals are set as classwork and projects.

## 2 CHALLENGING THE DESIGN PROCESS FROM COMPLEXITY

Since the design process is becoming more complex, several attempts to understand it better have been made, including those that aim to standardize or categorize its language (Roschuni et al. 2015), to those that reflect upon its scope and changing role within organizations, that usually move its focus away from producing objects (Muratovsky, 2015). For instance, in the emerging field of Experience Design, there are studies aiming to identify how to produce value through designing experiences (Kalbach, 2016; Mclellan 2000; Press & Cooper 2009). Camere and Bordegoni (2016) even proposed a model that could aid the designer in prototyping experiences as part of the experience-design process. Similarly, research on Service Design has gained notoriety, emphasizing, among other things, the need to consider intangible products as a result of the process (Secomandi & Snelders, 2011; Shostack, 1982; Stickdorn y Jakob, 2011). Moreover, Cascini (2012, p. 29) discusses the Theory of Inventive Problem Solving as a "reference methodology to support design activities driven by the forecasted evolution of technical systems," in order to enrich the design process of anticipatory design of future products and processes. Also, new collaborative practices have emerged, such as participatory or co-design (Lee, 2008) that aim to include in the design process, different people, and perspectives. These studies are only a few of the vast literature about different practices of the design process, highlighting the need for thinking and approaching differently to problems, requiring a systems perspective, and consequently, ecoliterate professionals.

A study of the design process could be traced back to the arts and crafts movement, and a line of evolution could be mapped to see how the definition of design has been changing, and its process has become more and more complex. The Design Council (TDC) sought to understand better the design process and how businesses are performing by it. By consulting eleven world-leading companies about how and when they bring designers into the development process of products and services, TDC identified common practices about the design process, synthesizing it into four stages: Discover, Define, Develop and Deliver. Moreover, the report presents a historical perspective of the design process, highlighting how it evolves when a new element is integrated. For instance, the transformation was evident when it changed from being a production centered to a user-centered process. Furthermore, this study argues that if ecoliteracy's principles were considered, the design process could evolve another step, from user-centered to become a systems-centered process. The role of the designer also stands out in the report, highlighting the need to evolve from individual to collaborative practices (TDC, 2007: 8), which concurs to a systems-centered approach.

According to Morin and Le Moigne (2000, 128), complexity has three principles that have been considered to develop a systems-based design process presented in this document:

Unite the opposing forces.

Consider internal movements in which the products and effects are themselves considered in the causal relationship.

Not only a part is in the whole, but the whole is in the part.

Ghislene and Franzato (2017), identified six constituents from complexity theory that can also be taken to ensure that the design process is holistic and impacting in the long-term.

A plurality of actors in cooperative and collaborative action. This is already perceivable from design practices such as co-design and participatory design. To incorporate this constituent, the design process has to maintain plurality throughout every stage, and not only when identifying needs.

Tangibility/intangibility in the creative paths. A systems-based design process should deliver tangible and intangible products. For instance, increasing interconnectedness is an intangible part of the process.

Inclusivity in the exogenous and endogenous circumstances. Ecoliterate designers become leaders that can understand out and inside organization-systems and how they affect each other. These readings must be considered in every stage of the design process.

Multi-diversity in the proposed relations. This constituent refers to the possibility of constructing different relations. Not only the resulting product of the design process should increase interconnectedness in existing relations, but promote establishing new ones as well.

Openness for the development of a given element, or lack thereof. Nowadays, there is certain tendency to perceive that design as a process does not mean to have an ideal methodology, instead, is to think about a flexible infrastructure "with the foresight and intelligence to respond quickly and appropriately to creative change" (TDC 2007, 10). This capacity of adaptation could be promoted through ecoliteracy because it enables the designer to understand how systems are changing and organizing themselves.

Systematicity in the relationships between all the terms involved. A systems-based approach does not mean disorganized. The challenge is maintaining structures while enabling flexibility at the same time. Systematicity means defining fixed nodes and layers for the design process, allowing its implementation, and adaptation to change and evolve. Finally, a system should resemble a network blurring hierarchies, pyramidal structures, and step-by-step models.

Two more constituents have been identified through other research:

Capability of Self Eco-organization. Autopoiesis is "the capability to change and adapt in light of changes by the environment without the system losing its identity" (De Mello 2017, p.91). However, this will only be achievable if it is incorporated in every stage of the design process. Resulting products should provide self-regulating strategies or become a means to rebuilt and maintain a living system.

Culture of Strategy. A culture within an organization is the critical thinking and constructive attitude that proposes new values and visions (Manzini 2015, 241). The culture of ecoliteracy emphasizes a problematic vision of the world, which allows to interpret contexts and recognize imbalances in society, culture, market, economy, and politics (De Mello 2017). This constituent encourages strengthening relations between the different actors of the ecosystem of innovation, thereby achieving collective systemic awareness.

The constituents mentioned above were considered to develop a systems-based design process presented in the next section.

## 3 BOUNDING ECOLITERACY: A SYSTEMS-BASED DESIGN PROCESS

The first stage of the design process from The Design Council (TDC) is the discovery stage, which, in a user-centered approach the designer would identify the needs of the user. From an ecoliteracy, systems-centered, approach, four nodes have to be addressed: interconnectedness, collaboration, endo/exogenous systems, and autopoiesis (Figure 1).

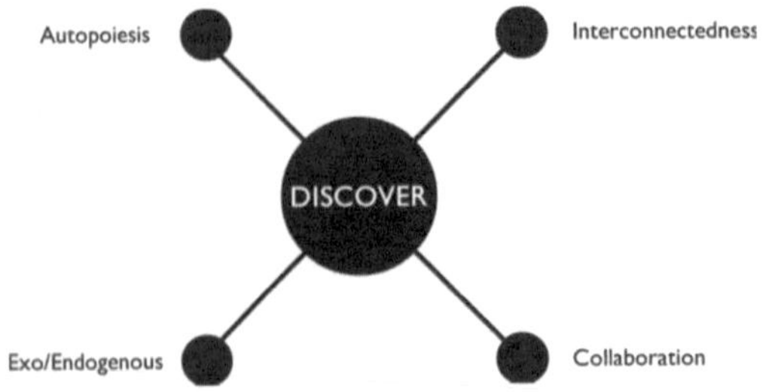


Figure 1. Four nodes at the discover stage of the design process from an ecoliteracy Perspective.

Instead of focusing on user needs, the design process should attempt discovering interconnectedness, systemic tensions and relations in all the terms or actors involved; discovering collaboration, aims identifying the level of plurality of actors in organization processes; discovering exogenous and endogenous circumstances that impact and are impacted by the organization; and finally, discovering autopoiesis or the capability to adapt to changes and context. This perspective does not only apply to design management, or design practices that are naturally systemic such as service design, experience design or so on; it also applies to identify innovation opportunities for products and services. However, instead of concentrating on the object attributes or the user, focuses on the system's attributes that will allow or constrain its development. Considering these systemic nodes will help the design team projecting a broader scope, enhancing the value, and improving the lifespan of the result.

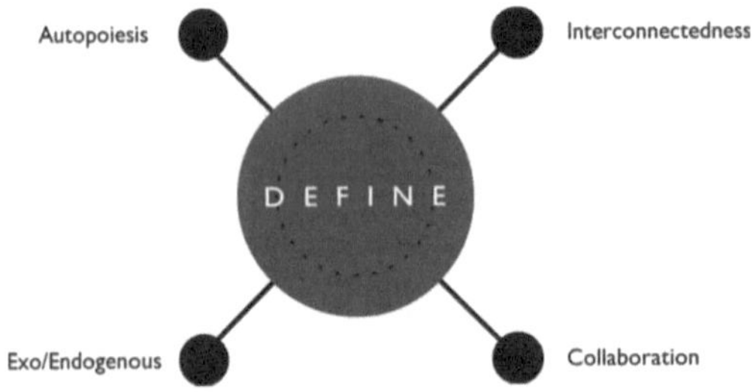


Figure 2. A broader scope of the define stage considering the four nodes of ecoliteracy.

Define is the second stage, according to TDC, which, from an ecoliteracy perspective would attain a broader scope than the previous stage, considering the same four nodes but interpreting them differently. This stage is about defining the aims and frames of the design project. The past stage, discovery, identified the current status of the system, and this stage defines the desired status. Thus, this process will develop a design to get from system A to system B (Findeli 2001). Defining interconnectedness should aim to promote multiverse relations, enabling systematicity, which means decentralizing decisions, hierarchy and opening multichannels of communication. A design project should define how and when to communicate with stakeholders, enabling channels to facilitate it. Defining collaboration also implies designing ways to involve stakeholders in the development process. Defining impact in exogenous and endogenous circumstances means determining how the design will affect communities and environments while producing it (endogenous) and while selling it (exogenous). This node also applies for organizational aspects within and outside institutional structures. Defining autopoiesis involves developing ways for the design to adapt to changes. A product or service could delay its expiration date if it is designed for adaptation, considering tangible and intangible aspects.

Develop, as the third stage (Figure 3), considers designing prototypes and processes to attain the goals set in the previous stage. Developing interconnectedness should promote communication with the brand, users, and producers; in other words, opening channels for communicating in every direction between stakeholders. Developing collaboration implies designing ways to involve a plurality of actors, either by consulting them or directly involving them in design workshops. Exo and endogenous impact consider developing ways to reduce environmental impact and amplify the social value of products and services, without compromising organizational resources. Developing autopoiesis requires a culture of adaptation, between the design team that can consider flexibility and long-lasting products and services in times of an everchanging world.

Products, services, and processes should deliver interconnectedness (Figure 4). In other words, design products should be connectivity facilitators. By doing so, collaboration will be promoted between people. Interconnectedness is about establishing relations between people of different positions, either they are inside or outside the organization, impacting on the third node exo/endogenous circumstances. From this point of view, the fourth node, autopoiesis, would mean delivering assessment strategies, updating processes of products and services, and creating feedback loops. This openness of the last node implies going back to the first stage and discovering new paths of interconnectedness, either delivered through the design process itself or as a collateral consequence. Thus, the design process from an ecoliteracy perspective builds an iterative process, maintaining systematicity, collaboration, and internal and external dialogue.

Since the delivery stage is also a way to discovering new interconnectedness, systematicity in the design process from a systems-centered perspective, should allow the process to begin at stage four, deliver, at stage one, or any stage. Figure 4 does not represent a step by step guide but a scope attribute of every stage. For instance, an organization could begin by looking at how they are currently delivering interconnectedness and collaboration which will lead to the discovery stage. Also, there is research reporting on the benefits of quick prototyping, which means starting at the development stage (Innella et al. 2016). By assessing any stage from

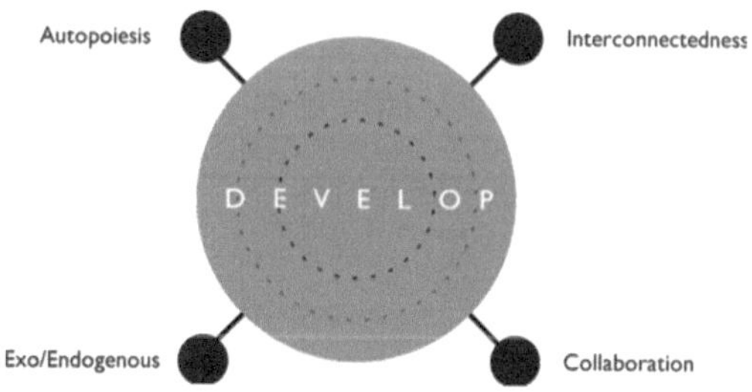


Figure 3. The develop stage broader than the past two.

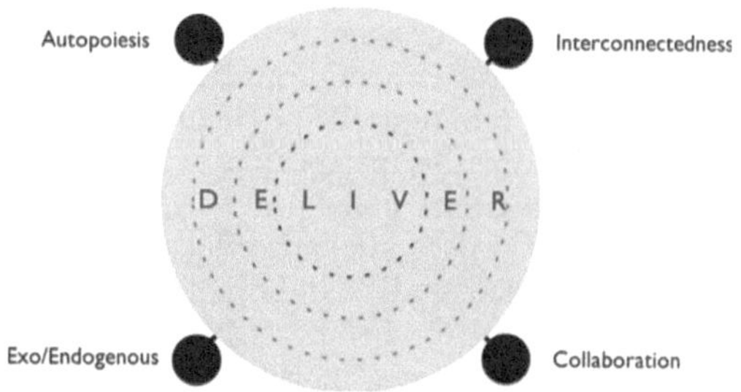


Figure 4. Deliver as the final stage of the design process.

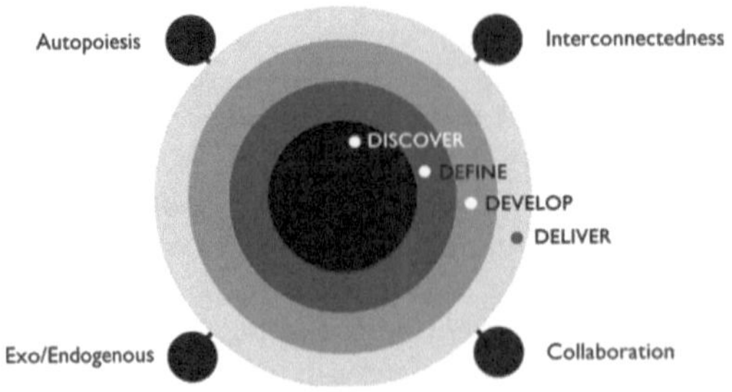


Figure 5. Four stages and nodes a system centered design process.

a systems-centered point of view, the design process would shift implicating the other three stages through reviewing each node of the system. Also, these changes in the design process imply changes in design education, as well.

## 4 TEACHING A SYSTEMS-BASED DESIGN PROCESS: DEVELOPING ECOLITERACY SKILLS

A systems-centered design process will require new tools for each node and each stage. Object-centered design processes entailed knowledge about shape, perception, materials, among others, while user-centered looked for tools to understand user behaviors and needs. Thus, a systems-centered will need tools and methods to understand systems and tensions that could give birth to pathways of innovation and fluent interconnectedness.

From an ecoliteracy perspective, the design process could become a nonlinear, layered, and multidimensional process, instead of a step by step process differencing one stage from the other. The designer should be able to see the whole process as a living system, visualizing every layer simultaneously, and its link to business, environment, and society.

Through the four nodes of every stage, designers will develop ecoliteracy competencies, activating cognitive, emotional, active, and connectional skills. Through the exercise of these skills, they will be able to reach high levels of awareness, sensitivity, and empathy. Design education should promote the iterative practice of these skills and a systems-based design process, training "leaders fluent in ecoliteracy [that will facilitate the] emergence of novelty, hence promoting conditions for supporting the creative tension even if the structure would have been temporarily destabilized" (Semetsky, 2010, p. 40)".

Identifying system tensions and turning points is a fundamental practice to develop ecoliteracy skills (Capra in Semetsky, 2010, p. 2). Discovering interconnectedness, collaboration, endo, and exogenous circumstances, and autopoiesis could begin with the conversation of tensions between members of a community. This approach represents a challenge in teaching practices and demands new teaching materials to encourage students to identify system tensions, which also implies redefining the starting points of design projects and assignments.

A paradigm shift towards a systems-based design process should avoid dualisms (Kelso and Engstrom, in Semetsky, 2010, p. 34). User/designer, natural/artificial, practice/theory are dualisms that disrupt the connection and cooperation between both parts. Ecoliteracy is a perspective that would decentralize, economic, or political interests balancing the practice of design. An educational outcome from this perspective would imply blurring the barriers between student/teacher, and other dualisms, collaboratively engaging learning.

Capra's claim about systems also implies systemic education: "Systemic school reform is based on, essentially, two insights: a new understanding of the process of learning and a new understanding of leadership" (1999, p. 6). From an ecoliteracy perspective, leadership is required throughout the whole design process; it will allow developing projects that correspond to their context and, at the same time, are innovation-driven.

Finally, feedback is also a fundamental aspect of the process: "because of feedback loops, a system tends to become creative of its own novel modes of existence that therefore may emerge at some critical [turning] points in the process of the system's evolution and learning" (Semetsky, 2010, p. 33). Hence, it implies the ability of students to identify feedback sources while developing a project and how the design outcome will gather feedback once implemented. Both activities require incorporating people and different informants in communities (as co-design has insisted on) but also feedback from peers, observation and performance data. A feedback cluster could help to construct feedback loops. This focus also demands developing teaching materials that help to understand and practicing feedback in and out the design studio.

## 5 DISCUSSION: THE SYSTEMIC TURN

Becoming and training ecoliterate professionals requires changing teaching practices, teaching methods, contents, and new models that allow the design process to expand and broaden its scope. Since a systems-based design process was presented in this document to highlight what ecoliteracy can bring to design practice, and how it can nurture its results, three central ideas are highlighted when applying ecoliteracy's lenses to design teaching practices:

Since there is no agreement on the ideal practice in the design process, schools need to teach flexibility, adaptation, connectedness, collaboration, self-learning, among others, so that students become capable of designing different routes to address constraints and challenges, instead of teaching particular design methods.

Interconnectedness also means that students have to recognize the system that they conform as students, their connection to peers and teachers, and how they are linked to design practice and the needs of society: "developing the ability to think systemically gives learners of all ages the potential to maximize the application of their diverse learning experiences and contribute to how we can better understand our complex interconnected world (Scalabrino & Oliva, 2013, p. 3). In other words, designers need to understand everything as a system, linking the user to society, the product to the environment, the business to the global economy.

Creating collaboration and feedback loops mean that education cannot be seen as a one-way channel anymore. Students have a lot to say about their education, and they can collaborate and become more motivated when participating actively in planning strategies for their own learning. Also, collaboration is open-ended, and at the same time, an iterative activity; this may be a challenge for educational institutions since they are traditionally framed in contents and timetables.

This paper is presented as an invitation to adopt ecoliteracy as a paradigm shifter in both design practice and design education towards a systemic turn. The findings discussed so far suggest a new role for designers. They have important implications for developing new strategies for Design education. It is not only a matter of understanding design disciplines in a different way, but also, it is about teaching design based on the fact that ecoliteracy should be addressed as a central perspective if the design discipline is seen as a meaningful and powerful practice to make substantial changes in the world. In order to develop active designers capable of transforming systems, it is necessary to transform what is taught and how it is taught in design education, considering the renovation of the educational system that would enable the evolution of students into social-ecoliterate innovators.

## REFERENCES


Cascini, G. (2012). TRIZ-based Anticipatory Design of Future Products and Processes, *Journal of Integrated Design and Process Science 16* (3), 29–63.

Camere, S. & Bordegoni, M. (2016). Unfolding the Notion of Experience (Virtual) Prototyping: A Framework for Prototyping in an Experience-Driven Design Process, *Journal of Integrated Design and Process Science 20* (2), 17–30.

Capra, Fritjof (2005). Speaking Nature's Language: Principles for Sustainability in *Ecological Literacy*. Stone, M. & Barlow, Z. (eds). San Francisco: Sierra Club Books.
Capra, F. (1977). *Creativity and Leadership in Learning Communities*, Mill Valley School District.
Capra, Fritjof. (1996). The Web of Life. A New Scientific Understanding of Living Systems. New York: Anchor Books.
Capra, Fritjof. (1999). Ecoliteracy, the challenge for education in the next century. California: Center for Ecoliteracy.
Ghislene, I. M., & Franzato, C. (2017). The relationship between Strategic Design and Metadesign as defined by the levels of knowledge of design. *Strategic Design Research Journal, 10*(2): 134–143.
De Mello, K. (2017). From strategic planning to the designing of strategies: A change in favor of strategic design. *Strategic Design Research Journal, 10*(2): 91–96.
Findeli, A. (2001) Rethinking Design Education for the 21st Century: Theoretical, Methodological, and Ethical Discussion. *Design Issues*, 17 (1), pp. 5–17.
Innella, G., Rodgers, P., Higuchi, T., Muchel, A. & Paskevicius, A. (2016). The benefits of prototyping early in design workshops. *Bulletin of JSSD, 63*(4), 21–28.
Kalbach, J. (2016). *Mapping Experiences. A Complete Guide to Creating Value through Journeys, Blueprints, and Diagrams*. Sebastopol: O'Reilly Media.
Kelso, J. A. S. & Engstrom, D. (2006). *The complementary nature*. Cambridge, Mass.; London: MIT.
Lee, Y. (2008). Design participation tactics: the challenges and new roles for designers in the co-design process. *Co-Design*, 4(1), 31–50.
Manzini, E. (2015). *Design, when everybody designs*. Cambridge: MIT Press.
McBride, B. B., Brewer, C. A., Berkowitz, A. R. & W. T. Borrie. (2013). Environmental literacy, ecological literacy, ecoliteracy: What do we mean and how did we get here? *Ecosphere 4*(5), 1–19. http://dx.doi.org/10.1890/ES13-00075.1
Muratovsky, G. (2015). Paradigm Shift: Report on the New Role of Design in Business and Society. *She ji* 1(2), 118–139.
Mclellan, H. (2000). Experience Design. *Cyberpsychology & Behavior*, 3(1), 59–69.
Morin, E. & Le Moigne, J.L. (2000). *A inteligência da complexidade*, São Paulo: Petrópolis.
Press, M. & Cooper, R. (2016). *The Design Experience*. New York: Routledge.
Roschuni, C., Kramer, J., Zhang, Q., Zakskorn, L., & Agogino, A. (2015.) Design talking: an ontology of design methods to support a common language of design. Retrieved from https://www.researchgate.net/publication/280558770_Design_Talking_An_Ontology_of_Design_Methods_to_Support_a_Common_Language_of_Design
Scalabrino, C. & Oliva, J.M. (2013). La imagen acerca de la sostenibilidad/insostenibilidad en empresas de economía social. Retreived from https://www.researchgate.net/publication/321278269_LA_IMAGEN_ACERCA_DE_LA_SOSTENIBILIDADINSOSTENIBILIDAD_EN_EMPRESAS_DE_ECONOMIA_SOCIAL_-_The_image_of_SustainabilityUnsustainability_in_Social_Economy_Enterprises_-_Coauthor_Jose_M_Oliva
Secomandi, F. & Snelders, D. (2011). The object of Service Design. *Design Issues*, 27(3), 20–34.
Semetsky, I. (2008). Developing ECOLITERACY as a sub-discipline of educational philosophy. Conference Presentation, Philosophy of Education Society of Australasia. Retreived from https://philpapers.org/rec/SEMDEA
Semetsky, I. (2010). Ecoliteracy and Dewey's educational philosophy: implications for future leaders. *Foresight, 12(1)*, 31–44.
Shostack, Lynn. (1982). How to Design a Service. *European Journal of Marketing*, 16(1), 49–63.
Stickdorn M. y Schneider, J. (2011). *This is Service Design Thinking*. New Jersey: John Wiley & Sons.
Stone, M. & Barlow, Z. (2005). Living Systems, sustainability, education and institutional change. Wals & Corcoran (Eds.) *Learning for Sustainability in times of accelerating change*, Wageningen Academic: Wageningen, pp. 381–384.
The Design Council, (2007). *Eleven lessons: managing design in eleven global companies*. Retreived from: https://www.designcouncil.org.uk/resources/report/11-lessons-managing-design-global-brands

*Research & Education in Design: People & Processes & Products & Philosophy – Almendra & Ferreira (eds)*
 *ISBN 978-0-367-49296-0*

# Exploring question asking practices in a design pre-jury

E.S. Himaki
*Department of Industrial Product Design, Faculty of Architecture and Design, Bahcesehir University*

ABSTRACT: This study investigates question asking practices in a design pre-jury, mainly focusing on its facilitative function in students' development. It aims at providing an initial understanding of question asking practices in design juries and students' understanding and interpretation of the questions posed by jurors. To this end, an empirical study was conducted in a project-based design course of first year Industrial Product Design students that consists of two stages: observation and recording of design pre-jury reviews, and interviews conducted with students. Building on observations in the pre-jury, frequencies of particular question categories and interviews conducted with students, initial inferences about the factors lead to particular types of question asking patters were provided. By interpreting each student's case separately, the tailored nature of the reviews was explored; unique patterns of questioning were revealed, and an initial understanding of students' perspective on question asking practice in design pre-jury reviews was obtained.

## 1 INTRODUCTION

As acknowledged by many scholars, design studio is the backbone of design education, and design critiquing is the prevalent way of communicating design knowledge and expertise in the studio (Oh et al., 2013; Goldschmidt et al., 2010; Uluoglu, 2000). Design critiquing and educator-student communication is increasingly recognized as an important subject in design education research, which draws considerable interest to different aspects of design reviews (Oh et al., 2013; Goldschmidt et al., 2010; Uluoglu, 2000; Ochsner, 2000; Schön, 1982).

By acknowledging the importance of question asking in thinking and learning processes and considering the influence of question asking in shaping design communication, I have started to investigate questions in a design studio setting. Believing that investigating questions posed in a design studio can help us understand how design knowledge is created and communicated in the studio setting, this particular study explores question asking practices in interim design reviews (pre-juries), and students' understanding and interpretation of the posed questions.

I begin by presenting a brief overview on related research regarding question asking in design and explain Eris's (2004) conceptual framework for questions in design, which forms the basis of the study. After providing the theoretical background, I continue with the empirical study, the analysis and interpretation of the analysis of data. At the end, I conclude with a summary of findings, further studies and possible implications.

## 2 QUESTION ASKING IN DESIGN

Questions and question asking practices have been a topic of investigation in many different research areas including philosophy, logic, linguistics, artificial intelligence and cognitive psychology (Graesser and Black, 1985), and the role of question asking in thinking and learning processes is widely acknowledged (Wilen et al., 2000).

The role of question asking is found important in design research as well. It has been recognized as "a particularly important lens for examining design discourse" (Cardoso et.al., 2014, p.60). Furthermore, designing is defined as a "question intensive" activity and the influence of questions formulated during design activity on design thinking is acknowledged (Eris, 2004).

In "Effective inquiry for innovative engineering design", Eris (2004) investigated the role of question asking in design and developed a "question-centric design thinking model". After reviewing existing taxonomies in the literature and testing them in design settings, Eris (2004) proposed a specific framework for design –that "characterizes and differentiates questions according to their conceptual meaning" (p.130)- by building on Lehnert's (1978) taxonomy, adopting 4 of Graesser's (1994) 5 additions and adding 5 new question categories that are unique to design situations.

Eris's (2004) conceptual framework for questions formed the basis of this study and utilized during the analysis as the coding framework. Eris (2004) classifies three main question categories: low level questions, deep reasoning questions (DRQ), and generative design questions (GDQ).[1] Low-level questions are asked in order to, "verify and clarify facts, identify and acquire relevant information, form the necessary communication base, and mediate social interaction" (Eris, 2004, p.131). Deep-reasoning questions are asked in order to "elicit patterns of reasoning in logical, causal or goal-oriented systems" (Graesser and Person, 1994). Generative Design Questions are asked in order to "disclose the alternative known answers, and to generate the possible unknown ones- regardless of their truth value" (Eris, 2004, p.37). DRQs and GDQs are also referred as high-level questions, that "relate to higher levels of reasoning" (Cardoso et al., 2014).

Premise behind low-level questions and deep-reasoning questions is that there is a knowable answer –not necessarily by the questioner or answerer- which is "expected to hold truth value". On the other hand, premise behind generative design questions is that "for any given question, there exists, regardless of being true or false, multiple alternative known answers as well as multiple possible unknown answers" (Eris, 2004, p.37). While DRQs are associated with convergent thinking, GDQs are associated with divergent thinking in design (Eris, 2004).

Also, in design education, question asking appears as a fundamental practice by the very nature of design reviews and critiquing. Building on Eris's model, Cardoso, Eris and Badke-Schaub (2014) explored the role of question asking in design reviews in a project-based design course. As a result of the study, they define 6 functions of design reviews and suggest specific inquiry approaches for each function. They also provide a model that explains the relations among "three question mechanisms" (low- level, DRQs, GDQs), and "cognitive moves" and "cognitive outcomes" enabled by each mechanism, and they argue that a comprehensive reflection is only possible in the iterative coexistence of three modes.

## 3 FIELD STUDY: QUESTIONS IN DESIGN STUDIO

With the aim of investigating question-asking practices in design pre-jury reviews and exploring students' understanding and interpretations of the questions posed by jurors, the study addresses three main questions:

1. What are the frequencies of low-level, deep reasoning and generative design questions in jurors' and students' utterances in a design pre-jury?
2. What type of questions attracts students' attention as being important and useful?
3. How do design students interpret questions posed by jurors?

[1] Eris adopts low-level and deep reasoning question categories from existing reviewed taxonomies. Graesser and McMahen (1993) introduced the concept of deep reasoning questions, questions in which the questioner is seeking to establish causality (Cardoso et al., 2014). Eris has proposed the Generative Design Questions category.

These questions are approached in a two-staged study: design pre-jury reviews of first year industrial design students were observed and recorded, and then semi-structured interviews were realized with participant students to explore their understanding of the reviews with a particular focus on questions posed by jurors.

### 3.1 *The pre-jury*

Critiques can be conducted in several different settings during a studio course, such as, desk crit, group crit, interim review, final (formal) review and informal interaction (Bailey, 2004). In this study, the area of inquiry is a pre-jury, also called as an interim review.

Conventionally, a design jury is conceptualized as a practice aims at assessing student performance through presented design work (Peterson, 1979; Anthony, 1991; Webster, 2006). It is widely embraced as one of the most important practices of design education. Interim reviews in particular, are reviews realized at "key milestones during a studio project" with the participation of the entire class (Oh et al., 2012). It is conceptualized as a constructive environment contributing to learning, with its focus on the process. Giving students opportunities to present their work while in-progress is found to be important and advised, since when students are presenting while in-progress, they also articulate their thinking besides presenting the product, and make reasoning behind their actions more transparent (Cennamo and Brandt, 2012).

This study was carried out in an undergraduate level industrial product design program in Turkey, which is a full-time 4-year long program. Observed design pre-jury was part of a first-year design studio course.

Studio course was carried out by four faculty members: an assistant professor, an instructor, a visiting instructor and, a research assistant. At the time of the study, researcher was part of the studio team as the aforementioned research assistant. In the observed pre-jury, jury members consisted of the studio team. The researcher adopted an observer stance and didn't participate the jury as a juror.

The design pre-jury was part of the final project, which was one of three projects conducted in the second half of the 1st year. Students were responsible from designing a 'chess set', and they undertook design projects individually. The duration of the project was six and a half weeks and the pre-jury was realized at the beginning of the fifth week. For the pre-jury, students were expected to bring a full set of 32 pieces, from designated material in real dimensions and a concept statement written in one sentence.

Six undergraduate industrial design students (three females, three males) participated to the study. Five of the participants were Turkish students and one was a foreign student. Language of the studio course was English and so the pre-jury was conducted in English as well. Since none of the participants are native English speakers, language barrier was at issue. In previously observed cases, it is seen that most of the students are having difficulties in explaining themselves in English. Thus, ability to speak English at a certain level was a criterion during the selection of participants.

Duration of the reviews changed between five to ten minutes. In most cases a 1 or 2 minutes presentation was followed by a discussion lead by jurors, and the average duration of the discussion part was approximately six minutes.

Before the pre-jury, written informed consent was obtained from all participants, and reviews were video and audio recorded with the permission of the participants to facilitate the interviews and the analysis.

### 3.2 *Interviews*

In the first studio day after the pre-jury, and the following day, semi-structured interviews were realized with participant students to explore their understanding of the review. Interviews were supported by a task, which is designed to provide an initial understanding of students' focus of attention in jurors' comments and questions. A brief explanation about the task was provided to participant students at the beginning of the interviews, and transcripts of the reviews were provided to them. For the task, participant students were asked to underline parts that they find important and useful for the development of the particular design project they were designing, and parts that help them build a deeper understanding about design and design process. To help students recall their reviews, interviews were accompanied by video records.

After completing the task, semi-structured interviews were realized to explore students' perspective. While going through participant students' remarks, they were asked to elaborate on the reasons why these comments and/or questions are found important and useful. Then, remaining questions they have received were brought forth for discussion. At the end, several questions were asked, such as, 'are there any specific type of questions that you find useful?' and 'are there any specific type of questions that you find disturbing and/or distressing?'

## 4 ANALYSIS

The outcomes of the empirical study were analysed by applying qualitative content analysis. A concept-driven strategy was adopted while categorizing questions posed in the design pre-jury, which uses a deductive strategy by basing the study on previous knowledge. Eris's (2004) conceptual framework for questions formed the basis of the analysis and utilized as the coding frame. Therefore, questions posed in the design pre-jury review were categorized under three main categories: low-level questions, deep-reasoning questions, and generative design questions. Categories of questions are exemplified with sample questions provided from the study (Table 1).

In order to conduct content analysis, all questions are extracted from the discussions. During the analysis Eris's (2004) definition of a question was utilized: "In a design context, a question is a verbal utterance related to the design tasks at hand that demands an explicit verbal and/or nonverbal response" (p.36).

Table 1. Main question categories and sample questions.

| Question categories | Sample questions |
|---|---|
| Low-level questions | Are there two versions? *(verification)*<br>Is it supposed to be like that, or is it supposed to be other way? *(disjunctive)*<br>What is holding them? *(concept complementation)*<br>How many [lines] are here? *(quantification)*<br>Which way you want to go? *(judgemental)* |
| Deep-reasoning questions | What do you mean by "as small as possible"? *(interpretation)*<br>Is that the reason why the base on the king is both red and black? *(causal antecedent)*<br>What happens if it [dimensions of the base] is 3 by 3? *(causal consequence)*<br>Why is everything the same height? *(rationale)* |
| Generative design questions | For example, how I would hold anything like this? *(method generation)*<br>Why this is glued here where there could be some other intersection details? *(proposal/negotiation)*<br>And if you want to use your hand, how is it to interact with it? *(scenario creation)*<br>For example, you can change the order, that is one thing, what else can you do because you are using layers? *(ideation)* |

## 5 INTERPRETATION OF THE ANALYSIS OF DATA

### 5.1 *Frequencies of question categories*

In the analysed reviews, the average duration of a design review was 7,2 minutes and the average duration of the discussion part of the review was 5,87 minutes. The average number of questions posed by jurors in a design review was 17,83 (7,3 low-level; 7,1 deep-reasoning; 3,3 generative design) whereas average number of questions posed by students was 1,33 (0,83 low-level; 0,5 deep-reasoning; 0 generative design).

As also suggested by a previous study (Cardoso et al. 2014), students generally pose a small number of questions during design reviews. Total number of questions directed in the analysed design reviews were 115 and only 8 of the questions were directed by students. When looked at the questions directed by the jurors: 44 questions categorized under low-level questions, 43 questions categorized under deep-reasoning questions, and 20 questions were categorized under generative design questions. When looked at the questions directed by the students: 5 questions categorized under low-level questions, and 3 questions categorized under deep-reasoning questions (Figure 1).

### 5.2 *Students' focus of attention in jurors' questions*

When looked at the students' remarks, it is seen that 20 questions posed by jurors were underlined or highlighted, and 16 of them were categorized under high-level question category (Figure 2). Based on these results, it might be suggested that participant students place importance to high-level questions they receive.

Additionally, when looked at the low-level questions remarked by students, it is realized that some of the questions categorized as low-level, are generative in the sense that they bring out something new from the perspective of students. For example, P1 underlined the question,

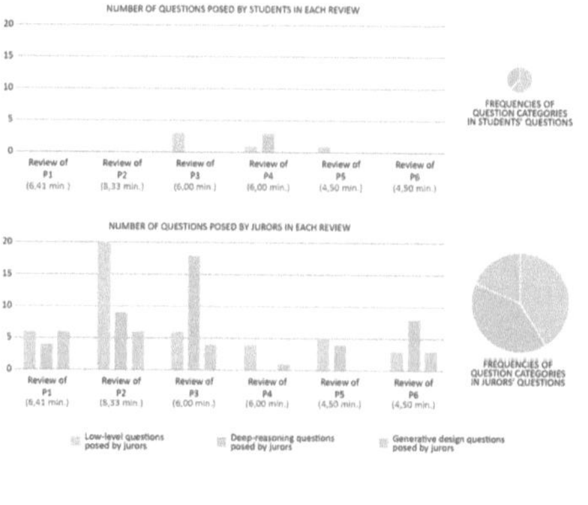


Figure 1. Graphs showing number of questions posed by students and jurors in each review, the duration of the reviews, and the frequencies of question categories in students' and jurors' questions.

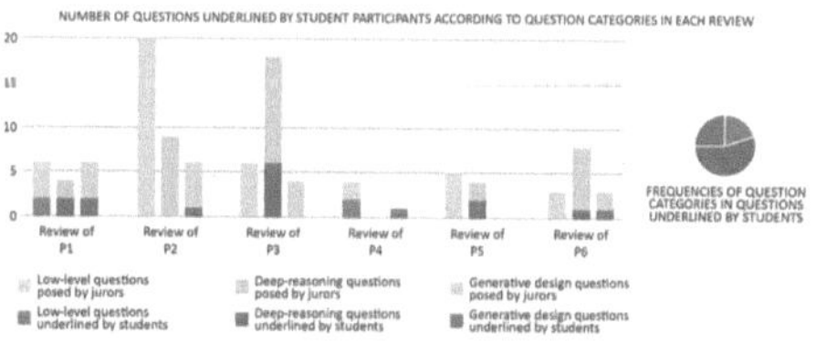


Figure 2. Number of questions underlined by student participants according to question categories in each review.

"Like for example this one is rolling and this one okay stays the way you do it, so I don't let it drop on the sides ha?" and stated that he was not aware of mentioned situation before he received the question.

### 5.3 *Patterns of questioning and students' interpretation of jurors' questions*

Analysis of the data showed that, question asking patterns and dominant question categories vary in design pre-jury reviews, and it is not very likely to make generalizations about question asking patterns. Nevertheless, it might be possible to draw initial inferences about the factors lead to particular types of question asking patterns by looking at each students' case separately, based on the observations in the pre-jury, frequencies of particular question categories and interviews conducted with students.

In case of P1, the discussion has started with DRQs and Low-level questions and continued with GDQs. Overall, he has received 6 Low-level questions, 4 DRQs, and 6 GDQs. When jurors were questioning the rationale behind his decisions through DRQ's like, "How do I understand [the difference among the pieces]?", "What about this little accessories?", "Are they different enough, to be able to distinguish?", P1 expressed that he was not really satisfied with the design and he discussed his hesitations about his design decisions openly. It is observed that throughout his review he actively participated to the discussion, and in his interview, he told that he was confident about understanding jurors' comments and questions. Compared to other students, he received a good number of GDQs, and this might had been triggered by his openness to criticism, ability to enter professional discourse and ability to reflect on the state of design. The frequency of GDQs leads his review into a divergent mode of thinking and it is observed that the student and jurors behaved like equal partners in the process.

Furthermore, when P1 is asked if there are particular comment and question types he found useful, he stated that even though it is easier to be directed, it is better not to hear instructions or directly applicable ideas: "Instead of a comment like 'you should do this', I prefer something like 'you can go for it, but there is also this... you have two options, but you choose it'". The fact that he is willing to take the responsibility of making design decisions on his own can be interpreted as a sign of positioning himself as an equal partner, which might have facilitated active participation in discussions.

A strong contrast appears in question asking patterns observed in the reviews of the students who presented more developed concepts with rationalized design decisions by adopting a designerly manner and students who presented poorly communicated, less developed concepts and couldn't enter the professional discourse. When looked at the case of P4, even though the students were expected to bring a full set of 32 pieces -from designated material in real dimensions- he only presented several initial mock-ups from modal making materials and received comments about the fact that he fall behind schedule, and he is not putting enough effort. In his review, jurors tend to make directive comments and posed only 5 questions, in which 4 of them belong to low-level question category. Thus, it is hypothesized that the state of project and the perceived competence of the student might play an important role in shaping the question asking frequency and pattern. The more primitive the project is -for the scheduled phase- and the more incompetent the student is in communicating in a designerly manner (both orally and visually), the more directive the comments become and questions tend to be low-level. The directiveness of the comments might be also caused by the time pressure as also suggested by Cardoso et al. (2014).

The project of P2 was probably one of the most developed projects that was close to finalization. Her review was the longest and she received highest number of questions. Overall in her review, jurors posed 35 questions: 20 Low-level questions, 9 DRQs and 6 GDQs. She presented a creative concept, yet according to jurors' comments there were several deficiencies and confusions in the logic of the concept concretization. Therefore, at the beginning of the discussion, P2 received several Deep Reasoning questions that are mostly questioning the rationale behind design decisions and quite a few low-level questions that are mostly asked to completely understand the concept. Later on, she also received a relatively good number of GDQs, similar to P1, and just like P1, P2 was also one of the participants who could discuss her ideas openly with a critical mindset and was able to reflect on the state of design. During her review she actively participated to discussions by negotiating ideas and decisions and it is observed that jurors "spent considerable time and energy working with the student's ideas" (Webster, 2006, p.291).

While interpreting her review, I came to realize that there are also some low-level questions posed by jurors with an intention to form a basis to discuss rationale behind design decisions. In a way, there are hidden rationale questions asked in the form of a low-level question such as quantification. For example, one jury member asked the dimensions of the bases of the chess pieces during P2's review to start exploring the reasoning behind decisions about dimensioning. There were some cases that students interpreted this kind of questions in a different way than intended, and only answered the question on the surface even though they received fallow-up rationale questions.

Similarly, in the review of P3 question asking occurred frequently. Jurors posed 6 low-level questions, 18 DRQs and 4 GDQs in the 6 minutes discussion time. DRQs were considerably dominant in the discussion and this might be caused by the perceived incompetence in design decision-making. Her decisions were found groundless, "random", and "emotional". Thus, she has received a number of rationale questions -in order to discuss the reasoning behind- such as, "Why are they (bases) hexagons?", "And why red and black?", "Why is it shaped like this?", "Why these pieces are, like that one, sticking out?" When talking about these questions she told that she build emotional bonds with some ideas and some particular aspects of her design, and she formulated this situation as something that needs to be changed, in her words, "so I figured here that I should keep the design ahead of my own feelings."

After talking about rationale questions she has received for a while, she expressed that on one hand she finds 'why questions' very useful, on the other hand she finds them demoralizing and restricting, in her words:

> *I feel as if I'm compromising myself according to the request of the instructors. I'm compromising my own design... Sometimes, I know they will criticize so I just leave the thing behind because of that, and it hurts me a little...I cannot take risk...They are not directly blocking you but, they hinder on the sly... When you are not sure about the thing you have done, why questions helps you to think, but when you are up with something you really like and confident about and you receive many 'why's, then you become like, 'I just wanted and did', you know.*

In the case of P5, the review was comparatively short and after a very brief explanation of the student, jurors started asking questions. She received 5 low-level questions and 4 DRQs. She was the only student who did not receive any GDQs. The concept she presented was clear and solid, so in order to help the student to transform 'concepts into design decisions and specifications' DRQs were posed by jurors (Eris, 2004). Thus, convergent mode of thinking was dominant in her review. When talking about the interpretation questions (in DRQ category) she received, she mentioned that aspects highlighted through these questions were mostly things that she didn't considered to be important, and made her discover new viewpoints.

In the case of P6, 3 low-level questions, 8 DRQs, and 3GDQs were directed to the student. Most of the questions posed were rationale questions as some deficiencies were perceived in reasoning behind design decisions. When discussing these rationale questions during the interviews, the student also mentioned that he made some of his decisions instinctively, without thinking comprehensively. Additionally, in his review, one of the jurors posed a scenario generation question in the role of a user and this question was marked by the students as being important in reminding him to look from user's perspective while designing.

## 6 CONCLUSION

This study can be considered as an initial step to explore question asking practices in a design pre-jury setting and students' understanding and interpretation of the questions raised by the jurors. Through the analysis of (1) frequencies of question categories, (2) students' focus of attention in jurors' questions, (3) patterns of questioning and students' interpretation of jurors' questions, the tailored nature of the reviews was investigated.

When looked at the frequencies of question categories, it is observed that high-level questions constitute 59% of the total questions. This percentage is found important, as high-level question asking has considered to be an effective practice in helping students to engage in higher level thinking processes (Cardoso et al., 2014). For the analyzed case, there might be several explanations for the dominancy of high level questioning such as, the structure of the pre-jury, timing of the pre-jury, the phase of design process, time pressure, level of students (e.g. first year), the individual characteristics of student participants, quality of the project presented, and the jury members' approach to design and teaching of design. Analyzing factors facilitating high-level question asking in a design jury setting lies beyond the scope of this study. Hitherto, it is presumed that the nature of pre-jury with its focus on the process might have an important role in facilitating high-level question asking. Furthermore, jury members' approach towards emphasizing the importance of the design process might be part of the cause. On the other hand, overall dominancy of DRQs presumed to be caused by the fact that participant students were first year students and jurors were trying to build an understanding of design decision-making.

When looked at participant students' focus of attention in jurors' questions, it is observed that students place importance to high-level questions they receive. Furthermore, when looked at patterns of questioning and students' interpretation of jurors' questions, unique and tailored nature of each review revealed different facades of the dynamics of question asking. It is suggested that there were some prominent factors that might play a part in shaping the question asking frequency and pattern such as, the state of design, perceived competence of the student, identified deficiencies in the project and process, etc.

Additionally, throughout the analysis it is realized that there are some questions categorized as low-level that are generative in the sense that they bring out something new from the perspective of the students, or that are posed by jurors with an intention to form a basis to discuss rationale behind design decisions. Thus, it is thought that for further studies new ways of adopting Eriş's (2004) framework can be searched with the aim of incorporating intentions and perspectives of the people asking and receiving the questions.

To conclude, considering the magnitude of question asking in design reviews, inquiries related to understanding and systematizing the practice of question asking and ways to improve its effectiveness is crucial. Thus, further studies can be conducted in order to better understand how knowledge is created and communicated through question asking practices in design reviews, to explore how different questions raised in design reviews are interpreted by students, how receiving some particular types of questions influence students' further decision-making process, how these influences manifest in the final product (if they manifest), and role of the questions in facilitating students' development.

## REFERENCES

Anthony, K. H. (1991). *Design Juries on Trial: The Renaissance of the Design Studio*. New York: Van Nostrand Reinhold.

Bailey, R. O. N. (2004). *The digital design coach: Enhancing design conversations in architecture education* (Doctoral Dissertation). Victoria University of Wellington, New Zealand.

Cardoso, C., Badke-Schaub, P., Eris, O., & Aurisicchio, M. (2014). Question asking in design reviews: how does inquiry facilitate the learning interaction? DTRS 10: Design Thinking Research Symposium. Purdue University.

Cennamo, K.& Brandt, C. (2012). The 'right kind of telling': knowledge building in the academic design studio. *Education Tech. Research Dev.*, *60*, 839–858.

Eris, O. (2004). *Effective Inquiry for Innovative Engineering Design.* Boston, MA: Kluwer Academic Publishers.
Goldschmidt, G., Hochman, H., & Dafni, I. (2010). The design studio 'crit': Teacher student communication. Artificial Intelligence for Engineering Design, *Analysis and Manufacturing, 24*(3), 285–302.
Graesser, A. C., & Black, J. B. (1985). *The Psychology of Questions.* Lawrence Erlbaum Assoc Inc.
Graesser, A., & Person, N. (1994). Question asking during tutoring. *American Educational Research Journal, 31*(1), 104–137.
Lehnert G. W. (1978). *The Process of Question Answering.* New Jersey: Lawrence Erlbaum Associates.
Oh, Y., Ishizaki, S., Gross, M. D., & Yi-Luen Do, E. (2013). A theoretical framework of design critiquing in architecture studios. *Design Studies, 34*(3), 302–325. doi:10.1016/j.destud.2012.08.004
Ochsner, J. K. (2000). Behind the mask: A psychoanalytic perspective on interaction in the design studio. *Journal of Architectural Education, 53*(4), 194–206.
Peterson, J. M. (1979). Me and My Critics: Students' Responses to Architectural Jury Criticism. *Studies in Art Education*, 64–67.
Schön, D. A. (1982). Design as a reflective conversation with the situation. In *The reflective practitioner: How professionals think in action* (pp. 76–104). New York: Basic Books.
Uluoglu, B. (2000). Design knowledge communicated in studio critiques. *Design Studies, 21*(1), 33–58.
Webster, H. (2006). Power, freedom and resistance: Excavating the design jury. *International Journal of Art & Design Education, 25*(3), 286–296. doi:10.1111/j. 1476-8070.2006.00495.x.
Wilen, W. & Ishier, M., Hutchison, J., & Kindsvatter, R. (2000). *Dynamics of Effective Teaching* (Fourth Edition). New York: Addison Wesley Longman.

*Design Studio*

*Research & Education in Design: People & Processes & Products & Philosophy – Almendra & Ferreira (eds)*
*© 2020 Taylor & Francis Group, London, ISBN 978-0-367-49296-0*

# Immersive behaviour setting in architectural education

H. Sopher & D. Fisher-Gewirtzman
*Faculty of Architecture and Town Planning Technion – Israel Institute of Technology*

ABSTRACT: Immersive Virtual Environments are acknowledged to support fundamental values of the Architectural Studio course. However, little is known about the way these environments are experienced by their users. This shortcoming creates difficulties in evaluating their educational sufficiency for different learners' needs. Immersive virtual environments are often expensive and require adequate staff and may create curricular changes, which emphasize the need to discerningly integrate the setting into future syllabi. This paper uses the theory of "Place" to identify the emerging relationship within immersive environments. We conducted observations and interviews upon a Studio course that used a traditional studio classroom and an immersive setting. The results provide significant insights on how the different educational settings are experienced and used. Students spend a significant amount of time in the studio. As such, Studio pedagogy ought to address learners' needs, actions and experiences, in order to promote learning processes.

## 1 THE ARCHITECTURE STUDIO SETTING

Central to architectural education, the Studio "is a challenging learning environment. Students learn by developing a design problem through reflective practice with an expert tutor, in what is known as "crits" (Schön, 1987). Design encompasses many forms of knowledge, including the design-process activities, constructive thinking modes, representation knowledge, and technological knowledge (Cross, 2006; Lawson, 2004). The process requires iterative cycles of structuring sub-problems that relate various building's features. While in practice the process evolves upon diverse knowledge sources, arriving from many professionals, Studio projects are often designed by a single student. To cope with this challenge, the pedagogical model follows the Constructivist learning approach (Lave & Wenger, 1991). The Studio thus strongly encourages active learning and diverse learning sources, originating from peer participation and professional guest reviewers, as well as adaptability to social changes (Eigbeonan, 2013). Studio pedagogy was criticized with reference made to hidden hierarchies and competitiveness among students that hinder shared knowledge (Wang 2010; Ward, 1990) and communication gaps (Webster, 2004), indicating that Studio research ought to address qualitative socio-cultural relationships between learners and educational settings, in order to better support learning.

Designed to support these challenges, traditional studio spaces offer a static open setting for learner-centred activities. They allow multiple class configurations and organization of personal learning stations and group crits. The walls are usually geared to enable hanging design outputs, in order to stimulate discussions and share knowledge. Studio spaces are commonly accessible beyond crit hours, allowing independent learning and interaction with peers.

## 2 IMMERSIVE VIRTUAL SETTING

In recent years, higher education has been increasingly exploring Immersive Virtual Environments (IVEs). IVEs enable their users to experience a sense of presence in a dynamic display, mediated through a virtual walk-through (Slater, 2009). IVEs co-depend on both physical and virtual properties. These room type spaces often possess a large-scale screen, allowing multiple attendees to simultaneously experience presence in the display in a content-focused manner. Blascovich et al., (2002) describe IVEs' social affect through self-criteria and target response. Accordingly, users' collective experience of presence is provided by the quality of the display and interaction with the virtual objects. Loomis, Blascovich, & Beall (1999) outline users' experience during and post an IVE session, referring to symptoms as enhanced explorative activities, sickness and imbalance.

Pedagogically speaking, immersive settings were found advantageous due to their potential in supporting the Constructivist approach (Mikropoulos & Natsis, 2011) and in supporting co-presence experience (Gandolfi, 2018). Research demonstrates IVEs' support for the fundamental goals of the architectural studio, such as spatial comprehension (Paes, Arantes, & Irizarry, 2017), stress and flow experiences (Boudhraa et al., 2019), decision support (Castronovo, et al., 2017), divergent-convergent developmental activities (Sopher, Fisher Gewirtzman, & Kalay, 2019) and active learning (Sopher, Kalay, & Fisher-Gewirtzman, 2017), particularly in advanced design stages (Sopher, Fisher-Gewirtzman, & Kalay, 2018), clearly indicating their eminent value to design pedagogy.

## 3 PROBLEM STATEMENT

The aforementioned developments afford Design pedagogy to integrate different environments in the syllabi, which creates a substantial need to assess the educational utility of each setting. Such decisions have additional implications. IVEs are often expensive and require adequate staff. Immersion supportive spaces create curricular changes (Fowler, 2015) that emphasize the need to utilize the setting discerningly. However, since the studio space was the sole educational context, most assessments have neglected to consider this component, thereby, do not provide a sufficient response to this objective. The literature describes the studio's crit models (Oh et al., 2013), project-based assessments (De la Harpe et al., 2009) and interaction over the various design tools (Goldschmidt & Smolkov, 2006; Lee & Yan, 2016).

Several studies depict the interaction with physical elements within the educational context. Learners' perceptions of a comfortable setting were different from those who used a setting that possessed extreme physical conditions (Marchand et al., 2014). A quiet space was favourable over a noisy one (Beckers, et al., 2016). Recent studies emphasize the emerging relationship between users and media enhanced spaces. Hod (2017) divides these spaces into a content-specific space, which affords specific, often teacher-directed, learning activities, and a content-flexible space, which affords a simultaneous emergence of various learning behaviours. Design literature describes the space's role in emerging patterns of privacy (Demirbas & Demirkan, 2000), learner engagement (Williams, 2017), evolving personal perception (Gray, 2013), and attitude towards the use of technology (Pektaş & Erkip, 2006), creating the ground for improved teaching styles and adequate design of the educational setting.

Since the use of IVEs in Architectural education is relatively new, little is known about the way these environments are experienced by learners. Survey studies have presented the recent topics of investigation of VR tools and IVEs in architectural pedagogy (Freitas & Ruschel, 2013; Milovanovic et al., 2017). These studies show that most existing knowledge focuses on technological developments and system application, while user experience is disregarded. Several IVEs require prior technological knowledge and training; factors that directly involve users' attitudes and abilities. This gap creates difficulties in estimating IVEs' educational utility to different learner needs. This paper aims to shed light on this deficiency, by adding theoretical perspectives that depict the emerging relationship between learners and an IVE during the studio course.

## 4 METHODOLOGICAL PERSPECTIVES ON LEARNERS' SITUATED INTERACTION

This study aims to demonstrate the way different learning environments are experienced by their users. To achieve this goal, the authors have grounded their investigation upon the founding basis of Environmental psychology. First and foremost, emerging interaction between users is considered inseparable from the physical attributes of a given space (Canter, 1977). Space in this notion, is seen as an active agent, affording or restricting certain behaviours. "Behaviour setting" characterizes a given space by its afforded activities (Barker, 1968). Research made use of this approach to describe individuals' attitudes towards spaces, termed as "place attachment" (Altman & Low, 1992). Spaces were socio-culturally described (Alexander, 1977). A recent study used this theoretical framework to assess higher education spaces (Beyraghi & Asl, 2018). Focusing on architecture studio spaces, Scupelli and Hanington (2016) found that learners perceived their study stations as personal territory, prioritizing them as a primary setting for meeting their needs.

An individual's experience encompasses an interaction with a setting that is external to the user, and an internal interaction with former beliefs and experiences (Wicker, 2002). Wicker addresses these interactions as "sense making," claiming that behaviours are co-affected by past experiences and present settings. Lewicka (2011) demonstrates how personal location-based experiences may be interpreted from observed processes and provide spaces' affordances.

Following the above notion, we consider the activities performed at the studio as situated experiences that emerge from the interaction between individuals and existing educational settings (Figure 1). A learner's internal interaction includes personal attitudes towards the physical, virtual and social settings, and possession of professional abilities. An external interaction describes the way one learner interacts with spatial and social components. It includes all observable types of learning encounters.

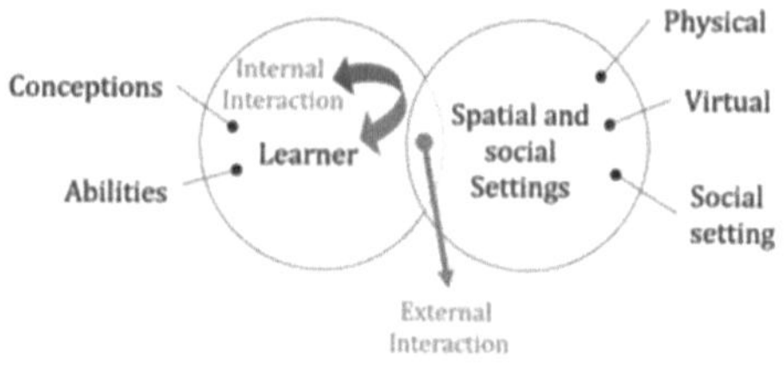


Figure 1. Emerging interaction with the educational setting.

The following sections describe the way internal and external interactions are identified through observable behaviours and expressions.

### 4.1 *Personal concepts as a source to determine a situated experience*

A personal concept delineates a space's existing or absent affordances with reference to learning performance. Research acknowledges the relationship between learner perception of the environment and learning attitude (Dart et al., 2000). Consequently, the way learners perceive the existence or absence of desired elements in a specific space may enhance (or decrease) their attitude towards learning. We distinguish between past concepts and those shaped by interaction with the environment throughout the semester. Past concepts include expressions that indicate the student's readiness for a new experience, previous learning habits, and prior knowledge. Shaped concepts describe changes in learning habits or attitude.

### 4.2 *Learner-space interaction*

The "Place-attachment" approach proposes useful principles for identifying learners' experiences. This aspect describes individual's feelings (positive or negative) that lead to a reaction towards the surrounding environment (Manzo, 2005). The habit of changing one's surrounding may serve as self-expression (Cooper, 1974). Observed acts of self-expression and statements of space-oriented feelings, are used to determine learners' external experience. For example, one's choice of a location for a personal learning station determines which crits by neighbouring peers they would be exposed to. In this sense, the ability to choose a tool to represent the architectural artefact serves as an act of self-expression, indicating individual abilities and will to learn. Observed acts of interaction with unfamiliar software are considered as one's positive will to learn.

## 5 CASE STUDY

The authors conducted a thorough study of a third-year studio course, taught by Dr. Dafna Fisher-Gewirtzman, during the spring of 2016. The course used an IVE and a conventional studio classroom (Figure 2). The IVE contains a 2.4 x 7.0-meter screen with a 75° field of view and three synchronized projectors enabling a dynamic 3D display (with the aid of 3D glasses). The space can host up to twenty people for a shared presence experience, exposed to the display of a single artifact. Due to maintenance requirements, the room's temperature is kept under 21°. Using the IVE requires the aid of a technical expert. Visits, therefore, have to be coordinated ahead of time.

Data were collected from eleven undergraduates (age 20-25). Table 1 depicts the sample's information. Each student is classified by a serial ID number, gender, ethnicity and previous technological knowledge. The first student, for example, is Arabic in origin and knows Sketchup software, therefore, his identification code is: 1MAS. All participants voluntarily signed a consent form, under the code of ethics in social science.

Semi-structured interviews were conducted with all of the students, except student 11MJRH, once the course was completed. The leading questions stimulated the discussion on different interaction types in each learning environment, use of personal learning stations, and learning habits during and outside course hours. Protocol analysis was used to retrieve themes common to each educational setting. One hundred and ten hours of non-interfering observations were used to collect evidence of interactions with the spatial and social settings.

Figure 2. A Studio crit in the immersive virtual environment (left) and in the traditional studio classroom (right).

Table 1. Participants information. Legend: F = Female; M = Male; A = Arab, J = Jewish; S = Sketchup; R = Revit; RH = Rhino; N = no previous technological knowledge.

| Learner ID | 1 | 2 | 3 | 4 | 5 | 6 | 7 | 8 | 9 | 10 | 11 |
|---|---|---|---|---|---|---|---|---|---|---|---|
| Gender | M | F | F | M | M | M | F | F | F | M | M |
| Ethnicity | A | J | J | J | J | A | A | A | A | A | J |
| Prior technological knowledge | S | R | RH | RH | N | N | N | S | S | S | RH |

## 6 RESULTS

Analysis revealed that each setting was correlated with different internal and external experiences. Data were aggregated into five thematic categories of behaviour setting that describe each space.

### 6.1 *Place attachment*

Personal learning stations were the main source to characterize acts of place attachment and self-expression. These stations were created in the traditional classroom at the start of the course. Ownership of a station was firstly marked by a simple note that contained a student's name, later enriched with personal equipment, such as sketching papers and modelling tools (Figure 3). A number of students brought headphones and used them while developing their project, which isolated them from the rest of the group and decreased the exposure to their peers' crits.

Personal stations were described in terms of territory and security. Having the space locked by the supervisor during breaks and lacking the possibility to develop personal learning stations, none of these expressions referred the IVE.

> *I left materials there (in the personal space). I saw student 1MAS leaves his bag there, but I don't leave things that are worth money.... I do not like it when people outside the studio come to work here. I always hang up a note to mark my place, my desk, a permanent place (Student 6MAN).*

Student 8FAS marked her territory by placing a note at her personal station after she had found out that someone else had used her position, stating: "*To the person who feels at home: Just don't feel at home. Don't move the models, don't leave food on the table, and don't leave your trash here*".

Learners' choices of presenting their outputs in digital formats revealed that due to the use of computers, the studio classroom and the IVE share similar activities. Although the studio classroom was enriched by some physical representations, these acts were mostly conducted to satisfy teaching demands rather than to fulfil an educational intent. Physical outputs were hung solely during the early stages of the semester, while advanced design outputs were temporarily exposed during crits. The IVE's screen was used to present digital outputs. Obviously, the exposure is restricted solely to presentation time. In that sense, it may be said that the virtual setting owns similar behaviour setting in both flexible and content-specific spaces. However, more acts of peer participation were observed during IVE crits, whereas at the studio space, students did not choose to join their peers' crits, making their learning more teacher-centered. Some interactions at the studio space referred to political conceptions. Student 8FAS placed in front of her personal desk a model that had been prepared for a different course, stating a political position (Figure 3, right). Such extracurricular activities did not emerge at the IVE. The setting was thus focused on the educational objectives of the course.

### 6.2 *Emerging Social structures*

The spaces were correlated with different social structures. The class population was divided into sub-groups that interacted frequently among themselves (Figure 4). The IVE was

Figure 3. A personal station at the start of the semester (left) and towards the end of the semester (right).

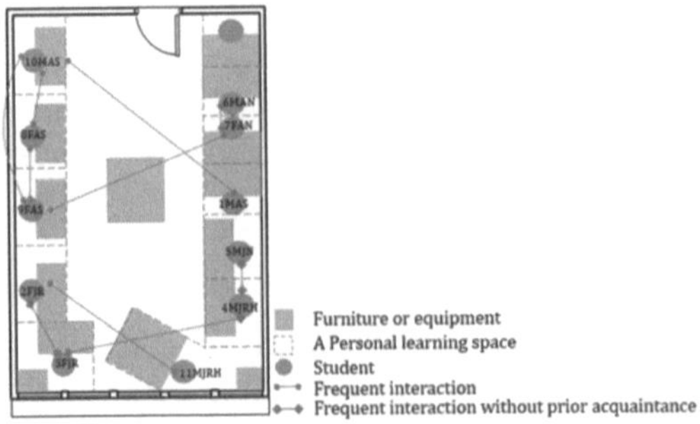


Figure 4. Emerging social structures at the traditional studio.

mentioned as an encouraging environment for peer participation. Peer assessments were perceived as a significant part of the learning process. However, most students were interested in assessments coming from their sub-group members' rather than other peer opinions.

Social factors, such as ethnic origin or prior acquaintance between students, were mentioned as a rationale for creating social structures, shared learning activities and competitive behaviours. These behaviours stood out more at the studio classroom and served as the reason for choosing the location of a personal station. Additionally, proximity between two stations gave rise to new interactions between students who did not know one another earlier, like students 2FJR and 3FJR. Since the IVE did not allow personal stations to be created, these hidden boundaries were broken, but as previously mentioned, these boundaries still affected peer assessment.

*Arab students are very competitive. We are all, like, friends, but highly competitive. So, when you are in a studio with many (Arab students), it makes you choose. You either don't invest at all, or you do everything you can. (Student 10MAS).*

Indirect learning interaction, obtained from viewing another student's output, or listening to a peer crit, should be considered as well. Having the IVE as a major component of the syllabus, students mainly worked with digital tools. They were, therefore, less exposed to peer progress at the studio classroom, which limited indirect learning sources. Since, in comparison with other studio courses, the course had more group crits (at the IVE), indirect learning was in fact promoted.

*(At the IVE) I felt that I was in some sort of delay compared to everyone else, and I stopped and prepared a physical model. (Student 2FJR).*

### 6.3 *Experiencing physical conditions*

All interviewees but one claimed that the studio classroom was more comfortable and flexible, due to the space's affordance of personal stations and the possibility of working while another student receives guidance. The IVE, on the other hand, was defined as unsuitable for a long stay, due to the extreme cold and lack of natural light. The space was experienced as inaccessible and inflexible, due to the need for prior coordination before attendance, and the inability to work while others present their work. Some positive expressions mentioned the IVE as a routine breaker, or a new space offering a new "event." Another advantage was the IVE's proximity to the faculty's computer lab, which allowed students to quickly make improvements in their digital models and re-evaluate them at the IVE.

### 6.4 *Experiencing the virtual conditions*

The IVE was used during formal reviews and group crits. Very few personal desk-crits were conducted there, allowing others to use the immersive screen at the same time. This opportunity was exploited only once, by students 2FJR and 8FAS, who did not interact at the studio space at all.

Figure 5. A spontaneous discussion between students 2FJR and 8FAS at the IVE.

The students used the IVE screen to present static and large-scale images and to perform virtual walkthroughs. Several animation films were presented towards the end of the course, at the instructor's request. Opposed to animations, the virtual walkthrough is not pre planned, making it possible to travel freely within the 3D display and discover unresolved spots. The students addressed the differences they experienced between these displays, referring to human scale, resolution qualities, the required level of detail, spatial comprehension and ability to evaluate design assumptions. Student 9FAS emphasized the sense of responsibility that she had experienced through use of the display at the IVE.'

*Every time I presented at the lab, I would look and recognize all my mistakes, because I saw it big (on a large scale). For example, when you present with a projector, you're fine. But when you present it (at the IVE), it requires responsibility. It therefore increases the chances that everyone will see the mistakes I made. (Student 9MAS).*

The motion during a virtual walk-through was observed as well. The movement had no gravity constraints and allowed users to fly or to pass through walls. Despite that, the users mostly imitated real-world movements, such as climbing stairs, or using doors to enter a space. A display on personal computers did not follow gravity constraints at all. These interactions demonstrate the degree of presence that was experienced by the users. Additionally, movements that did not obey real-world constraints enhanced symptoms such as imbalance and sickness and received negative comments from peers. Since at the studio, students learn by confronting real-world design problems, these gravity-bounded activities may indicate the IVE's advantage for fostering a situated learning approach.

### 6.5 *Emerging and shaped learning habits*

Analysis revealed a change in learners' working habits. Students experienced self-improvements in technological knowledge and time management. We examined the relationship between learners' technological abilities and willingness to perform an immersive walkthrough. The quality of movement during the walkthrough depends directly on the way the model is digitized. A cumbersome model, full of layers and details, will lead to slow movement, or none at all. Models that were designed on Sketchup software provided the smoothest walkthroughs, while Rhino and Revit models were lacking textures and had difficulties in movement. Students, therefore, had to either master in Sketchup or convert their files prior to IVE crits. We expected to see that students with high technological knowledge would demonstrate willingness to perform a virtual walkthrough. It was assumed that early acquaintance with modelling software would enhance a readiness for using new digital tools. Surprisingly, these students chose to use familiar software even if it did not allow a virtual walkthrough to be performed or supported a poor display. Others chose to convert their files only before the final review, where they were obliged to walk-through the model.

*I've learned to reach a level of graphics that helps people understand my project. I worked in a way that is different from previous Studios - mainly because I worked with Sketchup from the beginning. I also drew schemes and sections and sketches from the model. In previous projects, I had a computerized model only at the end of the semester. (Student 9FAS).*

Student 11MJRH expressed his unwillingness to try a different software: *"Rhino is my tool, I'm not leaving Rhino."* This expression may be considered as an experience of attachment to a familiar digital environment. Student 4MJRH experienced the emergence of a new learning habit. The student testified that the lack of prior knowledge in converting a file to a suitable software prevented him from performing walkthroughs. Once he overcame this challenge, he regularly converted files for additional walkthroughs. In order to hide unresolved areas in the model, some students specifically avoided walkthroughs during formal reviews. These results show that different concepts can interfere with one another: a) the concept of a review as setting to present coherent and resolved design decisions, and b) the virtual walkthrough concept as a setting that allows spontaneous and unplanned exploration of unresolved design problems.

## 7 CONCLUSIONS

This study uses commonalities and salience in learners' experiences to define patterns of two different educational environments. The IVE was found as a discouraging setting for place attachment. Students found the IVE to be a formal space, inaccessible for free entry – one that was a deterrent and not easy to use. Despite these difficulties, the immersive walkthrough was experienced as advantageous for self-assessments, peer assessments and spatial comprehension. It was positively perceived for informal crits rather than formal reviews. Personal attitudes were noted to affect the way students used the IVE. The IVE was found to restrict parallel learning activities and the emergence of permanent social structures, while enhancing the appearance of diverse social structures and shaped learning concepts. The studio space was considered as a flexible and comfortable setting. While these characteristics are highly appreciated, the results show that the space is no longer used according to its desired functions. Although the ability to reconfigure the setting was associated with a sense of attachment to one's personal space, students rarely stored personal equipment and mostly used the space during crits. The use of personal computers restricted knowledge sharing and made the studio setting more teacher-centered. The course's multi-setting syllabus may serve as an explanation for these findings. Considering that the course had multiple group crits at the IVE, students may have not felt the need to join additional peer crits at the conventional studio. This study expands recent understandings of similar multi-setting courses (Boudhraa et al., 2019; Rodriguez, Hudson, & Niblock, 2018), by adding personal traits to each setting.

The study has several limitations. The research sample is small and has no control group. The subject course alternately used both studio and the IVE. Consequently, some behaviour settings may have specifically emerged out of this criterion. Future work may include observation on a fully immersive studio course. Despite these limitations, the study presents a novel, place-based method to define learning spaces, and opens the gateway for further inquiry.

Using socio-spatial interactions to define the behaviour setting of a given environment lays the groundwork for promoting learning performance, particularly when dealing with Constructivist learning. A design of multi-setting syllabi can flourish by being attentive to different learner needs. Students spend a significant amount of time at the studio. As such, Studio pedagogy ought to address these objectives at times of critical technological developments.

## ACKNOWLEDGEMENTS

The research reported in the current article is kindly supported by the European Research Council grant (FP7 ADG 340753) and by the Jacobs scholarship. We thank Professor Yehuda E. Kalay and Dr. Efrat Eisenberg for enlightening insights. The students are dearly acknowledged for their participation.

## REFERENCES


Alexander, C., Ishikawa, S., & Silverstein, M. (1977). *A pattern language: towns, buildings, construction*. New York: Oxford University Press.

Barker, R. G. (1968). *Ecological Psychology: Concepts and Methods for Studying the Environment of Human Behavior*. Stanford University Press.

Beckers, R., van der Voordt, T., & Dewulf, G. (2016). Learning space preferences of higher education students. *Building and Environment*, *104*, 243–252. https://doi.org/10.1016/j.buildenv.2016.05.013

Beyraghi, S., & Asl, L. B. (2018). Semiotics of behavioral settings in educational spaces, emphasizing the social value of spaces (Case study of Islamic Art Faculty and Architecture and Art Faculty of Azad University of Tabriz. *Int. J. Architect. Eng. Urban Plan*, *28*(2), 1–24. https://doi.org/10.22068/ijaup.28.2.117

Blascovich, J., Loomis, J., Beall, A. C., Swinth, K. R., Hoyt, C. L., & Bailenson, J. N. (2002). Immersive Virtual Environment Technology as a Methodological Tool for Social Psychology. *Psychological Inquiry*, *13*(2), 103–124.

Boudhraa, S., Dorta, T., Milovanovic, J., & Pierini, D. (2019). Co-ideation critique unfolded: an exploratory study of a co-design studio 'crit' based on the students' experience. *CoDesign*. https://doi.org/10.1080/15710882.2019.1572765

Canter, D. (1977). *The Psychology of Place*. London: The Architectural Press LTD.

Castronovo, F., Oprean, D., Liu, Y., & Messner, J. (2017). Application of Immersive Virtual Reality Systems in an Interdisciplinary Studio Course. In *Lean & Computing in Construction Congress- Joint Conference on Computing in Construction* (Vol. 1). Heraklion.

Cooper, C. (1974). The House as a Symbol of Self. In J. Lang, C. Burnette, W. Moleski, & D. Vachon (Eds.), *Designing for Human Behavior: Architecture and the Behavioral Sciences* (pp. 168–172). Stroudsburg: Dowden, Hutchinson & Ross Inc.

Cross, N. (2006). *Designerly ways of knowing*. London: Springer.

Dart, B. C., Burnett, P. C., Purdie, N., Boulton-Lewis, G. M., Campbell, J., & Smith, D. (2000). Students' Conceptions of Learning, the Classroom Environment, and Approaches to Learning. *The Journal of Educational Research*, *93*(4), 262–270. https://doi.org/10.1080/00220670009598715

De la Harpe, B., Peterson, J. F., Frankham, N., Zehner, R., Neale, D., Musgrave, E., & McDermott, R. (2009). Assessment Focus in Studio: What is Most Prominent in Architecture, Art and Design? *International Journal of Art & Design Education*, *28*(1), 37–51. https://doi.org/10.1111/j.1476-8070.2009.01591.x

Demirbas, O. O., & Demirkan, H. (2000). Privacy dimensions: A case study in the interior architecture design studio. *Journal of Environmental Psychology*, *20*(1), 53–64. https://doi.org/10.1006/jevp.1999.0148

Eigbeonan, A. B. (2013). Effective Constructivism for the Arch-Design Studio. *International Journal of Architecture and Urban Development*, *3*(4), 5–12.

Fowler, C. (2015). Virtual reality and learning: Where is the pedagogy? *British Journal of Educational Technology*, *46*(2), 412–422. https://doi.org/10.1111/bjet.12135

Freitas, M. R.De, & Ruschel, R. C. (2013). What is happening to virtual and augmented reality applied to architecture? *Proceedings of the 18th International Conference on Computer-Aided Architectural Design Research in Asia (CAADRIA 2013)*, (Caadria), 407–416.

Gandolfi, E. (2018). Virtual reality and augmented reality for education. In K. Kennedy & R.. Ferdig (Eds.), *Handbook of Research on K-12 Online and Blended Learning* (2nd ed., pp. 545–561). ETC Press. https://doi.org/10.1145/2993363.3006041

Goldschmidt, G., & Smolkov, M. (2006). Variances in the impact of visual stimuli on design problem solving performance. *Design Studies*, *27*(5), 549–569. https://doi.org/10.1016/j.destud.2006.01.002

Gray, C. M. (2013). Informal peer critique and the negotiation of habitus in a design studio. *Art, Design & Communication in Higher Education*, *12*(2), 195–209. https://doi.org/10.1386/adch.12.2.195

Hod, Y. (2017). Future Learning Spaces in Schools: Concepts and Designs from the Learning Sciences. *Journal of Formative Design in Learning*, *1*(2), 99–109. https://doi.org/10.1007/s41686-017-0008-y

Lave, J., & Wenger, E. (1991). *Situated Learning: Legitimate Peripheral Participation* (24th ed.). Cambridge: Cambridge University Press.

Lawson, B. (2004). *What designers know*. Amsterdam: Elsevier.

Lee, S., & Yan, J. (2016). The impact of 3D CAD interfaces on user ideation: A comparative analysis using SketchUp and Silhouette Modeler. *Design Studies*, *44*, 52–73. https://doi.org/10.1016/j.destud.2016.02.001

Lewicka, M. (2011). Place attachment: How far have we come in the last 40 years? *Journal of Environmental Psychology*, *31*, 207–230.

Loomis, J. M., Blascovich, J. J., & Beall, A. C. (1999). Immersive virtual environment technology as a basic research tool in psychology. *Behavior Research Methods, Instruments, & Computers, 31*(4), 557–564. https://doi.org/10.3758/BF03200735
Manzo, L. C. (2005). For better or worse: Exploring multiple dimensions of place meaning. *Journal of Environmental Psychology, 25*(1), 67–86. https://doi.org/10.1016/j.jenvp.2005.01.002
Marchand, G. C., Nardi, N. M., Reynolds, D., & Pamoukov, S. (2014). The impact of the classroom built environment on student perceptions and learning. *Journal of Environmental Psychology, 40*, 187–197. https://doi.org/10.1016/j.jenvp.2014.06.009
Mikropoulos, T. A., & Natsis, A. (2011). Educational virtual environments: A ten-year review of empirical research (1999–2009). *Computers & Education, 56*(3), 769–780. https://doi.org/10.1016/j.compedu.2010.10.020
Milovanovic, J., Moreau, G., Siret, D., & Miguet, F. (2017). Virtual and Augmented Reality in Architectural Design and Education. An Immersive Multimodal Platform to Support Architectural Pedagogy. In *CAADFutures 17* (pp. 513–532).
Oh, Y., Ishizaki, S., Gross, M. D., & Yi-Luen Do, E. (2013). A theoretical framework of design critiquing in architecture studios. *Design Studies, 34*(3), 302–325. https://doi.org/10.1016/j.destud.2012.08.004
Paes, D., Arantes, E., & Irizarry, J. (2017). Immersive environment for improving the understanding of architectural 3D models: Comparing user spatial perception between immersive and traditional virtual reality systems. *Automation in Construction, 84*, 292–303. https://doi.org/10.1016/j.autcon.2017.09.016
Pektaş, Ş. T., & Erkip, F. (2006). Attitudes of design students toward computer usage in design. *International Journal of Technology and Design Education, 16*(1), 79–95. https://doi.org/10.1007/s10798-005-3175-0
Rodriguez, C., Hudson, R., & Niblock, C. (2018). Collaborative learning in architectural education: Benefits of combining conventional studio, virtual design studio and live projects. *British Journal of Educational Technology, 49*(3), 337–353. https://doi.org/10.1111/bjet.12535
Schön, D. A. (1987). *Educating the Reflective Practitioner. Toward a new design for teaching and learning in the professions*. (first). San Francisco: Jossey-Bass.
Scupelli, P., & Hanington, B. (2016). Design Studio Desk and Shared Place Attachments: A Study on Ownership, Personalization, and Agency. In *DRS2016: Future-Focused Thinking* (Vol. 2). https://doi.org/10.21606/drs.2016.259
Slater, M. (2009). Place illusion and plausibility can lead to realistic behaviour in immersive virtual environments. *Philosophical Transactions of the Royal Society B, 364*, 3549–3557. https://doi.org/10.1098/rstb.2009.0138
Sopher, H., Fisher-Gewirtzman, D., & Kalay, Y. E. (2018). Use of Immersive Virtual Environment in the Design Studio. In A. Kepczynska-Walczak & S. Bialkowski (Eds.), *the 36th eCAADe Conference* (Vol. 2, pp. 853–862).
Sopher, H., Fisher Gewirtzman, D., & Kalay, Y. E. (2019). Going immersive in a community of learners? Assessment of design processes in a multi-setting architecture studio. *British Journal of Educational Technology, 50*(5), 2109–2128. https://doi.org/10.1111/bjet.12857
Sopher, H., Kalay, Y. E., & Fisher-Gewirtzman, D. (2017). Why Immersive? - Using an Immersive Virtual Environment in Architectural Education. In A. Fioravanti, S. Cursi, S. Elahmar, S. Gargaro, G. Loffreda, G. Novembri, & A. Trento (Eds.), *The 35th eCAADe Conference* (pp. 313–322). Rome: CumInCAD.
Wang, T. (2010). A New Paradigm for Design and Technology Education? *International Journal of Art & Design Education, 29*(2), 173–183.
Ward, A. (1990). Ideology, culture and the design studio. *Design Studies, 11*(1), 10–16. https://doi.org/10.1016/0142-694X(90)90010-A
Webster, H. (2004). Facilitating critically reflective learning: excavating the role of the design tutor in architectural education. *Art, Design & Communication in Higher Education, 2*(3), 101–111. https://doi.org/10.1177/0267659109106701
Wicker, A. (2002). Ecological Psychology: Historical contexts, current conception, prospective directions. In A. Bechtel R, and Churchman (Ed.), *Handbook of environmental psychology* (pp. 114–126).
Williams, J. (2017). Charrette:essayDesign Studio: A Community of Practitioners? *Charrette, 4*(1), 2017.

*Research & Education in Design: People & Processes & Products & Philosophy – Almendra & Ferreira (eds)*
*© 2020 Taylor & Francis Group, London, ISBN 978-0-367-49296-0*

# On the notion of power in education and its presence in design studio

S.E. Karabulut
*Department of Industrial Design, Faculty of Architecture and Design, Istanbul Şehir Üniversitesi*

ABSTRACT: This paper aims to explore the notion of power in educational discourse and apply its theories to literature of design pedagogy, particularly on design studio. In the first section, it is introduced how design studio in this paper is taken as a living environment where theories of social sciences are applicable. Then the traditional and critical theories of pedagogy are explored and possible contributions of transitional pedagogy is discussed. In the following section, practices in the design studio are attempted to be examined with the gaining from the reading of educational theories. Corresponding weaknesses are tried to be pointed out. Lastly, it is discussed how power should be treated in order to generate a more inclusive design pedagogy.

## 1 INTRODUCTION

Today, *design studio* is commonly acknowledged as the core of undergraduate level design education in Turkey for not only being a space where students and teachers spend their time the most, but also for representing the department's pedagogical activities in general. Deans of the Consortium of Eastern Schools of Architecture has been argued that the design studio, similar to the internships of medicine students, is a unique model considering its intensity and involvement of its participants (1981). It is an exceptional learning environment in a university for being more than a room for lecturing. Design studio certainly involves more active intellectual and social participation than a typical classroom, with transitionary state of students between analytic, synthetic, and evaluative modes of thinking in different sets of activities (Dutton, 1987). It occasionally becomes a place, not dependent on the 'official' studio hours, where students work, share, chat, eat, relax, nap, or simply live (Anthony, 1991; Cuff, 1991; Oh, Ishizaki, Gross, & Yi Luen Do, 2013). Holding to this perspective, activities within it resembles its inhabitants' daily lives. Pedagogical activities in design studio, therefore, involve and are attached to the complex web of subtleties that are normally not as important considerations in regular classrooms. Consequently, the author's take is that the exploration of the activity of students and teachers in design studio in terms of comprehensive theories of social sciences would help to develop a deeper understanding of design pedagogy. For this purpose, a much-discussed issue of power is selected to be analyzed and utilized throughout the paper in contexts of both educational theories and design pedagogy. Doing so, arguments from educational theories are tried to be interpreted in the discussions on and practices of design pedagogy.

## 2 DISCUSSIONS ON POWER IN EDUCATIONAL THEORIES

Henry Giroux (2015) says that it would be inaccurate to try to understand education without first talking about much broader political, economic, cultural, and social structures in which it is embedded. The institutional burdens should be considered in order to understand the pedagogical activities executed under it. Educators find themselves surrounded or steered by some certain boundaries time to time, limited in their movements in educational domain (Giroux, 2015). Therefore, while exploring the micro-level activities in the design studio, it would be helpful to keep the issue of power in our minds, also as a more comprehensive institutional matter.

University education in general is highly involved in power-related issues according to Foucault (2017) for both being carried out under an institution and its application. He notes a threefold message that is directed from 'educator' to his students (educator here, also symbolizes the institution and the educational system he performs under) that constitute many elements about power relations in education. In the encounter of two agents, educator tells or implies the following: (1) "There are things that you do not know, and you should certainly know them"; (2) "There are things that you should know. I know them and I am going to teach you"; and (3) "After I teach you, then you should be knowing them. I am going to check if you learned or not" (Foucault, 2017). He calls these three statements (1) accusation, (2) obligation, and (3) control. These three statements are not only made their presence felt at the moment of student's first encounter with the educator, but also constantly reminded throughout student's education in the institute. It can be deduced from Foucault's description that in traditional pedagogy knowledge is a transferred phenomenon rather than a produced one. Accordingly, school is a place not for producing the knowledge, but for transferring it. In the constitution of this understanding, macro-structures are accountable for influencing institutions and their workers, however teachers' micro-level practices in classrooms (or design studios) are equally responsible for such an approach towards students. Nevertheless, responsibility brings the right of disposition. Responsibility on one level, gives the nature of teacher's individual pedagogical approach (including the distribution of power) to their own disposal. To this respect, trying to explore the issue of power, it would be misleading to underestimate the in-class activities of teachers.

Critical pedagogy, utilizing the concept of power, addresses traditional pedagogy in many aspects. Within traditional educational theory the focus is on how to transmit, teach, and systematically evaluate the 'legitimate' knowledge. Schools are treated as functional places for dominant society, in providing values, knowledge, and social relations to maintain the existing society (Fitzclarence & Giroux, 1984). So the main criticism towards the traditional educational theory is the fact that it disregards the function of power in nourishing the dominant interests of some groups in the expense of others (Fitzclarence & Giroux, 1984). Fitzclerence and Giroux (1984) pointed out that the pedagogical approach is there, not to support critical thinking but to make methodological improvements that are serviceable in the realm of dominant culture.

Bowles and Gintis (1976) argued that the inspection of how the social relationships in the workplace are reproduced in the classroom would benefit us to understand certain types of knowledge and control that characterized American schools. Agreeing with them Fitzclarence and Giroux (1984) acknowledges the differing pedagogical approach towards students from different social stratum e.g. working-class students are taught to be punctual and follow the rules, while upper-middle and ruling class students are trained in leadership skills and creative thinking. What is noteworthy here, is the assumption that while sustaining certain political, economic, and social interests, schools also function as a place for distinction. For Bourdieu and Passerson (1977), the culture that is transmitted in schools are not related only with the dominant culture in the society but appeals to a wider society for promoting the culture of dominant groups, and at the same time, invalidates the culture of subordinate groups of society. In other words, it creates excluded 'others'.

Critical pedagogy pointed out two main involvements of power: with knowledge in its production, organization, transmission, stratification, allocation, and evaluation (Fitzclarence and Giroux, 1984); and with social relations as in hierarchy, competition (Dutton, 1987), oppression, control, and domination. However, it fell short of providing alternative(s) to traditional pedagogy and answering some questions such as: why and how the 'legitimate' knowledge is produced in particular ways; or how all of the subordinate cultures are created and sustained. Fitzclarence and Giroux (1984) summarizes the problem of critical pedagogy clearly:

*[...] this position fails to treat power dialectically. Thus, power is primarily linked to forms of control that become synonymous with the logic of domination. There is no sense here of how power works in the positive sense as a form of opposition, or even as a form of affirmation for those groups engaged in defining the world and human experience in terms that move outside dominant discourses. (p. 467)*

As it can be understood from the argument above, power can be treated with an alternative thinking, which would help to examine pedagogical activities in a deeper sense and subsequently generate a more beneficial and inclusive pedagogical approach. In this regard, the issue of power cannot be framed with merely the conflict between the powerful and the powerless. Foucault's (1980) seminal take of the subject enlightens us about this view and includes some terms that are unconventional to the discourse of power, so it would be beneficial to quote in length:

*If power were never anything but repressive, if it never did anything but to say no, do you really think one would be brought to obey it? What makes power hold good, what makes it accepted, is simply the fact that it doesn't only weigh on us as a force that says no, but that it traverses and produces things, it induces pleasure, forms knowledge, produces discourse. It needs to be considered as a productive network which runs through the whole social body, much more than as a negative instance whose function is repression. (p. 119)*

On the basis of his annotation, adopting its relations with various terms, the notion of power is not approached as an absolute 'evil' concept in this paper. Correspondingly, social control could be utilized as a practice of liberation, establishing the settings for critical thinking and practice (Fitzclarence & Giroux, 1984).

In this respect, the empowerment of the teachers and students becomes a crucial factor in the nature of a desirable pedagogy. Foucault (1979) says that power is not something to be possessed but something exercised. Klaus Krippendorff (1995) has a similar approach to the issue. He states that power is not measurable by the amount of access to a resource of powerful and powerless. It is not something that some has and some other lack but "[it] manifests itself in the unwillingness to hold authorities accountable for what they say or do and ... in the refusal to provide accounts when requested".

Reserving the undeniable effects of the institutional administration and its direct executions, these above conceptions of power puts students' and teachers' actions in the classroom to a very significant position in our discussion of power. Not only for the executive impacts to the social relations of the pedagogical setting, but also for their undoubtedly political role that justifies or disapproves the mainstream modes of ideology and culture. Teachers have access to the relationships of power with knowledge and social relations, while holding the disposal of their interpretation and clarification (Dutton, 1987). If we are to see the declaration of legitimate forms and culture of empowerment as a great effort, then the form and content of the knowledge as well as the social practices within which it is adopted is a part of that effort (Freire, 1975). For that matter, it is significant to explore the performances of the actors that are utilized for manifesting power; and the forms of power which can operate in the environment of the design studio.

## 3 ISSUE OF POWER IN THE DESIGN STUDIO

In the respect of power in the design studio, firstly what is taught should be under consideration. What is the communicated knowledge in the design studio? What are the generalities (Uluoğlu, 2000) of design knowledge that are treated as commonsense? Perhaps the modernist understanding in product design education which can be still traced in the studio knowledge can set a good example for the design education in Turkey. A modernist discourse which directs design students in a certain way and preclude alternative thinking can be observed in many cases starting from students' first encounter with design in their first year of undergraduate education. We expect a novice student to be uninformed (or not acculturated) in design issues, therefore the knowledge communicated to him/her could easily be taken as legitimate knowledge. Since the knowledge they acquire in the first place will shape their cultural and ideological manner in their future lives, what is taught (the selection of legitimate knowledge) has a strategic role in what Dutton (1987) calls 'political socialization' of students.

Secondly, we should take the design studio itself as a place of social practice. As it has been mentioned above the social relations in classrooms show correspondence with the workplaces in the society. Dutton (1987) analyzes the mentioned social practice in two main headings: hierarchy and competition. Power being the subject, this paper's concern is the former. Hierarchy is the clear absence of equality in the distribution of power. According to Dutton, "hierarchy obviates the presence of dialog [. . .] as a fundamental precondition dialog requires an equality of participants – an equal distribution of power – which by definition is lacking in any system of hierarchy" (1987, p. 18). Dialogical inequalities degenerates the authentic conversation (Krippendorff, 2009) where knowledge is produced and therefore communicated. Gregory Baum (1977) says the agents should be equal in order to dialog to take place. Similarly, Paulo Freire (1973) highlights the significance of reciprocity, arguing that rather than the skill in persuasion, the ability to dialog with students indicates the successful educator. In respect to this, some conventional roles, e.g. master (in master-apprentice) or analyst (as in psychotherapy), of the teacher in the design studio should be questioned for promoting a hierarchical relationship with the student.

Chris Argyris (as cited in Dutton, 1987) points out significant observations in his extensive study of architectural program that encompasses different styles of teaching in different universities and different years, focusing on behaviors, teacher-student verbal communication, and course contents. One of his points is the observation that studio professors do not use the theories they claim to adopt. Which in my opinion creates a confusion in students' minds for where to attribute power, teacher or the discipline/curriculum. In relation to that, one other finding is that the studio setting is a teacher-centered experience and hence the design learning can be actualized only to the degree of students' capacity to understand and accept the professor's way of teaching. Considering the current situation in departments of design in Turkey, it is not unusual to witness students who (can) understand and accept the approach of a certain project facilitator, accordingly get better grades and also develop friendly relationship with the professor in some cases. Rest of the students here, experience a major unfavorable situation by becoming 'others'. In such an environment, students apply to the solution by 'playing along with the game' rather than 'feeling the game', from which *ritualistic mastery* arises (Webster, 2006). The term refers to unfavorable ways of behavior and thinking of students which prioritize developing societal skillsets over producing design knowledge in the pursuit of academic success.

Lastly, Argyris (as cited in Dutton, 1987) states that neither teachers nor students question enough the values and assumptions triggering their practices, which results in taking mystery as a symptom of mastery. He says that professors rarely "help the students recognize the ideas and theories that were embedded in their work or make explicit their own ideas, or reflect about their own work and thinking in a way that would help the students understand the discovery-invention-production process" (as cited in Dutton, 1987, p. 19). We can see writings in similar manner that could support this argument in design literature. For example, Schön (1982) in "The Reflective Practitioner" ends the chapter where he animates Petra (the student) and Quist (the teacher) with critiquing Quist's lack of reflectiveness on his own mastery performance under which a potential beneficial "fundamental structure of inquiry" for students lies. Goldschmidt et al. (2010) found that providing student with the rationale behind the instructor's teaching methods leads to better learning. As Schön's writing highlights the designer*ness* of the studio master and Goldschmidt et al.'s the educator*ness* they both agree in some extent that the agent must be more transparent in displaying his/her own process.

Perhaps the reasons for those mentioned behaviors of the teachers can be explained through the concept of power. Schön (1982) describes a comfort zone for teachers which he calls web of moves. In which teachers are moving freely without the obligation of see-*ing* something with a different point of view. They are swimming in the pool of familiar knowledge, and talk within a corresponding discourse that is in some cases difficult to get involved or intimidating for students. Klaus Krippendorff (1995) provides a good articulation of what design studio teachers might be experiencing while holding the (comfortable) power:

*First, we tend to see for the each of us obvious, the expected and the familiar. ... At that moment, you allowed yourself to be locked into one way of seeing, into one single uni-verse, and this excluded or blinded you from seeing any other uni-verse. Without the effort on your part to break out of this self-confinement, you would have neither experienced nor missed alternative ways of seeing, much less imagined that equally sane people could simultaneously be in altogether different uni-verses, without ever knowing this of each other. This is a quite remarkable fact that most common sense accounts of perception, of communication and, as we shall see, of power conveniently ignore. (p.103)*

Another problematical subject is the fact that studio teachers are also assessors. Even though some teachers find being supervisor and assessor as two separate and distinct roles, the fact that these roles are played by the same person causes a conflict in students' point of view. While pointing out the constructed-*ness* of the roles of teachers and students in the design studio, Belluigi (2016) stresses student's own identity in their conception of the assessor. She claims, for example, a 'hegemonic overlord' assessor would have less negative impact for the more 'acculturated' (by the dominant culture of the design studio) students compared to students who are considered 'weak'. The power differential between the assessor as 'critic' and student as 'pedagogised other' (Webster, 2006) is objectified and even reinforced by assessment. Feeling excluded by and because of their personal identities that are culturally constructed creates a deep fracture in the reciprocal relationship between teacher and student. Falchikov and Boud (2007) says that the worst mode of this dynamic is when teacher objectifies and student subjectifies all their judgements.

In sum, in a system of education where power is made an issue through its unequal distribution, a conflict between teacher and student (can be read as dominant and subordinate cultures) over the control of the education environment is inevitable. Although there must be and are many exceptional cases, it seems like the customary behavior and order in the design studio, in other words 'micro-technologies of power' (Foucault, 1979), strengthen the uneven distribution of power among teachers and students.

## 4 TO CONCLUDE: UNDERSTANDING THE ROLE OF POWER IN TRANSFORMATIVE DESIGN PEDAGOGY

Power can be a positive and a negative phenomenon at the same time, and can be utilized for an emancipatory practice. Teachers have their culturally constructed 'habitus'. Student side of the habitus should be recognized and explored so as to make use of them in the favor of a better pedagogy. Understanding their cultural experience would also help us to see what students want to learn besides and despite their background so they can cross the limits of mainstream culture and create their own attitudes. Fitzclarence and Giroux (1984) states that the notion of cultural power has to be taken seriously, "[a critical pedagogy] can do this by both legitimating and challenging the cultural experiences that make up the historical and social particulars that constitute the cultural forms and boundaries that give meaning to the lives of students and other learners" (p.473). They argue that two types of take in cultural power are necessary from teacher's perspective. First, is for teachers to develop forms of knowledge and social practices that can work with the cultural backgrounds that students bring to schools. Doing so, they will be confirming such experiences and putting students in a role of active participation. Second, is for teachers to work on the former experiences of the students so as to make progress in their strong and weak points. As Paulo Freire has long argued, such cases comprehend how students shape their view of the world.

Design education is a demanding field in which reflecting the knowledge back to the teacher is not sufficient enough for students. In final juries they are not only tested for their abilities in repeating back what they are taught, but expected to produce ideas, solutions, perspectives; in sum, knowledge. Accepting that, the essential focus should be ensuring a convenient environment to produce knowledge.

In conclusion, development of a better pedagogy as in distributing the power equally (or fair) among the inhabitants of the design studio is highly dependent on teachers' approach on what the content and nature of knowledge should be, how it should be communicated, and in which social relationships it should be appropriated. It is worthy to note that the design discourse is vulnerable to effects of broader economic, social, cultural considerations of institutions, however narration is on teachers' disposal through which the 'disciplinary truth' can be shaped and is communicated with students. The awareness of the access to the relationships between power and knowledge, and power and social practices would empower teachers in forming the future of design pedagogy.

## REFERENCES


Anthony, K. H. (1991). *Design juries on trial: The renaissance of the design studio*. New York: Van Nostrand Reinhold.

Baum, G. (1977). Truth beyond relativism: Karl Mannheim's sociology of knowledge. *The Père Marquette Lecture in Theology Series*, *9*, 1–81.

Belluigi, D. Z. (2016). Constructions of roles in studio teaching and learning. *International Journal of Art & Design Education*, *35*(1), 21–35.

Bourdieu, P., and J. Passeron. (1977) *Reproduction in Education, Society, and Culture*. Beverly Hills, CA: Sage Publishers.

Bowles, S., and Gintis, H. (1976) *Schooling in Capitalist America*. New York: Basic Books.

Cuff, D. (1991). *Architecture: The story of practice*. The MIT Press.

Deans of the Consortium of Eastern Schools of Architecture (1981) "The Challenge to Schools of Architecture" Architecture Education Study, Vol. 1: The Papers The Andrew Mellon Foundation and the Consortium of Eastern Schools of Architecture (distributed by the MIT Laboratory of Archi- tecture and Planning, Publications Program).

Dutton, T. (1987). Design and Studio Pedagogy. *Journal of Architectural Education (1984-)*, *41*(1), 16–25. doi:10.2307/1424904

Falchikov, N. & Boud, D. (2007) *Assessment and emotion: the impact of being assessed*, in D. Boud & N. Falchikov [Eds] Rethinking Assessment in Higher Education: Learning for the Longer Term. New York: Routledge, pp. 144–55.

Fitzclarence, L., & Giroux, H. A. (1984). The paradox of power in educational theory and practice. *Language Arts*, *61*(5), 462–477.

Foucault, M. (1979) *Discipline and Punish: The Birth of the Prison*. Harmondsworth: Penguin.

Foucault, M. (1980) *Power/Knowledge: Selected Interviews and Other Writings, 1972–1977* ed. C. Gordon. Brighton: Harvester Press.

Foucault, M. [Çeviri Konuşmalar]. (2017, October 24). Michel Foucault - Okul Hakkında [Video File]. Retrieved from https://www.youtube.com/watch?v=Px-JaP-7dQbY

Freire, P. (1975) *Pedagogy of the Oppressed*. New York: Seabury, 1975.

Giroux, H. A. [MacPherson Institute]. (2015, October 22). *Henry Giroux: Where is the Outrage? Critical Pedagogy in Dark Times* [Video File]. Retrieved from https://www.youtube.com/watch?v=CAxj87RRtsc&t=360s

Goldschmidt, G., Hochman, H., & Dafni, I. (2010). The design studio 'crit': Teacher student communication. *Artificial Intelligence for Engineering Design, Analysis and Manufacturing*, *24*(3), 285–302.

Goulet, D. (1973) "Introduction" Paulo Freire, *Education for Critical Consciousness*, The Seabury Press (New York), p. xiii.

Krippendorff, K. (1995). Undoing Power. *Critical Studies in Mass Communication*, *12* (2), 101–132. https://doi.org/10.1080/15295039509366927

Krippendorff, K. (2009). Conversation: Possibilities of its repair and descent into discourse and computation. *Constructivist Foundations*, *4*(3), 135.

Oh, Y., Ishizaki, S., Gross, M. D., & Yi-Luen Do, E. (2013). A theoretical framework of design critiquing in architecture studios. *Design Studies*, *34*(3), 302–325. doi:10.1016/j.destud.2012.08.004

Schön, D. A. (1982). *Design as a reflective conversation with the situation*. In The reflective practitioner: How professionals think in action (pp. 76–104). New York: Basic Books.

Uluoglu, B. (2000). Design knowledge communicated in studio critiques. *Design Studies*, *21*(1), 33–58.

Webster, H. (2006). Power, freedom and resistance: Excavating the design jury. *International Journal of Art & Design Education*, *25*(3), 286–296. doi:10.1111/j.1476-8070.2006.00495.x

*Research & Education in Design: People & Processes & Products & Philosophy – Almendra & Ferreira (eds)*


# In the midst of things: A spatial account of teaching in the design studio

J. Corazzo
*Sheffeild Hallam University*

ABSTRACT: The role of the material space of the studio is underarticulated in design pedagogy, despite the studio being distinctive and central to design education. This underarticulation represents challenges when defending space and its occupation, designing new studio spaces and for those interested in ways in which the material space of the studio can enhance learning experiences. In this paper, I argue that spatial accounts of learning and teaching can begin to address this underarticulated and under-researched area of design education. Specifically, this paper develops spatial accounts by focusing on design tutors' experiences and practices of teaching in the studio. Using ethnographic mapping and interviews with design tutors to show embodied and spatial accounts of teaching in the design studio offer new lenses with which to understand design education teaching practices.

## 1 INTRODUCTION

This short paper investigates the educational design studio, not as is typically encountered – a cultural ideal, rather it looks to the educational studio as a socio-spatial phenomenon. This argument for a socio-spatial lens is underpinned by two observations. First, despite the central position the studio occupies in design education (Orr and Shreeve 2017) the studio as a material resource is becoming increasingly precarious in an era of higher education managerialism and massification (UK, AUS, US). And, although the literature on studio-based design pedagogy continues to grow, little of it directly addresses the role of material space and its contribution to learning. Second, drawing on a sociomaterial perspective, I will argue for the ways socio-spatial accounts of design education can enrich our understanding of how teaching and learning happens.

To address both observations there is a need to better articulate the role of the material space of the studio. Noting such an articulation would be necessarily complex and beyond the scope of this short paper, I will instead address an aspect of this gap in the literature. Design tutors' experiences and practices of teaching in the studio (Corazzo 2019).

The specific aim of this study is to develop a spatial account of teaching in the studio. I will begin by defining the studio (a term that has multiple meanings in education). I will then move on to discuss the theoretical framework. Here I outline the necessity of a sociomaterial approach as the only means to get a purchase on the relations between material space, teaching and learning. I will then briefly describe the methodological challenges for gathering and analysing data in ways that keep the social and the material co-present. Finally, as this work-in-progress, I will share some preliminary results.

## 2 CONTEXT

### 2.1 *What is the studio?*

The studio is a space where students engage in the process of making alongside, or under the guidance of, an 'expert' tutor (Schon 1987). Although the studio takes many forms, core features can be identified: project-based learning, learning-by-doing, the use of material space, and a tendency to demand physical and temporal immersion. As Orr and Shreeve (2017: 15) note, "[s]tudio education is not delivered. Studio education is forged" and this serves to underline both the transformational intent of the studio and its core purpose; to develop proto-artists and proto-designers.

Although the concept of the studio is broadly shared in art, design and architecture education, the term 'studio' can mean more than one thing. Schon's (1987) learning constructs of the studio is a useful analytic lens for identifying the different, but overlapping and related meanings (see Figure 1). In this study, I predominantly focus on the studio as a physical space and the studio as a mode of teaching and learning.

### 2.2 *Precarious studio*

In its ideal form, studio education is predicated on small class sizes, large spaces for inhabitation, around the clock access, the permanent display of work-in-progress and specialist resources (Boiling, Siegel, Smith and Parrish 2013). Thus, studio education is often viewed as resource intensive and expensive when compared with many Higher Education (HE) disciplines. Consequently, in an increasingly marketised HE system (UK), studio education has come under greater scrutiny.

In parallel, the continuing expansion of HE has rendered the studio ideal of a distinct workspace for each student a mostly redundant concept in design education (Marshalsey 2015). Similar issues have been noted in the US (Boling, Schwier, Gray, Smith, & Campbell 2016). The twin pressures of financial efficiencies and expanding student numbers have led to sustained concerns about the impact on disciplinary teaching and learning models (Boddington & Boys 2011; Harrison & Hutton 2014; Rodgers & Jones 2017).

Despite these sustained concerns and perceived threats to studio-based education, educators and researchers have continued to struggle when articulating the contribution of studio space to teaching and learning. As critics have noted (Mewburn 2012; Vyas and Nijholt 2012; Corazzo 2019), existing accounts of the studio tend to underplay its role as a material and spatial

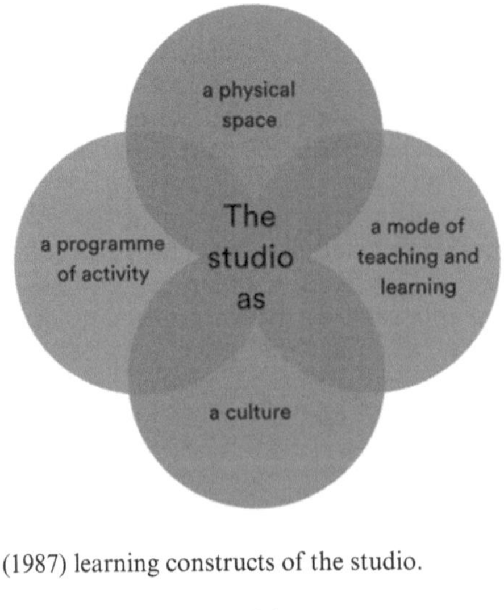


Figure 1. Based on Schon (1987) learning constructs of the studio.

entity. Although, it should be noted, this is typical of a broader 'spatial lacuna' in educational research (Elkington and Bligh 2019).

## 3 THEORETICAL FRAMEWORK

### 3.1 *Realist perspective*

Given the importance and centrality bestowed upon the studio in design education, it is perhaps even more important such a gap is addressed. The gap prevails, in part, because many of the social theories deployed by educational researchers attend to the human dimension of teaching and learning. As Orlikowski (2007:1436) succinctly puts it "[l]anguage matters. Discourse matters. Culture matters. But there is an important sense in which the only thing that does not seem to matter anymore is matter".

Where the spatial dimensions of social activity are considered, space is commonly treated as an environment in which social activity takes place, rather than being integral to its occurrence (Giddens 1979). As such, material space becomes an invisible backdrop for the complexity and vibrancy of social space. In turn, such a perspective renders material space as a 'passive container for social action' (McGregor 2004: 350).

The conception of space as a container tends to dominate discussions. In this 'realist perspective' patterns of space and material entities are seen to enable or constrain particular types of human activities. In this view, space has an essence. It can be designed as open, flexible and innovative, qualities which, when harnessed by those within it, can develop 'capacities in students for the twenty-first century' (Mulcahy, Cleveland & Aberton 2015: 578).

Mulcahy, Cleveland & Aberton (2015) argue such a view is problematic because it operates in a predominantly singular direction, and implies a causality that is difficult to prove. Furthermore, it leaves the questions of how change happens unanswered, and while purporting to account for the role of space in learning, it does so by actually separating teaching and learning practices from physical spaces (Mulcahy, Cleveland & Aberton 2015). This separation, in turn, frames the relations between space and learning as instrumental – predictable, causal and stable and sustains an unrealistic and ineffectual paradigm for thinking about space and learning (McGregor 2004; Mulcahy 2006; Cleveland & Aberton 2015; Acton 2017; Bligh 2019).

The separation of space and learning is, according to (Sorenson 2009), symptomatic of a general 'blindness' towards how educational practices are effected by materials. As is typified in the realist enabling/disabling perspective, materials are considered to be things in support of education. However, Sorenson (2009) argues educational practices actually get done by a combination of the social (humans) and the material (non-humans). In other words, humans may use materials and space, but the spatial and material may also 'use humans' and influence educational practices.

### 3.2 *Relational perspective*

Breaking with the dominant realist perspective on space, a relational perspective focuses simultaneously on the social and the material dimensions of educational settings. The social and the material are joined to form the sociomaterial. It is considered a relational perspective because the attention is not on the social or material as separate entities, but on the relations between the social and the material.

A sociomaterial approach has the potential to offer ways of understanding how space and learning are generated together (Decuypere & Simons 2016) and considers the ongoing mutual entailment of human and non-human (Sorenson 2009). As such, it offers ways to explore "encounters between space and its occupation" (Boys 2011: 51). In this sociomaterial approach, space doesn't happen in advance – simply waiting and ready to be used, rather space and the social are generated together (Mulcahy, Cleveland & Aberton 2015). Space is not "viewed as a container within which the world proceeds… space is seen as a co-product of those proceedings" (Thrift 2003: 96).

This view, argues Boys (2011), means space and the social are "inseparable and interlocked, dynamically informing and influencing each other" (Boys 2011: 50–51), space and its use is mutually constituting. As Mulcahy, Cleveland & Aberton (2015: 590) argue, the space should be thought of as a verb rather than a noun: "as something we do (a matter of encounter), rather than something we have (a new learning environment, a finished design) affords acknowledgement of the multiplicity and mutability of spatial and pedagogic practices".

So, could such a theoretical perspective provide us with the conceptual tools to attend to the relations between the spatial and social dimensions in the educational settings of the studio? Acton (2017: 1442) has argued for such an approach in education:

> *The benefit of taking a sociomaterial approach to spatial research is that it carefully illuminates the junctures, tensions and lived practice of spatial-social relationships. It allows attention to focus on embodied learning and teaching, the synergies between place and people, the relations between the imagined affordances implicit in infrastructure design and construction, and the experienced realities of the people who inhabit those spaces in practice.*

## 4 METHODOLOGY

Having outlined a theoretical framework that considers space and learning as mutually co-constituting, I now turn to the specifics of this study – design tutors' experiences and practices of teaching in the studio. This study presented a three-fold methodological challenge: i) the taken for granted and tacit nature of teaching in the studio; ii) the difficulties of articulating the relations between space and teaching/learning; iii) lacking the time and resources to undertake direct ethnography.

### 4.1 *Graphic elicitation*

These challenges prompted a decision to use graphic elicitation. In brief, graphic elicitation involves the use of drawings (created by researchers or participants) together with interviews to elicit participants experiences and understanding (Copeland and Agosto 2012). Graphic elicitation occupies a 'hinterland' between language and the graphical. This hinterland is essential because of the focus it places on the relations between the drawing/diagram and what is said. As Pink (2006) notes, this is not a purely visual process; the visual is interpreted within an interview context, the meaning is made in the relations between the visual and the verbal.

In addition to opening up a hinterland of meaning-making, proponents argue graphic elicitation can have several benefits over traditional qualitative interview processes. These include: memory recall, seeing the normal in new ways (Banks 2001), attending to the multiplicity of experiences that are easily expressed in spoken or written language (Pink 2006) and supporting participants to express complex and abstracts notions (Copeland and Agosto 2012).

This study draws directly on aspects of an approach used by Nolte-Yupari (2017) who deployed graphic elicitation and map making to research new art teachers use of their classrooms. For this study, participants were asked to draw a map of the studio they had taught in that day. Participants were then asked to re-enact the movements and journeys they made in the studio throughout the day by drawing lines – "a gestural re-enactment of journeys actually made" (Ingold, 2007: 84). Participants were interviewed while they drew. The map was used to ask questions and elicit experiences, stories and behaviours of teaching in a studio. This process was filmed with audio (see Figure 2).

This study focused on tutors working on a large undergraduate graphic design programme (approx. 100 student per year). Tutors were purposively sampled to include a mix of teaching experience, gender and different years (and therefore different studios). The hand-drawn maps and interviews were undertaken at the end of a teaching day and in the same studio where the teaching had taken place (see Figure 3).

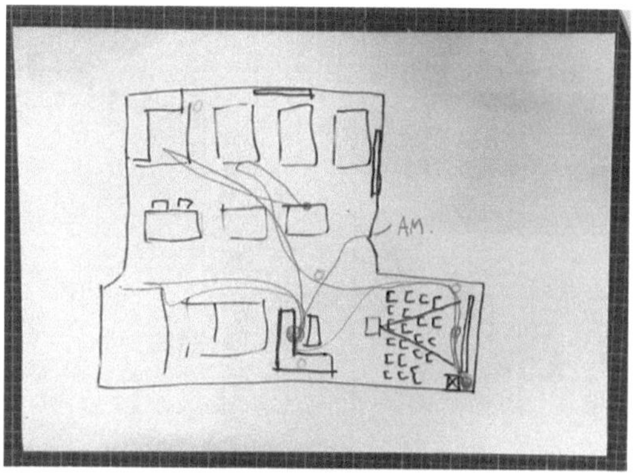


Figure 2. Participant map.

Figure 3. The interview in situ.

## 4.2 *Analysis*

Analysis of data was undertaken by watching the video with audio and annotating the participants map of the studio. Here, deliberate efforts were made to co-locate verbal extracts with space (see Figure 4). The verbal extracts were re-read a number of times and coded. Following this, tentative categories were generated by merging codes. All the while, attention was made to the body, tools, spaces, activities and encounters.

# 5 RESULTS + CONCLUSION

At the time of writing stage, five participants talk and draw interviews have been conducted and preliminary analysis on two of these has begun. Five nascent categories have emerged from the analysis: bodies-in-relation, visibilities, territories, informalities, and mutabilities. As way of a conclusion, I will briefly outline each.

## 5.1 *Bodies-in-relation*

The ethnographic mapping and drawing have worked to highlight a 'bodily pedagogy'; movement, body position and verticality (sitting or standing) are part of a repertoire of studio

Figure 4. Visual analysis.

teaching. The use of the body to signal and shape particular types of encounter; in relation to how 'autonomy' is developed, encouraged, nurtured and maintained.

### 5.2 *Visibilities (see and be seen)*

Analysis reveals a conscious 'signalling' by tutors of their presence in the studio – to 'see and be seen'. Particular 'vantage points' were adopted to enable this behaviour. Even when tutors were not engaged in direct dialogue, there was a tendency to notice, clock, and observe the presence and activities of others. Perhaps because of this omni-visibility, tutors saw partitions as problematic, and there seemed to be an ongoing, but subtle, tussle between tutors keeping the studio open plan to enable lines of sight in all directions and some students erecting partitions and attempting to partially or fully evade these sightlines.

### 5.3 *Territories*

A temporal dimension emerged; participants discussed the ways in which the year unfolded by 'literally ceding ground to them [students] in terms of less tutor-controlled spaces. This has been evoked in relation to 'their areas' and them coming into 'our areas', but also attempts to dispel these territories and make the studio a 'home from home'.

### 5.4 *Informalities*

Participants in both their use and accounts of space appeared to spend considerable time shifting 'formalities' into 'informalities.'. From choosing not to stand behind a lectern, to the deliberate use of a sofa to conduct informal tutorials.

### 5.5 *Mutabilities*

Specific material entities in studios take on highly mutable forms. For example, a sofa in the studio that is on one hand keeping with a 21st century learning space/and or a trope one might expect to see in a professional studio. In this example, the sofa took many forms – a staff space, a student space, a peripheral space enabling escape from an uninteresting brief, a way to signal availability and a place where 'different kinds of conversations' happen to those in a desk crit.

### 5.6 *Mess*

These are very tentative categories and should be considered with caution. They are also messy, lacking the sureties of distinct categories and suggestions for future 'best practice'. However, the methodology and preliminary analysis suggests there is conceptual potential to 'illuminate the junctures, tensions and lived practice of spatial-social relationships' (Acton 2017: 1442). Perhaps most importantly, they begin to illuminate the potential disjuncture between 'imagined affordances' of space (and its design) and the lived experiences of those that inhabit them.

## REFERENCES


Acton, R. (2017) 'Place-people-practice-process: Using sociomateriality in university physical spaces research' Educational Philosophy and Theory, Vol.49(14) pp.1441–1451.

Banks, M. (2001) Visual methods in social research, London: Sage.

Boddington, A. & Boys, J. (2011) Re-Shaping Learning: A Critical Reader – The Future of Learning Spaces in Post-Compulsory Education. Rotterdam: Sense.

Boling, E., Siegel, M., Smith, K. and Parrish, P. (2013) 'Student goes on a journey; stranger rides into the classroom: Narratives and the instructor in the design studio', Art, Design & Communication in Higher Education, Vol.12 (2), pp.179–194.

Boling, E. Schwier, R., Gray, C., Smith, K., & Campbell, K. (eds) (2016) Studio Teaching in Higher Education: Selected Design Cases, Florence: Taylor and Francis.

Boys, J. (2011) Towards Creative Learning Spaces: Re-Thinking the Architecture of Post-Compulsory Education, Abingdon: Routledge.

Copeland, A. and Agosto, D. (2012) 'Diagrams and Relational Maps: The Use of Graphic Elicitation Techniques with Interviewing for Data Collection, Analysis, and Display" International Journal of Qualitative Methods Vol.11(5), pp.513–533.

Corazzo, J. (2019) 'Materialising the Studio; a systematic review of the role of the material space of the studio in Art, Design and Architecture Education' In press.

Crowther, P. (2013) 'Understanding the signature pedagogy of the design studio and the opportunities for its technological enhancement', Journal of Learning Design, Vol.6 (3), pp.18–28.

Decuypere, M. & Simons, M. (2016) 'Relational thinking in education: topology, sociomaterial studies, and figures' Pedagogy, Culture & Society, Vol. 24:(3) pp.371–386.

Elkington, S. and Bligh, B. (2019 'Future Learning Spaces Space, Technology and Pedagogy: Case Study Series published by Advance HE.

Fenwick, T. & Landri, P. (2012) 'Materialities, textures and pedagogies: socio-material assemblages in education' Pedagogy, Culture & Society, Vol.20(1) pp. 1–7.

Giddens, A. (1979) Central Problems in Social Theory, London: Macmillian.

Harrison, A. & Hutton, L. (2014) Design for the Changing Educational Landscape: Space, Place and the Future of Learning, Abingdon: Routledge.

Heywood, I. (2009) 'Making and the teaching studio', Journal of Visual Art Practice, Vol. 8 (3) pp.195–204.

Ingold, T. (2007) Lines: A brief history, New York, NY: Routledge.

Jacklin, H. (2001) 'Locating Pedagogy', British Sociological Association annual conference, York University.

Marshalsey, L. (2015) 'Investigating the Experiential Impact of Sensory Affect in Contemporary Communication Design Studio Education', International Journal of Art and Design Education Vol.3 (3), pp.336–348.

McGregor, J. (2004) 'Spatiality and the Place of Material in School', Pedagogy, Culture and Society, Vol.12 (3), pp.347–372.

Mulcahy, D., Cleveland, B. & Aberton, H. (2015) 'Learning spaces and pedagogic change: envisioned, enacted and experienced' Pedagogy, Culture & Society, Vol. 23(4), pp.575–595.

Mulcahy, D. (2006) 'The Salience of Space for Pedagogy and Identity in Teacher Education: Problem-based Learning as a Case in Point.' Pedagogy, Culture and Society Vol.14 (1) pp.55–69.

Nolte-Yupari, S. (2017) 'Everything Into One Place: Beginning Teachers and the K-12 Art Room Taskscape' Studies in Art Education: A Journal of Issues and Research Vol.58(3) pp. 222–233.
Orlikowski, W. J. (2007) 'Sociomaterial practices: Exploring technology at work' *Organization Studies*, Vol.28, pp.1435–1448.
Orr, S. and Shreeve, A. (2017) Art and Design Pedagogy in Higher Education: Knowledge, Values and Ambiguity in the Creative Curriculum, London: Routledge.
Pink, S. 2006. Visual methods [dictionary entry]. In The Sage dictionary of social research methods, ed. V. Jupp, pp. 320–2. London: Sage.
Pink, S. (2008) 'Mobilising Visual Ethnography: Making Routes, Making Place and Making.
Images' Forum Qualitative Sozialforschung/Forum: Qualitative Social Research, Vol. 9(3), Art. 36.
Rodgers, P. & Jones, P. (2017) 'Comparing University Design Students' and Tutors' Perceptions of Creativity' The Design Journal, Vol.20 (4) pp. 435–457.
Schön, D. (1987) Educating the Reflective Practitioner: Toward a New Design for Teaching and Learning in the Professions, San Francisco, CA: Jossey-Bass.
Shreeve, A., Sims, E. & Trowler, P. (2010): 'A kind of exchange': learning from art and design teaching' Higher Education Research & Development, Vol.29 (2) pp.125–138.
Shulman, L. (2005) 'Signature Pedagogies in the Professions' Daedalus Vol. 134 (3) pp. 52–59.
Smyth, J., and McInerney, P. (2013) 'Making 'Space': Young People Put at a Disadvantage.
Re-engaging with Learning." British Journal of Sociology of Education Vol.34 (1) pp.39–55.
Sorenson, E. (2009) The Materiality of Learning: Technology and Knowledge in Educational Practice, New York, Cambridge University Press.
Thrift, N. (2003) "Space: The Fundamental Stuff of Human Geography." In Key Concepts in Geography, eds by Holloway, S., Rice, S. and Valentine, G. pp. 95–107. London: Sage.
Vyas, D. and Nijholt, A. (2012) 'Artful surfaces: an ethnographic study exploring the use of space in design studio', Digital Creativity, Vol.23 (3–4), pp.176–195.

*Research & Education in Design: People & Processes & Products & Philosophy – Almendra & Ferreira (eds)*
*© 2020 Taylor & Francis Group, London, ISBN 978-0-367-49296-0*

# Integrating immersive visualization laboratory into a design studio

O. Nezer & D. Fisher-Gewirtzman
*Faculty of Architecture & Town Planning, Technion-Israel Institute of Technology, Israel*

ABSTRACT: The Immersive Virtual Environment (IVE) Visualization Laboratory (VisLab), at the Faculty of Architecture and Town Planning at the Technion-Israel, facilitates immersion in a virtual, digital, three-dimensional space – a new working environment and tool. The Visualization Laboratory was presented in a design studio to a group of third- and fourth-year Architecture students. This was the first time that they had been exposed to the new environment and tool of the Visualization Laboratory. The main purpose of the research was to assess and describe the comprehension and evaluation of the students regarding their design process with the integration of the Visualization Laboratory into the design studio course, and in comparison, to the other design tools they use. The research presents the reader with the interpretation, the experience and the meaning that evolved in the observed setting, as described by the students and using a qualitative research approach.

## 1 INTRODUCTION

> *Maps make us feel safe. The idea of controlling space, gives us security... Or maybe it is the distance or altitude that makes us feel superior. Or maybe because we are not seen, we are invisible, and that is why we feel safe.* (Rechmaoui, Artist, 2010)

The emergence of Virtual Reality (VR) technology in the architectural design environment underlines the need to research the integration of VR into the pedagogy of architectural design.

Study methods in the studio are intended to simulate the practical experience of the architect undertaking a design challenge (Schon, 1985). This process of knowledge acquisition occurs through resolving design assignments, carried out under conditions of doubt and uncertainty (Cross, 2001, 2006; Dewey, 1933; Schon, 1983). Engaged in an experiential learning process, the student must employ reflective thinking to introduce clarity to the design challenge, (Cross, 2011; Dewey, 1934, 1938; Kolb, 1984; Salama, 2016; Schon, 1987) - as the design problems are not always clearly defined (Buchanan, 1992; Rittel & Webber, 1973).

The traditional design studio exercises a formidable influence on the future architect. Whilst the student is free from real-world pressure and constraints during the simulated design process, they must nevertheless contend with the emotional and psychological stress that accompanies simulations (Austerlitz & Aravot, 2007; Ochsner, 2000; Willenbrock, 1991).

The Immersive Virtual Environment (IVE) Visualization Laboratory (VisLab), at the Faculty of Architecture and Town Planning at the Technion-Israel Institute of Technology, facilitates immersion within a virtual, digital, three-dimensional space. The Visualization Laboratory was introduced, as a new pedagogical tool, to a group of architecture students taking a design studio course focusing on the adaptive reuse of existing buildings. The research presented in this paper describes and assesses the design work process in the studio over the course of an academic semester, detailing the integration of VR into architectural design pedagogic practice.

### 1.1 *Research question and objective*

A new working environment endowed with contemporary technological tools; the Visualization Laboratory was introduced to a group of Architecture students in a design studio course. The research described in this paper observed the students' design processes, over the course of a semester.

The research describes the first-time experience of students using the Visualization Laboratory in their design process in a studio, with and in addition to design tools which they had previously been exposed to and had worked with. Emphasis was placed on detailing the thought processes manifested by the students during their design assignment, using both the new tool and the other tools and in the context of the design theme of the studio course: the adaptive reuse of existing buildings.

The main objectives of the research were:

To assess and describe the comprehension and evaluation of the students regarding their design process with the integration of the Visualization Laboratory into the design studio course, and in comparison, to the other design tools they use.

To present the reader with the interpretation, the experience and the meaning that evolved in the observed setting, as described by the students.

A secondary objective of the research was:

To examine how the design studio course, focusing on the adaptive reuse of existing buildings, contributed to the work process of the students in the design studio.

## 2 RESEARCH METHODS

A qualitative research approach was deemed best suited for this study, given the objective of interpreting phenomena according the meanings that people attach to them. In line with the orientation of this approach, the researcher tries to interpret the phenomenon by observing and listening to the information without preconceptions. After data collection and classification, the former in the research's natural environment, the researcher attempts to establish interpretative concepts through an inductive thought process. (Shkedi, 2003). In this sense, the current research is not attempting to validate a theory or an assumption, but rather to reflect the process that the students experienced and from their own point of view. From the variety of methods available within the qualitative research approach, the Case Study tradition, (Creswell, 1998) was selected.

Data collection took place over the course of a full semester of the design studio class. Present at each of the twice-weekly classes and group discussions, the researcher assumed the role of a silent non-participant observer. Activities connected to the observer role included taking notes, recording discussions, and collecting other relevant material. Reviewing the data collected during the observation process, the researcher identified a number of concepts, which were then used to establish the interview protocol.

At the end of the semester, the researcher conducted in-depth one-on-one and structured (but open-ended) interviews with 22 students. The interviews were recorded and subsequently transcribed. Gaining access to the participants involved a gradual process of building mutual trust and reciprocal closeness. The content of the insight shared by the students was encoded by the researcher. A Student I.D. system, consists of a serial number and a letter, was used to maintain the privacy of the students.

### 2.1 *Research environment and participants*

The design studio course took place in two working environments, the studio classroom and the Visualization Laboratory. The studio classroom was an open space studio hall, facing outwards with individual work and display stations. A small area to the fore of the hall, with several large layout tables at its centre, was available for gatherings and group discussions. (See Figure 1).

Figure 1. Research environment – the studio classroom.

The lab contained a concave screen, 2.40 metres high by 7.0 metres high, with a 75-degree field of vision. This was supplemented by three Projectiondesign® projectors, a high-quality sound system, seven infra-red cameras, 3D capabilities and motion sensors for controlling the on-screen display. The Visualization Laboratory facility room was equipped to present the experience of 3D space to between twenty and thirty people, all wearing 3D glasses; the primary user, wearing 3D glasses with sensors, navigated the model by means of a joystick. The lab was kept dark and at a regulated temperature during use, to better present the projected model and to maintain the equipment. A trained technician – required by protocol to be present at all times whilst the lab was in use – was available for support. (See Figure 2).

The research setting: The setting of the case study for this research was the observation of a design-thematic studio course focusing on the adaptive reuse of existing buildings. The main advantage that this design theme presented to the researcher was the access, early in the semester, to an existing building which the students could exercise with and engage using the capabilities of the Visualization Laboratory. The selected buildings for intervention were in Brutalist Style, and located in Tel-Aviv, Israel.

*Participants:* The participants in the study were 23 third- and fourth-year Architecture students (17 females and 6 males). This was the first time that they had been exposed to the new environment and tools of the Visualization Laboratory. The class had two instructors, both experienced architects and teachers. One of the instructors was, in addition, experienced in the use of the Visualization Laboratory.

*The studio course*: As customary with the faculty's syllabus, the course consisted of twice-weekly meetings of at least five hours in duration. During the meeting, the students would work on their assignments, receiving one-on-on instruction from either one of the two teachers leading the studio. The students brought to the class sketches, physical models and personal laptop computers, depending upon the specific assignment assigned and their own preferences.

Figure 2. Research environment – The visualization laboratory – A review session.

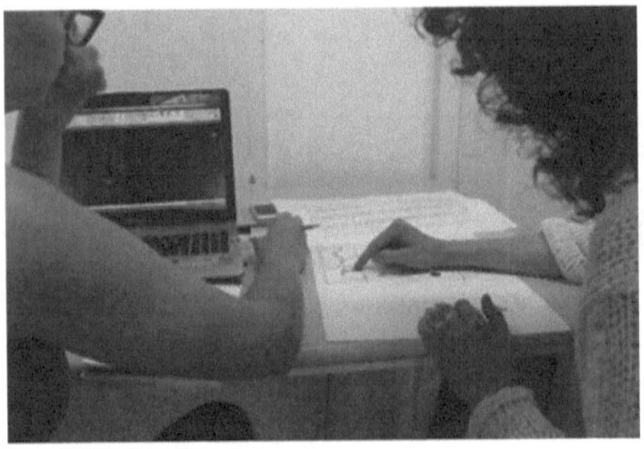

Figure 3. Non-participant observations.

During the one-on-one sessions, held in the studio classroom, student and instructor sat at the student's workstation, with other students occasionally joining to listen. The researcher habitually sat at the back and to one side, close enough to watch and listen but not in a participating mode. (See Figure 3). The students were free to choose where to spend their time working on their project, either in the studio classroom or in the Visualization Laboratory; they were, however, required to complete assignments given in each of the environments. The classroom and the Visualization Laboratory were both available for the students' use outside the scheduled hours of the course.

## 3 DATA ANALYSIS

Data analysis was conducted using qualitative methods, employing a spiral process until the account was clarified. The information was classified into "families" of themes, each "family" consisting of sub-themes named "codes". The different codes presented a variety of perspectives regarding its parent family (Creswell, 1998).

Reading through the data collected, the researcher used a graphic mark-up process indicated by colours. Codes in the analysed text were identified and marked in different colours, and were subsequently grouped. For the process of establishing the "families" and "codes", the researcher developed a work method using an index card system. (See Figure 4). Selected identified quotes from the text were copied onto separate index cards, and classified under a different code. Codes were given names which encapsulated their meaning. The index card system allowed for distinguishing, by colour and number, between individual students, and added an additional layer of colour differentiation by "family". A subjective evaluation ruler was placed on each card, allowing the researcher, during the process of classification, to indicate in graphic form the specific code's "strength of presence" in the text. (See Figure 5).

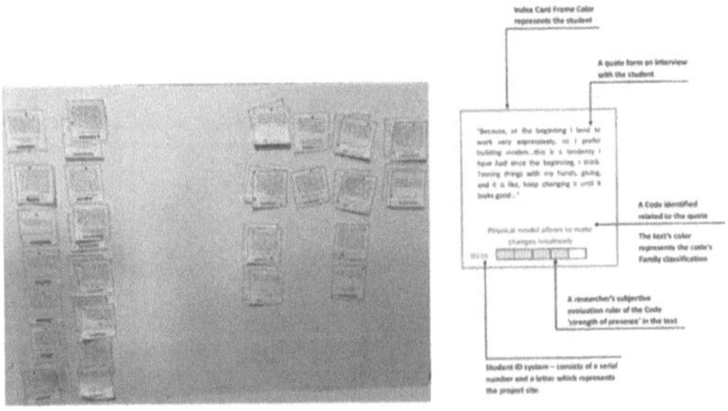


Figure 4. Using index cards (left) index card diagram (right).

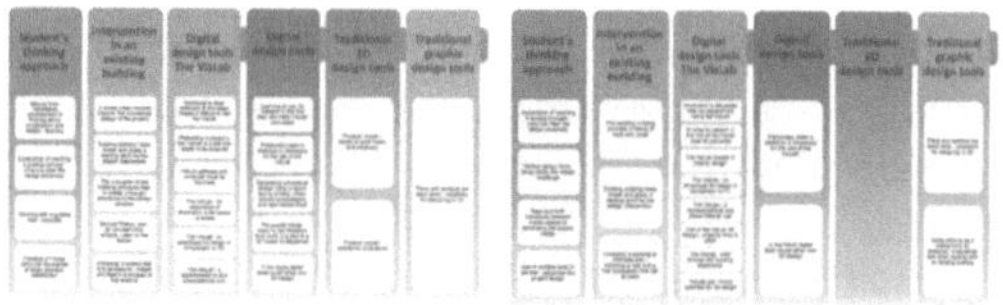

Figure 5. Grouped codes diagram – Student B006 (left) grouped codes diagram – Student B007 (right).

Another graphic display method which assisted the researcher in the data analysis process was the use of grouped codes diagrams, presenting the codes that emerged in each of the students' interviews. The codes were classified according to the following six families:

Traditional graphic design tools – represented by the colour **light blue**
Traditional three-dimensional design tools – represented by the colour **purple**
Digital design tools – represented by the colour **brown**
Digital design tools – The Visualization Laboratory – represented by the colour **orange**
Intervention in an existing building – represented by the colour **green**
Student's thinking approach – represented by the colour **red**

Within each family, the codes were ranked by their placement on the subjective evaluation ruler, appearing on the index cards. This form of graphic display presented the researcher with an overview of the student, and the relative value attributed to individual codes and families – by the student and in comparison, to other students. Figures 5 present family codes for two different students.

The use of an index card system and the grouped codes diagrams at the data analysis phase of the study allowed the researcher to extend the spiral process of repeating and redefining the families and codes until their final clarification. The data was inductively developed, through working out findings over the course of the research.

## 4 FINDINGS

As noted above, the information was classified during the data analysis process into six "families" of themes, each "family" consisting of sub-themes named "codes". The order in which the families are presented in the text and diagrams was for purposes of clarity of thought and readability, and does not suggest any hierarchy. Following is a brief summary of the findings for each of the six families:

***Traditional graphic design tools:*** This identified the students' approach to hand sketches, photomontages, and drawings of plans and sections. Those tools were within the usage comfort zone of the students. However, the students did not perceive these – with exception of the section – as design generators for a three-dimensional evolution of the project. The students also experienced difficulties in creating two-dimensional plans from the conceptual model that they have created with a traditional three-dimensional design tool.

***Traditional three-dimensional design tools:*** This identified the students' approach to physical models, created in the traditional way customary to the design studio prior to the digital era. The students demonstrated a strong connection with the physical model tool. Most chose to present and develop their concepts using a traditional three-dimensional model, even when they are not specifically asked to do so. The physical models allowed them the capacity to work freely and intuitively, to engage fully with a structure and its composition, and to consider design alternatives via an overall out-to-in view of the project. The process involved with this tool, of creation and the use of a variety of materials, brought out a marked enthusiasm and willingness to engage. The students were not always aware of all the meanings built into the models. The instruction process at the studio assisted in revealing these meanings, and helped the students to consolidate a design generating idea.

***Digital design tools:*** This identified the students' approach to three-dimensional digital models, generated on the students' personal computers with software like SketchUp, Revit and Rhino. Responses to using the digital model tool were mixed. Many of the students started the semester with little or no experience at all of it, and voiced this discomfort. Even for those with prior experience, significant effort was required to create the comprehensive digital model, taking up many working hours – an investment unlikely to have been made if it were not a specific requirement of the course. Use of the digital software did not act as a design tool in developing a concept. The students found the process slower and less intuitive, in comparison with the traditional design tools. Nevertheless, they appreciated the inherent potential of the digital tool for accurate and detailed design processes.

***Digital design tools – The Visualization Laboratory:*** This identified the students' approach to the new design tool, the Visualization Laboratory, which they were introduced to for the first time over the course of the semester observed by the researcher. With regards to the Visualization Laboratory, two layers of findings emerged, a visible layer and a concealed layer. The students demonstrated many challenges in their use of the Laboratory. They practiced using the three-dimensional software required for building the digital model during the semester; it may have been a more effective process if the skills had been acquired at an earlier stage. They also encountered a number of technical difficulties while trying to use digital models generated by software other than the system's default. Time and effort was required to generate the required digital model. In order to present this in the Visualization Laboratory, the model must be complete. The students found this requirement unnecessary and on the account of time spent on the design. They did not identify any advantage to using of the Visualization Laboratory for design as compared to their personal computer – aside from interior design, where the magnification enabled by the Visualization Laboratory assisted in identifying missing details, materials and textures. The experience of the Visualization Laboratory was more of a representational and presentational tool than a design tool. Few were able to experience immersion in the space while using the Visualization Laboratory. One surmises that the use of the Visualization Laboratory does not yet present as a design tool within the usage comfort zone of the students. Beyond the described visible layer, some also revealed a deeper concealed layer of perception. Aside from the lack of privacy, working on the project in the Visualization Laboratory, there was the inevitable – and potentially discomfiting – exposure at the presentation stage. The fact that the model is designed to be accessible to all in the Visualization Laboratory did not allow the students to hide weaknesses, or to demonstrate strong aspects of the design according to their own choice.

***Intervention in an existing building:*** This identified the students' approach to the design theme of the studio course, the adaptive reuse of existing buildings. The students noted the uniqueness of working with an existing building. The project site was selected by a majority of the students following a visit and a first impression that left its mark. Most of the students emphasized that the designer must engage with the work of the preceding architect(s). A system of planning principles, informed by the existing design, had to be defined by the intervention designer. Understanding this tended to assist the student in creating a flow of thought for the design process and to establish design principles for the interference with the existing building. The students found this process challenging, but also helpful as it provided a framework and boundaries, and helped them conceptualize a starting point for their design. The students considered the experience of visiting the existing site incomparable to the virtual "trip" presented by the Visualization Laboratory. The virtual reality demonstrated through the models they had created failed to convey the complexity of the existing site; the discovery process occurring through exploring the existing reality.

***Student's thinking approach:*** This identified the students' approach to various topics that emerged with regard to their conceptualization and development of the project. The students placed high emphasis on the search for a design generator – a guiding concept for the project. Some described the search as complex; however once it was clarified, the design process became much easier and enjoyable, based as it was upon the guidelines established in the design-guiding concept. The back-and-forth transitions between scales, development of building details and exploration of the building structure assisted the students in progressing with the project design. The availability of multiple design tools introduced an element of flexibility to the design process. The use of the various tools made it possible for the students to develop different aspects of the projects without being bound to a predetermined sequence, only later aggregating the parts to a single whole. However, the students found the overall external view of the project, and appreciation of the relationship between the parts and the whole, as crucial to their design process. For this, the traditional tools and the "small screen" held an advantage over the Visualization Laboratory. Many of the students encountered difficulties at some point during the design process; however, the supportive atmosphere of the studio course, together with excellent feedback from the instructors, was helpful. For some of the students, reviews from external guests invited to critique the projects at various points during the work process helped in focusing their work. The students appreciated the opportunity to have choices available to them within the limits of the course guidelines, affording them a sense of independence.

## 5 DISCUSSION, CONCLUSIONS AND FUTURE RESEARCH

The codes and families utilized in analysing the course experience translate into a rich and textured portrayal of the observed setting, as described by the students. The description assesses the relative merits of traditional design tools, such as varied graphic tools and physical models, alongside use of the new Visualization Laboratory. The research also presents the students' experience of the studio course, focusing on the adaptive reuse of existing buildings, and how this contributed to their work process in the design studio. Study and observation of the studio work process and the traditional design tools used by the students, were the context in which the new Laboratory tool and environment were observed.

The architectural studio, as noted by existing studies, (Austerlitz & Aravot, 2007; Ochsner, 2000; Willenbrock, 1991), presents to the student both a powerful emotional experience and some degree of psychological stress. As revealed in the findings, the students attempted to control their exposure and privacy with regard to the design processes and presentation. The Visualization Laboratory system created a new level of exposure, unfamiliar to the students. More consideration should be given to restructuring the work environment at the Visualization Laboratory, and to the design review and presentation process. Creating an intimate environment for the use of the Visualization Laboratory would afford the students some degree of privacy and control during work sessions, as they have with other design tools.

The use of scale, in design and as a means of communication with the construction world, is a fundamental aspect of an architect's education. During the design process, the designer looks on the project from the outside, shaping a comprehensive view (Dorta, Kinayoglu, & Boudhraa, 2016; Kan, Tsai, & Wang, 2011; Scheer, 2014). As the findings show, the students' familiarity with small-scale models translated into a design advantage, due to the overall out-to-in view it afforded them. With the Visualization Laboratory's view being from the in-to-out (due to presented digital model being larger than the designer), the students struggled to achieve a comprehensive view during the work process, and sought to return to their comfort zone where a sense of safety and control could be maintained (Rechmaoui, 2010). This challenge can be addressed though the development of new courses for first-year students, offering them practice opportunities to develop a sense of control within the immersion environment of the Visualization Laboratory.

The naturally occurring verbal language, of the traditional design studio education process (Avidan & Goldschmidt, 2013), was not sufficiently nuanced for the new design education setting of the Visualization Laboratory environment. A new vocabulary and body language adapted for this purpose (the use of large gestures, for example) should be developed for this environment.

Merging a project encompassing the adaptive reuse of existing buildings, in a studio and a Visualization Laboratory-based course, will allow for the simplification of the design process for at least some students. Through this, students can be directed to the tasks, while gradually introducing new components and considerations, differentiated accorded to experience and

Figure 6. Integration of tools in the evolving design environment: physical models on the floor and virtual representation on the screen.

levels of reflective thinking. In addition, an emotional relationship with the building, which the students started to develop from the first tour on site, highlights the advantages of this type of project for enriching architectural design education.

The students reported difficulties –slowness and complexity, specifically – in using the Visualization Laboratory. This, in addition to the fact that the design was effected on the students' personal computers with other design tools – all located outside of the Visualization Laboratory – did not allow for full expression of the possibilities of experiential learning in the Visualization Laboratory, as a whole, dynamic and recurring process (Dewey, 1934). The feedback received by the student was absorbed in a passive state and could only be reviewed and implemented later, as a reflective process, when all the required tools were available.

The students reported the significant benefit of using a variety of tools in their design process. Thus, one concludes that it would be beneficial if a new evolving design environment, integrated the tools available in the Visualization Laboratory (perhaps with additional new forms of digital tools) with the studio activity, which is based on traditional tools and methods (See Figure 6). In such a setting, the students would be able to choose the desired tool on the basis of current needs and the demands of the project. Presumably, new tools could influence the work process in the studio, and perhaps the outcome of the design.

The observation of the student design methods and processes in the new design environment has helped in identifying a range of topics and questions for discussion and further study, highlighting the importance of the continued integration of the Visualization Laboratory and other state-of-the-art design and visualization tools into studio design courses. Thus, this research hopes to contribute to the development of architectural design education.

## REFERENCES


Austerlitz, N., & Aravot, I. (2007). Emotions of architecture students: A new perspective for the design studio. In A. M. Salama, & N. Wilkinson (Eds.), *Design studio pedagogy: Horizons for the future* (pp. 233–245). Gateshead, United Kingdom: The Urban International Press.

Avidan, Y., & Goldschmidt, G. (2013). Talking architecture: Language and its roles in the architectural design process. *ICoRD'13*, 1139–1150.

Broadbent, G. (1995). Architectural education. In M. Pearce, & M. Toy (Eds.), *Educating architects* (pp. 10–23). London: Academy Editions.

Buchanan, R. (1992). Wicked problems in design thinking. *Design Issues 8*(2), 5–21.

Creswell, J. W. (1998). *Qualitative inquiry and research design choosing among five traditions*. Thousand Oaks, CA: Sage Publications, Inc.

Cross, N. (2001). Designerly ways of knowing: Design discipline versus design science. *Design Issuses 17* (3), 49–55.

Cross, N. (2006). *Designerly ways of knowing*. London: Springer-Verlag.

Cross, N. (2011). *Design thinking*. Oxford: Berg.

Dewey, J. (1933). *How we think*. Boston: D.C. Heath and Company.

Dewey, J. (1934). *Art as experience*. New York: Capricorn Books.
Dewey, J. (1938). *Experience and education*. New York: Macmillan.
Dorta, T., Kinayoglu, G., & Boudhraa, S. (2016). A new representational ecosystem for design teaching in the studio. *Design Studies 47*, 164–186.
Kan, J. W., Tsai, J. J.-H., & Wang, X. (2011). "Scales" Affecting design communication in collaborative virtual environments. In X. Wang, & J. J.-H. Tsai, *Collaborative design in virtual environments, ISCA 48* (pp. 77–87). Springer Science & Business Media B.V.
Kolb, D. A. (1984). *Experiential learning: Experience as the source of learning and development*. Englewood Cliffs, N.J: Prentice-Hall.
Ochsner, J. K. (2000). Behind the mask: A psychoanalytic perspective on iteraction in the design studio. *Journal of Architectural Education 53*(4), 194–206.
Rechmaoui, M. (2010). *Lecture at the Map Marathon*. (Artist, Performer) The Serpentine Gallery, London.
Rittel, H. W., & Webber, M. M. (1973). Dilemmas in a general theory of planning. *Policy Sciences 4*, 155–169.
Salama, A. M. (2016). *Spatial design education*. New York: Routledge.
Scheer, R. D. (2014). *The death of drawing architecture in the age of simulation*. London: Routledge.
Schon, D. A. (1983). *The reflective practitioner*. New York: Basic Books.
Schon, D. A. (1985). *The design studio, an exploration of its traditions and potential*. London: RIBA Publications Limited.
Schon, D. A. (1987). *Educating the reflective practitioner: Toward a new design for teaching and learning in the professions*. San Francisco: Jossey-Bass Publishers.
Shkedi, A. (2003). *Words of meaning: Qualitative research – theory and practice*. Tel-Aviv: Ramot.
Willenbrock, L. L. (1991). An undergraduate voice in architectural education. In T. A. Dutton (Ed.), *Voices in architectural education: Cultural politics and pedagogy* (pp. 97–119). New York: Bergin & Garvey.

*Research & Education in Design: People & Processes & Products & Philosophy – Almendra & Ferreira (eds)*
 *ISBN 978-0-367-49296-0*

# In quest of a successful design studio course: A course evaluation template

K. Gelmez
*Department of Industrial Product Design, Istanbul Technical University*

ABSTRACT: It is hard to uncover design studio courses because it includes complex, intricate and rich learning experiences. Therefore, it needs a special attention in terms of both instructional strategies and academic studies. This paper addresses developing a qualitative way of evaluating a design studio course. This is an attempt to shed a light into this complex issue together with the students. To achieve this, senior industrial design students participated a two-part session to generate a design studio course evaluation template. Based on the reviews and suggestions from the students, the evaluation template was developed including these three main dimensions, which are *communication*, *content and process,* and *overall.*

## 1 INTRODUCTION

What makes a design studio course successful? How does a design teacher evaluate his/her design course's success? Does the quality of project outcomes show that the design studio course has a success? If so, to what extent? Is it a great sign of success if a project is manufactured and commercialized in the scope of a design studio course? Do we need an external evaluator to measure the success? Is it possible to create an ideal design studio course? What do we need to achieve this?

These are the common questions that the author has come across during his eleven-year design teaching experience. This becomes his motivation to dive into this intricate issue through an academic study.

From a perspective of constructivism, learners are in the focal point of the learning process. Evaluating a course is a critical issue that design teacher can consult to the design students, who are the most important shareholders in a design studio course, actually the owners of the learning process.

As a course already includes an immense complexity all by itself, it is unthinkable to evaluate without the contribution of the students. This study specifically addresses a writing activity to evaluate a design studio course together with and by the design students.

### 1.1 *Course evaluation and design studio*

Course evaluation is considered as a critical and intricate issue in education domain (Spooren et al., 2013). In the literature, it is possible to find studies regarding preparation, analysis and interpretation of the tutor and course evaluation systems. Basically, focusing on the validity issue of student evaluation, there are numerous academic studies on course evaluation systems based on relationships between student evaluation of teaching and student, teacher and course characteristics (Spooren et al., 2013). More specifically, we can come across studies revealing the relationship between student evaluation and the age of the student (see Centra, 1993; McKeachie, 1979); and gender of the student (see Centra and Gaubatz, 2000; Feldman, 1993; Santhanam & Hicks, 2001); and the success of the student (see Davis 2009; Abrami, 2001; Brashkamp & Ory, 1994; Spooren, 2010); and the personal trait of the student (see Abrami,

Perry, & Leventhal, 1982); and the student interest (see Olivares, 2001); and teaching experience (see Centra, 2009; Feldmen, 1983; McPherson et al., 2009); and the gender of the teacher (see Basow & Montgomery, 2005; Feldmen 1993; Centra 2009); and the race of the teacher (see Centra 1993); and personal trait of the teacher (see Feeley, 2002; Murray et al., 1990; Patrick, 2011); and research productivity of the teacher (see Centra, 1993; Stack 2003); reputation of the teacher (see Griffin, 2001); and title of the teacher (see McPherson et al., 2009); and quality of teaching (see Barth 2008; Pan et al. 2009); and class size (see Bedard & Kuhn, 2008; McPherson et al., 2009); and class attendance (see Ting, 2000); and year level of the course (see McKeachie, 1979; Santhanam & Hicks, 2001); and the workload (see Centra, 2003; Dee 2007); and the course discipline (see Basow & Montgomery, 2005).

Course evaluation is not a mere assessment tool. Besides, it can be used in order to increase teachers' pedagogical qualities (Platt, 1993; Spooren et al., 2013) even though there are no clear signs that course evaluations enhance the quality of teaching (Kember et al., 2002). In addition, course evaluation methods can be utilized as a tool to enhance the learning processes of learners (Erikson et al., 2016). In other words, Erikson et al. (2016) added a new objective which is to leverage the learner's learning process based on the theory of student involvement by Astin (1999). It is also asserted that current course evaluations are teaching- and teacher-centered (Edström, 2008), and conventional course surveys are not sufficient to correspond to the complexity level of learning (Erikson et al., 2016). Platt (1993) states that the open-ended questions are allocated in a very little part of the surveys, which implies "I do not want to read anything from you or I am not interested what you write."

It is hard to cover all of these objectives in one type of evaluation method (Erikson et al., 2016). The literature suggests developing original methods pertaining to the nature of the courses. In this context, we can see that there are qualitative methods for student course evaluations (Erikson et al., 2016; Wulff & Nyquist, 1986; Lattuca & Domagal-Goldman, 2007). In order to get a feedback for effective teaching, it is necessary to use more inclusive ways besides end-of-course surveys (Lattuca & Domagal-Goldman, 2007).

Similarly, one of the noteworthy problems regarding student evaluation of teaching is related to anonymous way of delivering surveys with standard questions (Spooren et al., 2013). Thus, the use of course evaluation surveys can be considered as "impersonal" attitude towards the students (Platt, 1993; Spooren et al., 2013).

It is believed that course evaluation is beneficial when it is well-prepared and supported with different methods (Stark & Freishtat, 2014). There is a need for more comprehensive methods in addition to course evaluation surveys to get feedbacks for effective teaching (Lattuca & Domagal-Goldman, 2007). In a similar vein, Lattuca & Domagal-Goldman (2009) advocated to use various sources of data in course evaluation.

Spooren et al. (2013) claim that there are serious problems regarding anonymous evaluation surveys, and thus "humanist" methods need to be developed (Platt, 1993; Spooren et al., 2013). Svinicki (2002) also advocated to find proper ways to support and motivate the students to give feedback. In this sense, it is important to offer a context to the students that the teachers and the students can meet.

In this respect, it is possible to find qualitative methods to evaluate a course (see Erikson et al., 2016; Lattuca & Domagal-Goldman, 2009; Wulff & Nyquist, 1986). Wulff and Nyquist (1986) strongly recommended qualitative ways to study instructional improvement including course evaluation in order to correspond the complexity of learning and teaching. In their study, they utilized interviews and observations to evaluate the lecture and laboratory components of a course. Thanks to this, they asserted that they were able to grasp contextual, multivariate and complete perspective of the students.

In their study, Erikson et al. (2016) asked the following question to evaluate the course: "What could have been done in this course in order to better support your learning?" They identified four major themes based on their analysis e.g. teaching/learning activities, communication, teachers as agents, students as agents. They suggested that qualitative course evaluations can enlighten us about students' viewpoints of their own learning processes, that is demanding to reach in a conventional evaluation form.

There is an invitation in the literature to find a way of humanizing course evaluations to value the learners and the lively experience of learning process. In the context of this study, course evaluations are regarded as a tool for feedback of teaching and a learning instrument for learners.

The pedagogical intention of this paper is best described and summarized in the following paragraph by Platt (1993):

> *Let us glance at a typical student evaluation. It consists of a series of questions that the student is to answer by checking boxes, circling numbers on a scale, or marking phrases on a continuum, with a soft- lead pencil, whose marks can be easily scanned and tabulated by a machine. Clearly, very little, in effort or time, is required to fill out such a form. It does not ask the students for examples, questions, thoughts, or a story. Little even in penmanship is required. Little is also insisted upon. If a student wrote a letter instead, it would not be welcome.*

This study is a proposal for an alternative way to evaluate a design studio course qualitatively from students' perspectives. As course evaluation is a part of learning process, it should not be thought as a discrete process from the learning. In other words, course evaluations should be given importance as well as the other teaching activities (Svinicki, 2002).

From a constructivist learning perspective, learning continues after the semester is over. That is, the end of the course or the semester should mean being open to new learning experiences in order to make a new connection with. Evaluation is one of the most activities held in a design studio, especially in design critiques and juries. However, design research literature has been silent on course evaluations.

Design studios, which lie at the heart of industrial design education (Crowther 2013; Green & Bonollo 2003), are the practice-based courses where students apply different knowledge and skills (Demirbas & Demirkan 2007). They offer a lively and rich learning experience that seems to be tougher to grasp.

Although it is hard to find an academic study directly related to course evaluation in design education literature, we can see limited number of studies getting feedback from the design students.

Cartier (2011) examined expectations of first-year and fourth-year design students via face-to-face interviews. Attoe and Mugerauer (2006) studied excellent studio teaching, where they conducted interviews with university teachers who have won awards for teaching excellence. They put forward 14 elements that affect efficient studio environment.

Chen (2016) investigated the industrial design students' learning problems and the means that the students employed to tackle these problems in undergraduate setting. Most of the difficulties is related to concept generation, design presentation, and design research. "People, object, method, and environment" are the resources that students get help while overcoming the problems.

Chen and You (2010) explored how students responded towards an Internet-mediated design studio by conducting surveys and focus group interviews. In their study, there were four dimensions to evaluate namely "course, internet, learner, and instructor." "Course" contains objective, content, activity and delivery, "internet" refers to capability, limitation and cost. Whereas "learner" includes motivation, participation, performance and process, "instructor" embraces strategy, pedagogy, evaluation and organization. While the "instructor" and "student" are the key users, the "course" includes the content. On the other hand, "internet" refers to the specific communication path utilized in the studio. The results of revealed that the students are positive to Internet-mediated design studio course in general. However, there are some certain limitations stemming from the use of internet.

In design education literature, various type of writing activities can be found that might change according to the objective and context of the writing task (Gelmez and Bağlı, 2018b). In this respect, writing can be a tool to use to evaluate a learning process and course evaluation.

Kurt and Kurt (2017) revealed that students keeping "reflective design journal" can reflect on "awareness," "organising and planning," "monitoring," and "evaluating" their learning processes. In addition to these metacognitive processes, reflective writing can be utilized in order to grasp affective processes in a design studio (see Gelmez and Bağlı 2018a). Regarding affective domain, course content, pedagogical moves, and extra-curricular activities in a design studio can influence design students emotionally (see Gelmez and Bağlı 2018a).

Reflective writing can be utilized as "a communication tool/mediator" between the students and the teachers in addition to design critiques, collective discussions, design juries, face-to-face meetings and e-mails. As design students can convey their opinions, emotions, needs and wishes in the scope of a design studio, the teachers are able to check how the students react the course content and pedagogy (see Gelmez and Bağlı 2018b).

From this point forth, a design studio course evaluation template was generated with the participant-design students in this study.

## 2 DESIGN STUDIO COURSE EVALUATION TEMPLATE

A qualitative method was developed together with the students in order to evaluate an industrial design studio course. In this context, a study was conducted in the middle of the semester in the context of course titled EUT 419E Industrial Design Studio IV in the Department of Industrial Product Design at Istanbul Technical University in 2017-2018 Spring Semester.

EUT 419E Industrial Design Studio IV is a compulsory course with 9 ECTS for 4$^{th}$ grade students. Students enrolled to this course are expected to exhibit professional level designer skills and to develop comprehensive design concepts. In the first half of the semester, a packaging design project, called as "Unpackageables," was assigned to the students. The project brief started with a story given in the following (Gelmez, 2018):

> *Last week, I went to a supermarket to buy a packet of gluten-free pasta. I had to read the information written on the package in tiny font size to see the ingredients in it. Then, I wanted to buy a bottle of deodorant. Again, I had to open its lid to smell and choose the fragrance. In this short shopping experience, I came to realize that there were some elements such as odors, flavors, textures, sounds, emotions, experiences, identities that I called as "unpackageables."*

In this project, the students were expected to design a package system for "unpackageables" in 5 different states and in 3 different sizes.

After this project, 13 senior industrial design students enrolled to the course participated voluntarily to this present study. They were given a written informed consent. Since they were senior students, they experienced several design studio classes, which made them as experts to evaluate a design studio course.

The study has two main parts. In the first part, the students were asked to write a review of answering this question: Could you write a review to evaluate the half of the semester? 8 students submitted their reviews.

In the second part, the students split up two groups and they were asked to generate suggestions to evaluate a design studio course by answering this question: How can an industrial design studio course be evaluated?

Reviews and suggestions were both analyzed by the studio tutors, who were an assistant professor (author of the paper) and a research assistant. Based on this analysis, a design studio course evaluation template was generated and implemented to the students at the end of the semester.

In their reviews and suggestions, the students' expressions are clustered under three main dimensions, which are *communication, content and process,* and *overall.* In this section, these dimensions will be revealed by exemplifying with the students' reviews and suggestions[1]. The reviews and suggestions were written in Turkish to prevent any loss resulting from language problems.

The communication dimension of evaluation template refers to the contact points among the actors in the design studio. In this respect, actions occurred among student(s)-student(s) and teacher(s)-student(s) are the common topics that the participant-students mentioned in their reviews. These are some highlights from their reviews that underline the importance of communication between teacher(s) and student(s):

[1] Students' reviews and suggestions given were translated from Turkish to English by the researcher.

"Being listened sentence by sentence increased both my attention and my motivation."

"I can say that course instructors are quite respectful."

When we checked the suggestions on design studio course evaluation, we can come across several statements regarding the teacher's roles and tasks:

"The attendance of the teacher"

"The authority of the teacher to the class and the ability to transfer what he/she knows"

"Equal time distribution of the teacher to each student"

"The awareness of the teacher about the project and the critiques that he/she gives"

"The mastery of teacher to the project"

"The quality and the quantity of design critiques"

"The control of the teacher over students' projects"

"The motivation and effort of the teacher to students' attendances"

Design critiques and design juries are the prominent issues specific to a design studio course. Therefore, the students often referred to the collective design critique sessions and the design juries:

". . . I learned not only while I was presenting, but also while I was listening my friends. In fact, some critiques that my friends have taken were more useful than those that I have taken."

"Everyone has found opportunities to say their opinions, in this sense it was a quite interactive class."

"While we were talking about everyone's projects, I felt that my project had been settled down."

Also, the physical environment that the classes took place is another influential factor in a design studio course:

"If I rated physical conditions out of 10, I would give 6-7."

Assessment can be another issue that was raised in some reviews:

"As a result, I would like to add I thought that I understood the process of the project before the assessment and the jury, however, I realized that I didn't understand in fact."

The second dimension of the design studio course evaluation is about the content and the process of a design studio course. In students' reviews, it was possible to find various examples where they narrated their design processes, and commented the project topic and the brief:

"My brain has expanded while I was thinking the project at home and on my way."

"The process of this project has activated all of unusual our brain muscles."

"The project brief has provided an opportunity to take a breath among all other standardized project briefs."

"I was very tired during the project of 'Unpackageables':)"

"In the first weeks, what would show up as an outcome was very unclear."

"I think that the choice of the project topic has a complimentary effect to all these things."

In addition to these reviews, we can see students' suggestions related to the project brief and topic and the process:

"The plan of the course content, its implementations and evaluation"

"The notification of the project to the students"

"The clarity of the implementations' contributions"

"The quality and complexity of the project to the class level and student progress"

"The benefit of the project topic to the students"

"The interactivity of the class"

"The allocation of time to develop project in class"

"The effect of teamwork if any"

"The benefit of project to students' future design careers"

"The inclusion of the project to students' design portfolios"

Additionally, there were expressions that the students established connections among other classes. Regarding this, they often compared the other design studio courses that they experienced with the existing one:

"I haven't taken a course that a student could illuminate other students with his/her ideas."

"I think that this project is the most philosophical and open-to-comment project among the project courses that I have taken before."

"As a result, this is the class that I have entertained and learned most during my last 4-year [period of study]."

The last dimension is related to an overall evaluation. This covers statements about the relationship among classes, and overall semester evaluation:

"The project is a very nice test of developing arguments."

"Other than this, it was a nice project."

Benefitting from the reviews and suggestions, an evaluation template was generated. This template consisted of stickers with incomplete sentences and a blank folded A4 format paper into A5 (see Figure 1 and Table 1). The participant-students could use the stickers and

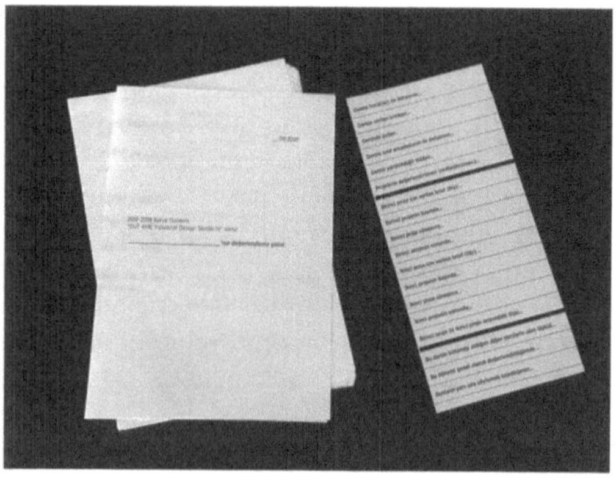

Figure 1. Design studio course evaluation template (left: A5 paper, right: stickers with incomplete sentences).

Table 1. Course evaluation template.

| Dimensions | Incomplete sentences | Objectives |
|---|---|---|
| Communication | My dialogue with the teacher(s)...<br>The design critiques in the class...<br>The design juries...<br>My dialogue with my classmates...<br>The studio environment...<br>The assessment (grading) of the projects... | To get feedback on communication dimension<br>To get insights about specific actions such as design critiques and design juries |
| Content and process | The brief of the first project...<br>In the beginning of the first project...<br>During the first project...<br>At the end of the first project...<br>The brief of the second project...<br>In the beginning of the second project...<br>During the second project...<br>At the end of the second project... | To find connections between projects<br>To make comparisons within the term<br>To get specific insights on project topics and briefs<br>To get clues on project process |
| Overall | The relationship between the first project and the second project...<br>The relationship between this course and the other courses in the department...<br>When I evaluated this semester in general...<br>Beside this, I would like to say... | To grasp an overall evaluation<br>To give students an opportunity to say something they want |

complete the sentences by writing. The aim of the incomplete sentences is to trigger students to write on something specific. One of the participant-students in her review declared that "I see writing activity as a more consistent way to express ourselves and our projects." As reflective writing has a function of "a communication tool/mediator" between the students and the teachers in a design studio (Gelmez and Bağlı 2018b), the evaluation template has been designed as a reflective writing format.

## 3 CONCLUSION

This study is an attempt to develop a specific method to evaluate a design studio course. As design studio courses are complex, intricate and rich learning environments, this type of qualitative evaluation method may correspond to grasp more natural insights from design students.

The template developed together with the students is expected to help design students to give responses on design studio course in terms of *communication*, *content and process* and *overall*. The qualitative nature of the template can give opportunities to understand a design studio course from students' aspects. Incomplete sentences given as stickers can function as triggers and reminders for design students to write down something specific about the course.

According to Wagner (1999), employing different sources and techniques while evaluating a course is helpful for curriculum improvement and student achievement. Reflective writing activity presented in this study is not a substitute method instead of conventional course evaluation, face-to-face meetings, interviews or other communication types among students and teachers. Rather, it is recommended as a supplementary evaluation method to the current forms of evaluation systems.

The benefit of the template developed in this study has two-folded. The first one refers to learning aspect. It may help learners to review their learning processes. The second benefit is expected on pedagogical aspect. Design teachers may improve their teaching strategies by considering the reviews.

In further studies, it is planned to analyse the design studio course evaluation template gathered.

## ACKNOWLEDGEMENTS

I would like to express my gratitude to the students participated in this current study. I would also like to thank Res. Assist. Enver Tatlısu with whom I conducted the aforementioned course. This study (Project ID: 41620) was supported by Scientific Research Projects (BAP) of Istanbul Technical University (ITU).

## REFERENCES

Abrami, P. C. (2001). Improving judgments about teaching effectiveness using teacher rating forms. In M. Theall, P. C. Abrami, & L. A. Mets (Eds.), *The student ratings debate: Are they valid? How can we best use them?* New Directions for Institutional Research, No. 109 (pp. 59–87). San Francisco: Jossey-Bass.

Abrami, P. C., Leventhal, L., & Perry, R. P. (1982). Educational seduction. *Review of Educational Research*, *52*, 446–464.

Astin, A. W. (1999). Student involvement: A developmental theory for higher education. *Journal of College Student Development*, *40*, 518–529.

Attoe, W., & Mugerauer, R. (1991). Excellent studio teaching in architecture. *Studies in Higher Education*, *16*(1), 41–50. doi: 10.1080/03075079112331383081.

Barth, M. M. (2008). Deciphering student evaluations of teaching: A factor analysis approach. *Journal of Education for Business*, *84*, 40–46. doi: 10.3200/JOEB.84.1.

Basow, S. A., & Montgomery, S. (2005). Student ratings and professor self-ratings of college teaching: Effects of gender and divisional affiliation. *Journal of Personnel Evaluation in Education, 18*, 91–106. doi:10.1007/s11092-006-9001-8

Bedard, K., & Kuhn, P. (2008). Where class size really matters: Class size and student ratings of instructor effectiveness. *Economics of Education Review, 27*, 253–265. doi: 10.1016/j.econedurev.2006.08.007

Braskamp, L. A, & Ory, J. C. (1994). *Assessing faculty work: Enhancing individual and institutional performance*. San Francisco: Jossey-Bass.

Cartier, P. (2011). Most valuable aspects of educational expectations of the students in design education. *Procedia—Social and Behavioral Sciences, 15*, 2187–2191. doi: 10.1016/j.sbspro.2011.04.077

Centra, J. A. (1993). *Reflective faculty evaluation: Enhancing teaching and determining faculty effectiveness*. San Francisco: Jossey-Bass.

Centra, J. A. (2003). Will teachers receive higher student evaluations by giving higher grades and less course work? *Research in Higher Education, 44*, 495–518.

Centra, J. A. (2009). *Differences in responses to the student instructional report: Is it bias?* Princeton, NJ: Educational Testing Service.

Centra, J. A., & Gaubatz, N. B. (2000). Is there gender bias in student evaluations of teaching? *The Journal of Higher Education, 71*, 17–33.

Chen, W. (2016). Exploring the learning problems and resource usage of undergraduate industrial design students in design studio courses. *International Journal of Technology and Design Education, 26*(3), 461–487. doi: 10.1007/s10798-015-9315-2

Chen, W., & You, M. (2010). Student response to an Internet-mediated industrial design studio course. *International Journal of Technology and Design Education, 20*(2), 151–174. doi: 10.1007/s10798-008-9068-2

Crowther, P. (2013). Understanding the signature pedagogy of the design studio and the opportunities for its technological enhancement. *Journal of Learning Design, 6*(3), 18–28.

Davis, B. G. (2009). *Tools for teaching*, (2nd ed.). San Francisco: Jossey-Bass.

Dee, K. C. (2007). Student perceptions of high course workloads are not associated with poor student evaluations of instructor performance. *Journal of Engineering Education, 96*, 69–78.

Demirbas, O. O., & Demirkan, H. (2007). Learning styles of design students and the relationship of academic performance and gender in design education. *Learning and Instruction, 17*(3), 345–359. doi: 10.1016/j.learninstruc.2007.02.007

Edström, K. (2008). Doing course evaluation as if learning matters most. *Higher Education Research & Development, 27*(2), 95–106. doi: 10.1080/07294360701805234

Erikson, M., Erikson, M. G., & Punzi, E. (2016). Student responses to a reflexive course evaluation. *Reflective Practice, 17*(6), 663–675. doi: 10.1080/14623943.2016.1206877

Feeley, T. H. F. (2002). Evidence of halo effects in student evaluations of communication instruction. *Communication Education, 51*, 225–236. doi: 10.1080/03634520216519

Feldman, K. A. (1993). College students' views of male and female college teachers: Part II: Evidence from students' evaluations of their classroom teachers. *Research in Higher Education, 34*(2), 151–211.

Gelmez (2018). Ambalajlanamayanlar [Unpackageables]. *Ambalaj Dünyası [Packaging World], 32*(170), pp. 70–72.

Gelmez, K., & Bağlı, H. (2018a). Tracing design students' affective journeys through reflective writing. *International Journal of Technology and Design Education, 28*(4). pp. 1061–1081. doi: 10.1007/s10798-017-9424-1

Gelmez, K., & Bağlı, H. (2018b). Exploring the functions of reflective writing in the design studio: A study from the point of view of students. *Art, Design & Communication in Higher Education, 17*(2). pp. 177–197. doi: 10.1386/adch.17.2.177_1

Green, L. N., & Bonollo, E. (2003). Studio-based teaching: History and advantages in the teaching of design. *World Transactions on Engineering and Technology Education, 2*(2), 269–272.

Griffin, B. (2001). Instructor reputation and student ratings of instruction. *Contemporary Educational Psychology, 26*, 534–552. doi: 10.1006/ceps.2000.1075

Kember, D., leung, D. Y. P., & Kwan, K. P. (2002). Does the use of student feedback questionnaires improve the overall quality of teaching? *Assessment & Evaluation in Higher Education, 27*, 411–425. doi: 10.1080/0260293022000009294

Kurt, M., & Kurt, S. (2017). Improving design understandings and skills through enhanced metacognition: Reflective design journals. *International Journal of art &* Design Education, 36(2). doi: 10.1111/jade.12094

Lattuca, L., & Domagal-Goldman, J. (2007). Using qualitative methods to assess teaching effectiveness. *New Directions for Institutional Research, 136*, 81–93. doi: 0.1002/ir.233

McKeachie, W. J. (1979). Student ratings of faculty: A reprise. *Academe, 65*, 384–397.

McPherson, M. A., Todd Jewell, R., & Kim, M. (2009). What determines student evaluation scores? A random effects analysis of undergraduate economics classes. *Eastern Economic Journal*, *35*, 37–51. doi: 10.1057/palgrave.eej.9050042

Murray, H. G., Rushton, P. 1., & Paunonen, S. V. (1990). Teacher personality traits and student instructional ratings in six types of university courses. *Journal of Educational Psychology*, *82*(2), 250–261.

Olivares, o. J. (2003). A conceptual and analytic critique of student ratings of teachers in the USA with implications for teacher effectiveness and student learning. *Teaching in Higher Education*, *8*, 233–245. doi: 10.1080/1356251032000052465

Pan, D., Tan, G. S. H., Ragupathi, K., Booluck, K., Roop, R., & Ip, Y. K. (2009). Profiling teacher/teaching using descriptors derived from qualitative feedback: Formative and summative applications. *Research in Higher Education*, *50*, 73–100. doi:10.1007/s11162-008-9109-4

Patrick, C. L. (2011). Student evaluations of teaching: Effects of the Big Five personality traits, grades and the validity hypothesis. *Assessment & Evaluation in Higher Education*, *36*, 239–249. doi:10.1080/02602930903308258

Platt, M. (1993). What student evaluations teach. *Perspectives on Political Science*, *22*(1), 29–40. doi: 10.1080/10457097.1993.9944516

Santhanam, E., & Hicks, O. (2001). Disciplinary, gender and course year influences on student perceptions of teaching: Explorations and implications. *Teaching in Higher Education*, *7*, 17–31. doi:10.1080/13562510120100364

Spooren, P. (2010). On the credibility of the judge. A cross-classified multilevel analysis on student evaluations of teaching. *Studies in Educational Evaluation*, *36*, 121–131. doi: 10.1016/j.stueduc.2011.02.001

Spooren, P., Brockx, B., & Mortelmans, D. (2013). *On the validity of student evaluation of teaching: The State of the Art. Review of Educational Research*, *83*. doi: 10.3102/0034654313496870

Stack, S. (2003). Research productivity and student evaluation of teaching in social science classes. *Research in Higher Education*, *44*, 539–556. doi:10.1023/A:1025439224590

Stark, P. B., & Freishtat, R. (2014). An evaluation of course evaluations. *ScienceOpen*, (September), 1–26. https://doi.org/10.14293/S2199-1006.1.SOR-EDU.AOFRQA.v1

Svinicki, M. D. (2002). Encouraging your students to give feedback. *New Directions for Teaching and Learning*, *2001*(87), 17–24. doi:10.1002/tl.24

Ting, K. (2000). A multilevel perspective on student ratings of instruction: Lessons from the Chinese experience. *Research in Higher Education*, *41*, 637–661. doi:10.1023/A:1007075516271

Wagner, Z. T. (1999). Using student journals for course evaluation. *Assessment & Evaluation in Higher Education*, *24*(3), 261–272. doi: 10.1080/0260293990240301

Wulff, D. H., & Nyquist, J. D. (1986). Using qualitative methods to generate data for instructional development. *To Improve the Academy Professional and Organizational Development Network in Higher Education*, 119, 37–46.

*Visual Design Representations*

*Research & Education in Design: People & Processes & Products & Philosophy – Almendra & Ferreira (eds)*
 *ISBN 978-0-367-49296-0*

# Mental imagery as a tool in visualization: A participant observation study

T.E. Tüfek
*Istanbul Bilgi University*

ABSTRACT: Designers are required to produce 2-dimensional representations of objects in both design and production phases. To create right 2-dimensional representations, designers mostly relies on their ability to create visual mental images of these objects. Constructing a right visual mental image is important for visualization process, as designers can work on these images and reason visual transformations. The ability of generating accurate mental images is not strong in design students as they lack of necessary knowledge and experience. In visualization courses, students are exposed to the employment of this ability by their instructors. This paper discusses the tacit characteristics of the utilization process of visual mental imagery in a visualization course. The tactics that experienced designers and educators use for demonstrating students the ways of constructing detailed and accurate visual mental images are also explored through the study.

## 1 INTRODUCTION

When they first start university, design students face a very different kind of education system. This system contains a more hand on approach with the notions like juries, presentations and visualization. Consequently, students are required to adopt a new set of skills in the freshman year of their design education. One of these skills is very essential for designers; 2-dimensional visualization. Designers are required to have a certain level of visualization ability to be able to conduct design and production processes properly. However, just like the whole design education, 2-dimensional visualization with conventional tools is also not a skill which can be taught by just writing down or verbalizing. Although, there are rules and instructions in some parts –especially in technical drawing- the whole visualization process requires a skill principally. That is the ability of 3-dimensional thinking. 3 dimensional thinking is closely related with the notion of mental imagery. Correct perception and correct creation of mental images are crucial abilities for designer to make accurate 2-dimensional representations of the objects they are designing.

This paper presents a study, which is conducted in a visualization course. This design visualization course introduces both technical drawing and quick sketching to first year industrial product design students. The course contains one-on-one critique sessions in which instructors comment on, criticize and correct the drawings of the students. The objective of this study to examine the students' utilization of visual mental imagery in this course. Furthermore, to understand how the interaction between student and instructor helps the improvement of 3-dimensional thinking ability of students.

## 2 PERCEPTION AND MENTAL IMAGERY

Over the years, the notions of imagery and perception were continuously draw attention of researchers from a variety of fields like, philosophy, cognition and psychology. To understand the notion of mental imagery, the notion of perception should be indicated at first. In APA Dictionary of Psychology perception defined as follows;

> *"The process or result of becoming aware of objects, relationships, and events by means of the senses, which includes such activities as recognizing, observing, and discriminating. These activities enable organisms to organize and interpret the stimuli received into meaningful knowledge and to act in a coordinated manner". (2019)*

In some way, perception could be referred as a comprehension and accumulation process of human in existence of environmental stimuli. Whether this stimulus is visual, auditory or tactile, it is perceived and stored in human's memory as a procedure. The experience through the perception process enables the existence of the imagery, since the mental imagery actually contains similar structures of perception. (Finke, Ward, & Smith, 1992; Kosslyn, 1995; Oxman, 2002). In Stanford Encyclopaedia of Philosophy mental imagery is described as a "quasi-perceptual experience; it resembles perceptual experience, but occurs in the absence of the appropriate external stimuli." (2019). In other words, mental imagery is a kind of reanimation of the experience occurred in the course of perception.

Many people consciously or unconsciously experience mental imagery in their daily lives (Betts, 1909; Doob, 1972; Marks, 1999). Visual mental imagery is a sub-category of mental imagery, which initially relies on visual inputs stored in memory. Kosslyn, defines visual mental imagery as "seeing in the absence of the appropriate immediate sensory input; imagery is a perception of remembered information, not new input" (1995, p. 267). Mental images are loyal to the characteristics of the things they are representing (Arnheim, 1969). Correspondingly, previous studies show that mental images perform similar to their real counterparts (Finke, 1980, 1986; Shepard & Cooper, 1982; Kavakli & Gero, 2001). That is to say, as a result of visual mental imagery process, people can imagine not only objects but also their reciprocal and spatial relationships as if they are seeing them in real.

Sketching and drawing are processes that results in generation of 2-dimensional depictions, concepts and impressions stored in the memory of creator is utilized in these processes (Goldschmidt, 1991). Visualizing an object is a common practice in design. Designers often make variety of drawings as representations of both realized and non-realized products. Designers also reflect on their designs by using drawings in the process (Schön, 1992). Imagery enables people to understand and visualize objects and their attributes by using the data in their minds. This procedure also facilitates to envision the possible visual outcome when alterations made on depicted image (Kavakli & Gero, 2001). Presumably, product designers also rely on their visual mental imagery, while they are creating 2-D visualizations products yet to be realized. This mental process of design is usually accompanied with 2-D representation tools like sketching and drawings. Meaning that, designers picture the product while making drawings of it. Whether the designer is making drawings of an existing product or in the process of creating a novel entity, he regularly depends on his perception and imagery.

## 3 CREATING MENTAL IMAGES AS TACIT KNOWLEDGE

The concept of tacit knowledge first introduced by Polanyi; he states that "we can know than we can tell." (1966, p. 4). The term of tacit knowledge is repeatedly used to indicate the information that is acquired by doing and experiencing. This information is hard to transfer by just verbalizing, because of its individual and self-evident nature. Thus, it is the opposite of the codified knowledge (Chergui, Zidat, & Marir, 2018; Dampney, Busch, & Richards, 2002; O'Brien, 1995). Tacit knowledge is also existent in design, Schön's idea of "knowing-in-action" explains professionals from the fields like music, sports and design utilize tacit knowledge in their everyday lives (1983). Alexander states that the tacit knowing is apparent when designer intuitively knows the correctness of a "form" according to its "context defines the problem" (1973, pp. 15–16).

As it has been mentioned before, visual mental imagery is experienced by almost everyone in their daily lives. At first glance, creating a drawing (technical, isometric or perspective) of an object may not seem connected with tacit knowledge, as there are strict rules, methods and instruments for creating drawings. However, are these rules sufficient to create a correct visualization of an object? Polanyi's game of chess example shows similar characteristics with the action of creating a drawing. Polanyi explains that there are "rules" of playing chess however "the principles controlling the game cannot be derived from the rules of chess" (1966, p. 34). Similarly, there are rules varies according the type of visualization. However, the process of creating and manipulating mental images of objects in order to create 2-D representations of it, in other words conscious utilization of visual mental imagery can be considered as a tacit knowledge.

## 4 THE STUDY

### 4.1 *The course*

This study was conducted in a first-year visualization course of an industrial product design department. The course was mainly taken by first year industrial design students with exceptions of some failed second and third years and carried by three instructors. The instructors are professional designers from different backgrounds (one is industrial designer, an interior designer and another industrial designer with an engineering background). There were 64 students registered to the course, 42 of them were in their freshman year. The lectures were started at 10:00 am, ended at 13:00 pm on Mondays, and continued for 13 weeks.

In most of the weeks, students were given assignments. These assignments were divided between the instructors for grading. In each course session, after a short lecture, students were given in-class exercises. While in-class exercise was taking place, students met the instructor who graded their assignments, and review their previous homework. Review sessions were conducted face to face with each student, and created an appropriate environment for conveying critical feedback. Through the sessions, instructors mainly corrected and commented on the drawings of the students. In these sessions, the problems and mistakes in the drawings of the students were discussed and the approach of each instructor differs.

### 4.2 *Methodology*

The main aim of this study was to observe how the instructors help to improve the students' ability of creating right mental images through the one-on-one interactions. For that reason, participant observation was one of the main methods used. One-on-one critique sessions were on the focus. Through the study, more than 20 one-on-one critique sessions were observed. 4 of these critique sessions were further analysed.

Qualitative methods are employed in the study, since the nature of the interaction between student and instructor is rich in terms of narrative. A number of methods were used to collect and analyse data. In the first part of the study, four critique sessions of two instructors (two sessions of each instructors) were observed. Participant observation method was employed, which means the observer/researcher took notes about the behavioural responses of the actors through the sessions. The sessions also were recorded and transcribed into texts for further analysing. In the second part of the study, a semi-structured interview was conducted with the two instructors. The overall stance of the instructors towards the course and their opinions and approaches about the 2-D visualization were inspected. This interview was also recorded and transcribed.

After the completion of transcription of the critique sessions, in the light of field notes of the observer, these sessions were inspected in terms of possible interactions relating building of correct mental images.

## 4.3 *Utilization of mental imagery in the course*

As it is stated before, there are similarities between the operations of perceived real objects and the mental images of these objects. This connection often utilized by the designers while they are working on creating new or improved products. Presumably, these mental processes are employed also in the process of creating 2-D representations. In the case of design communication course, students are frequently asked to create 2-D representations of various objects. In some cases, students were given real objects, conversely, in some cases, several drawings of an object were given and another version of these drawings were demanded. For example, in the isometric drawing phase of the course, orthographic drawings of objects were given to the students, then they are required to make isometric drawings of the same objects according to the drawings given prior. This kind of exercises mostly require students to work on mental images of these objects.

The importance of visual mental imagery was emphasized numerous times by the instructors in semi-structured interview. Instructor A explains how some students have less control over their mental imagery when they try to visualize objects; "Actually, this course requires information. There are some rules ... However, we (students) have difficulties in 3-dimensional thinking. Especially when we start to orthographic (drawings), there are some "intuitive" things. Some students can see that (the object), some can really not and they say so. "I cannot visualize." they say."

In the course, there were instructions for students to follow in order to create accurate drawings, instructors' concern about the rules were apparent especially in the technical drawing section. However, applying the rules precisely was not a guarantee to an accurate drawing. If the student fails to create the right mental image of the product he is designing, he most likely fails to make correct drawings. There were instances that students were unable to create right mental images of the objects they are assigned to draw. The causes of this situation were varied; sometimes this inability was because the lack of information in other related fields like material and production, sometimes it was because of poor perception. But in most of the times, there was a combination of a couple of reasons. An exemplary instance took place in one-on-one critique session between Instructor A and Student 1. Students 1 showed her teacup section drawing she made accordingly to a real object. After a short inspection, Instructor A, identified a problematic point and the conversation below took place;

I: What is this line?

S: Material thickness in there.

I: Is that it only?

S: It is because this is empty inside. I mean, I thought that it is empty inside. Maybe it is filled, I am not sure.

I: It is actually filled. Let's think like this...

S: [Cuts her off] Actually that cup has a very thick material thickness. That is why I am not sure if it is filled or not.

...

I: It is probably like that, because this is glass production. [hatches over the students drawing]

...

S: ... I take it as empty but if it is filled, it should be hatched.

I: In any case, this should have a material thickness on its bottom. Here is filled with material, otherwise the tea would leak out.

Student in this example had the real teacup while creating the section drawing. Still, since, she could not cut the object in half, she had to rely on her version of the visual mental image of the object. On the other hand, even though the instructor did not have the actual product in her hands, she could visualize the divided teacup correctly as she has the necessary information for creating an accurate visual mental image. As it can be interpreted from the conversation above, the student's limited knowledge on materials and production methods lead to a deficient visualization of the object.

Even when the students are asked to draw an actual object without any further requirement of drawing unseen parts and views of it, the outcomes could be incorrect. This case was apparent especially in the proportions of the drawings. Some students had difficulties to represent the proportions of the given objects correctly into their drawings. Instructor B mentioned the issue by saying "I have always taken notes on the assignments about proportions. I mean, be reasonable right? This is not what you need to draw, this is not what you see. You need to draw rationally." The indication of Instructor B to the way of students' seeing, refers to the notion of perception. Some of the students had trouble to perceive the objects in a systematic and holistic way, consequently these students failed to represent the objects in right proportions. Instructor A categorizes correct perception of an object is an attribute of a designer. "...You are going to be a designer. If you ask a normal person, he would explain (the object) as; "there is a hole and a circle". But you need to describe (the object as;) "the size of the hole is one third of the circle and the distance between them is five times of the size of the circle". You need to look (at the objects) like this from now on." she explains.

During the course it has been observed that some students are naturally keen on utilizing their visual mental imagery. In other words, these students had a higher level of 3-dimensional thinking ability. Visualization of assigned objects were easier to these students from their peers. Although, these students may have a quick start to the course, it has also been observed that the ability of 3-dimensional thinking can be improved through experience and trying. Both of instructors stated that one-on-one critique sessions as a new implementation to the course was helpful for students to improve their 3-dimensional thinking ability.

### 4.4 *Approaches of instructors in one-on-one crits*

One-on-one critique is a specific method primarily employed in studio environment. In design studio, instructors and students come together and discuss over the project of the student (Goldschmidt, Hochman, & Dafni, 2010; Uluoglu, 2000). For this drawing course, a similar method was adapted. Meaning that, through the semester students discussed their drawings with their assigned instructors. In the interview session, instructors claimed this one-on-on approach allowed them to understand the troubles that each student has while visualizing objects.

In the one-on-one sessions, it is observed that instructors employed different approaches to explain students how to create mental visualizations of an object. These approaches varied according to the correctness of the mental process of the student and the accuracy of the visualization. Still, it is noticed that instructors generally used these approaches in combinations. The approaches observed in this specific course can be grouped under three categories; *intervention to an existent drawing, invention of a new drawing and indication to an existent object.*

*Intervention to an existent drawing* is an approach instructors employ when there are small problems in the drawings of the student. While Instructor A was pointing out a mishap in the button part of the assignment Student 2 made;

I: ... You should add more details. For example, there are juts that sticks out right? They are buttons. Then there should be thickness right? [Corrects the 2-D looking buttons on student's drawing]"

S: Yes, exactly

I: [Continues to correct the drawing by adding the button hole] "In the same way, if something enters (into the object), the surface should bend towards (inside of the object) and this (the button) sticks out."

In this example, student has an initial drawing with a little need for correction. Other than that she was able to perceive the overall object rather accurately. Thus, the instructor preferred to explain her own flow of reasoning a visual mental image of the object step-by-step. This way student could memorize and use these steps as a possible checklist for creating a detailed mental image.

Sometimes students could have perceptional problems. This issue shows itself as bigger mistakes in the drawings which cannot be fixed by just making small corrections and reminders. In these cases, instructors generally tried to direct students into a more conscious way of looking towards the objects. Instructors use two methods -mostly together- to show students how to create right image first in their minds and then on the paper. One is *to create the drawing from the scratch*. For example, while Instructor B and Student 3 was discussing over her 2-dimensional looking isometric drawing of a mouse, Student B states;

S: I don't understand this.
I: I can see that. ... Really, If I haven't given this as an assignment, there was no way that I guess this as a mouse.

After a while he inspected the drawing, Instructor B took a pencil and started to draw next to the Student B's drawing. While drawing he continued to explain the visualization process which he thinks acceptable in this situation;

I: Basically, you should have started with the base. You know, in car design they also do that. Generally, a boundary box is drawn at first. ... The middle points of the boundary box are marked; I mean (this is) the base area. ... Then you need to, kind of, take sections (of the object), then you should move up and down. ... After that you connect the lines, what you do is creating a geometry actually...

By constructing the object step-by-step and explaining each stage, the instructor demonstrated a systematic way to perceive and represent the object.

Along with creating a new drawing Instructors often preferred to indicate to the real object if it is available. For instance, in the first class of section drawing Instructor B used paper cups to explain the subject to a student. He cut the cups in half both vertically and diagonally and hatched the parts that touched to the cutting plane to show how the section drawing should be done. By showing the direct outcome of taking section, student was able to see directly the right mental image she should create and store this visual in her memory, thus, they could use this information to reason the visual mental images of other objects.

At times when an object is not available for helping to show the right visual, instructors commonly encourage students to create mental images of the object and work on them. The indications like "Let's think like this, lets image a ..." frequently used in the conversations between instructors and students.

## 5 CONCLUSIONS

Creating the right mental images of objects is an ability that is important for designers, especially when they are creating 2-dimensional representations of these objects. Although, this ability is existent in every human being as a default, to be able to create right and detailed images, designer should practice and improve this ability of hers.

Design educators and professionals utilize their 3-dimensional thinking ability while they are creating visual representations of the objects they are designing. Their process of creating right visual mental image can be categorized as tacit knowledge by its nature. Because of their limited knowledge and lack of practice design students may not be able to use their visual mental imagery successfully. Yet, design students have visualization courses that provide an appropriate environment for them to observe and internalize the processes of their instructors.

In the context of this study, it is observed that the instructors have different tactics to support students to improve their visual mental image construction ability. In one-on-one sessions, the instructors employed a variety of approaches like intervening to an existing drawing, explaining by creating a novel visualization or explaining over an existent object, according to the success level of the student. Naturally, this is a very limited study in terms of the number of the interactions inspected. However, the qualitative aspect of the data collected may give very initial insights about the possibilities and the effects of the communication between design student and instructor in terms of the usage of visual mental imagery. Expectedly, this study sparks interest for further studies in this subject.

## REFERENCES


APA Dictionary of Psychology. (2019). Retrieved from https://dictionary.apa.org/perception.

Alexander, C. (1973). Notes on Synthesis of Form. Harvard University Press.

Arnheim, R. (1969). Visual Thinking. University of California Press.

Betts, G.H. (1909). The Distribution and Functions of Mental Imagery. New York: Teachers College, Columbia University.

Chergui, W., Zidat, S., & Marir, F. (2018). An approach to the acquisition of tacit knowledge based on an ontological model. Journal of King Saud University - Computer and Information Sciences. https://doi.org/10.1016/j.jksuci.2018.09.012.

Dampney, K., Busch, P., & Richards, D. (2002). The Meaning of Tacit Knowledge. Australasian Journal of Information Systems, 10(1), 3–13.

Doob, L.W. (1972). The Ubiquitous Appearance of Images. In P.W. Sheehan (Ed.), The Function and Nature of Imagery (pp. 311–332). New York: Academic Press.

Finke, R. A. (1980). Levels of equivalence in imagery and perception. Psychological Review, 113–132. https://doi.org/10.1037/0033-295X.87.2.113.

Finke, R. A. (1986). Mental imagery and the visual system. Scientific American. https://doi.org/10.1038/scientificamerican0386-88.

Finke, R. A., Ward, T. B., & Smith, S. M. (1992). Creative Cognition: Theory, Research, and Applications. The MIT Press.

Goldschmidt, G. (1991). The Dialectics of Sketching. Creativity Research Journal. https://doi.org/10.1080/10400419109534381.

Goldschmidt, G., Hochman, H., & Dafni, I. (2010). The design studio crit: Teacher-student communication. Artificial Intelligence for Engineering Design, Analysis and Manufacturing: AIEDAM. https://doi.org/10.1017/S089006041000020X.

Kavakli, M., & Gero, J. S. (2001). Sketching as mental imagery processing. Design Studies. https://doi.org/10.1016/S0142-694X(01)00002-3.

Kosslyn, S. M. (1995). Mental Imagery. In S. M. Kosslyn & D. N. Osherson (Eds.), An Invitation to Cognitive Science, Visual Cognition (2nd ed., pp. 267–296). MIT Press.

Marks, D.F. (1999). Consciousness, Mental Imagery and Action. British Journal of Psychology (90) 567–585.

O'Brien, R. C. (1995). Employee involvement in performance improvement: a consideration of tacit knowledge, commitment and trust. Employee Relations, 17(3), 110–120.

Oxman, R. (2002). The thinking eye: Visual re-cognition in design emergence. Design Studies. https://doi.org/10.1016/S0142-694X(01)00026-6.

Polanyi, M. (1966). The Tacit Dimension. The University of Chicago Press.

Schön, D. A. (1983). The Reflective Practitioner: How Professionals Think in action. Basic Books. https://doi.org/10.1542/peds.2005-0209.

Schön, D.A. Research in Engineering Design (1992) 3: 131. https://doi.org/10.1007/BF01580516

Shepard, R. N., & Cooper, L. A. (1982). Mental Images and Their Transformations. Cambridge: MIT Press.

Uluoglu, B. (2000). Design knowledge communicated in studio critiques. Design Studies, (21), 33–58.

*Research & Education in Design: People & Processes & Products & Philosophy – Almendra & Ferreira (eds)*
*© 2020 Taylor & Francis Group, London, ISBN 978-0-367-49296-0*

# Pinpointing fragility through the act of drawing as a moment of embodiment

L. De Brabander, T. Lagrange & J. Van Den Berghe
*Faculty of Architecture, KU Leuven, Belgium*

ABSTRACT: Fragility is evident yet uncharted in architectural design. To fathom the term and develop ways of understanding it, we explore the potential of drawing. Through making drawings, and reflecting on their outcome, we probe a more effective implementation of fragility in architectural design processes. Lagrange (2016) previously argued that fragility is a subjective phenomenon that resides in both the material and mental space. To fathom fragility, we address the following threefold: material space (the drawing as an object), the moment of embodiment (the body: the drawing hand) and mental space (the mind). This paper elaborates on how the central moment of embodiment, installed as a mediator between material and mental space, might facilitate ways of understanding fragility by demonstrating how empathy can be activated through specific ways of drawing. We do so by explaining two research cases: a Belgian dune landscape and the investigation of the first author's drawing archive.

## 1 STATE OF THE ART

The research aims to develop ways of understanding the phenomenon of fragility within architecture through setting up different drawing processes. Fragility has not explicitly been defined yet in architecture. In a landscape, in this case a Belgian dune landscape, material and immaterial elements are present, but some only occur in the mind, adding up to an ever-accumulating archive of experienced material spaces (van Schaik, 2008). Our empathic abilities enable us to suspect that there is fragility in both the material and mental space, but we do not understand it yet. Fragility is intangible and unclear, and thus difficult to grasp, capture and implement in architectural design processes. Lagrange (2016) previously argued that fragility is a subjective and dynamic phenomenon, occurring in both the material and mental space (Otero-Pailos, 2016). In this research, we explore and apply the potential of drawing to facilitate an understanding. In sciences such as economics and medicines, fragility is understood in an applied way (Bertocci & Guerzoni, 2011; Tranquilli Leali et al., 2011). Independently, they determined sets of parameters or fragility indexes, to define how fragile the contexts they are investigating are. Opposed to these disciplines, fragility has been conceived as an aesthetic, e.g. in the history of an art form such as ballet, in which the formation of the ballerina's feet, and by expansion, her whole body, had to look and appear fragile while being physically strong (Carroll, 2010).

## 2 OBJECTIVES: THE MOMENT OF EMBODIMENT AS A MEANS TO UNDERSTAND FRAGILITY

From an architectural point of view, we aim to determine our own stance within the already existing theoretical framework concerning fragility. Our research case is a Belgian dune landscape and its close surroundings. In this paper, we demonstrate how the physical act of drawing constitutes a central moment between material and mental space, i.e. a moment of embodiment. Moreover, we discuss how the act of drawing enables us to develop an understanding of fragility through empathy, activated through specific ways of drawing.

First, we want to elaborate on the twofold of material and mental space to then instate the central moment of embodiment. Materiality is connected to substance; in this research, material space encapsulates the substance of the drawing as an object.

> *"With materiality (substance), the architectural drawing can evolve from a mere tool to a thing, and when executed with care this thing becomes a 'work'. [...] Once a 'work,' the verb 'to work' enters the discourse, and the process of making comes into the scope."* (Van Den Berghe, 2013, p. 666)

The drawing as an object needs a subject in order to be. A drawing depicts something. The mental space consists of an ever-growing archive of feelings, thoughts, memories, phantasms, words, word pictures, and thus of experienced material spaces (Verschaffel 2012). According to Verschaffel (2012) the mental space is hidden *"somewhere in the torso and the head"*, it *"hollows out the body and fills it with an 'inside,' an empty space that is surrounded by skin, but which seems boundless and thus infinite"* (p.8). In our view, this inside can be conceived as a landscape as well. However, it never becomes as clear as things perceived in the physical dune landscape. Due to its unclearness, mapping the topography of this mental space is challenging, but it is crucial to ever understand its fragility. Besse (2001) states that the map is the key instrument through which a (mental) landscape can be drawn, because every map depicts a draughtsman's personal perception of a landscape.

The physical act of drawing happens around the hinges and joints of the fingers, the wrist, the elbow and the shoulder. The hand is central, it receives impulses from the mental sphere, it holds the pencil, and marks the substance of the paper (Van Den Berghe, 2013). *"The term 'design' comes from the Italian 'disegno,' meaning drawing, suggesting both the drawing of a line on paper and the drawing forth of an idea to a thing"* (Hill, 2005, p.224). We argue that the act of drawing constitutes a central moment of embodiment that only occurs when drawings are being made (Figure 1). It takes place in the kairotic moment when the threefold of mind, drawing hand, and material are present (De Smet and Janssens, 2016). The moment of embodiment is temporal; it disappears when the draughtsman stops drawing. Material and mental space *are*, they are inevitable and moreover, *a given* (Birnbaum, 2008). In the moment of embodiment, drawings are in a stage of becoming (Glanville, 1990). Only when the drawings have become drawings through a moment of embodiment (the body: the drawing hand), they can be (part of) a material space. A unidirectional transfer of *things* takes place from the stage of *becoming* (moment of embodiment) to the stage of *being* (material and mental space) (Figure 1). The dune landscape that is a case in this research, is a given in the mental space. It was selected out of a range of possible sites through an extensive drawing process, based on on-site experiences. In that way, the realm of the material and mental space is continuously enlarged.

We want to remark that although the physical dune landscape is a material space, we do not conceive it as a material space that is the product of the embodied trajectory. We rather conceive it as (part of) a mental space. It leaves a mark in the mind, an experience that leads to emotion and, through an indispensible moment of embodiment, brings about a materialization that might involve knowledge about fragility. The mental space is the subject of the drawings. Material space is the drawing and by expansion what is made out of that drawing e.g. a building.

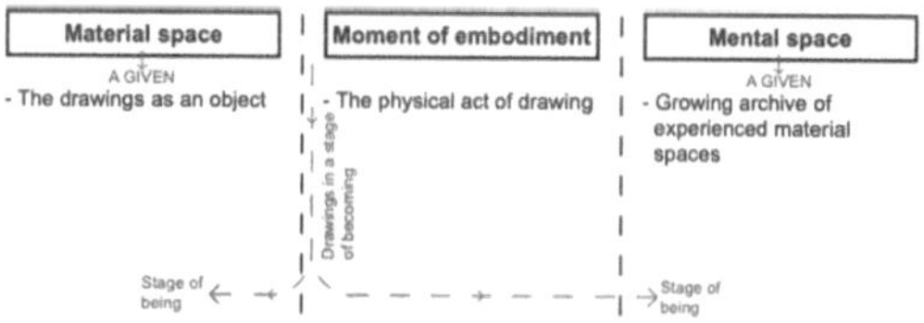


Figure 1. The moment of embodiment scheme (figure by the first author).

## 3 METHOD

In this experimental, design driven research, we use drawing to access and possibly understand fragility within the architectural discipline. *"Gaston Bachelard writes about the imagination of the hand: 'Even the hand has its dreams and assumptions. It helps us understand the innermost essence of matter. That is why it also helps us imagine (forms of) matter'"* (Bachelard in Pallasmaa, 2002, p.28). Opposed to the material and mental space, the moment of embodiment is evitable, and yet, it is the crucial hinge that might enable ways of understanding fragility. This is where empathy enters the discussion (Figure 2). Giambattista Vico (1725) stated in his Scienza Nuova:

> *"As rational metaphysics teaches that man becomes all things by understanding them, imaginative metaphysics shows that man becomes all things by not understanding them [...] but when he does not understand he [...] becomes them by transforming himself into them"* (Vico, 1725, p.405).

Empathy is a human quality that facilitates understandings of non-understood concepts through becoming that which is not understood. The draughtsman's empathic abilities enable to suspect that there is fragility in the material space, the moment of embodiment and the mental space that he or she does not understand, and that it is inevitable in architectural design processes. In order to understand, a moment of embodiment is created through drawing, wherein empathy can be activated. In this case, empathy is activated through the draughtsman's specific manner of drawing. Here we make a fundamental distinction between two kinds of drawings:

1. Drawings of facts ('Factual Drawing') referring to existing material spaces and the accompanying mental spheres.
2. Drawings of fictions ('Fictional Drawing') that present possible architectural interventions as design propositions.

If executed with care, both are able to activate empathy and establish understandings of phenomena such as fragility. Both ways of drawing are not merely representation, but demonstrate two sides of an ontological presence that are indispensable and needed to describe a fragile architectural reality as a whole. In this paper, emphasis lies on the first. Besse (2001) argues that the map, and by expansion every drawing representing an existing material space, has to be considered as an iconographic expression of the mental life of humanity. Brian Harley says that it *"helps the human mind to make sense of his universe at different scales"* (Harley & Woodward, 1987, p.1). In this case, imagination is not regulated by fantasy. Making a drawing means imagining the real, to give its image to the surface of the earth (Besse, 2001).

## 4 FIELDWORK

### 4.1 *Drawing a geological section as an architect*

First, a series of handmade analytical drawings were made of a Belgian dune landscape. The series comprises three kinds of handmade drawings: topographic sections, a geological section and plans. Here, we put emphasis on the geological section (Figure 3). Dune landscapes have

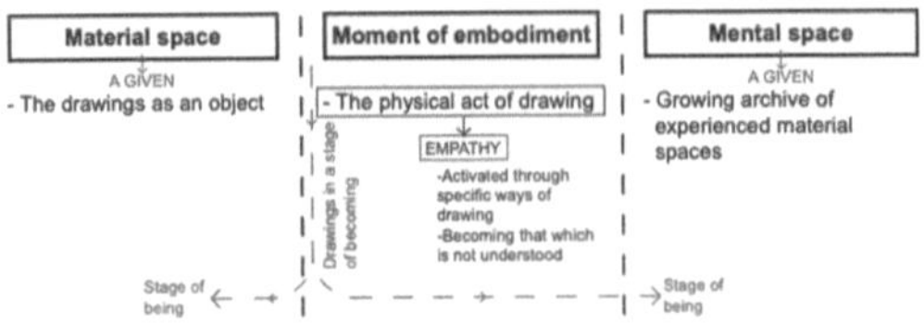


Figure 2. The moment of embodiment second scheme (figure by the first author).

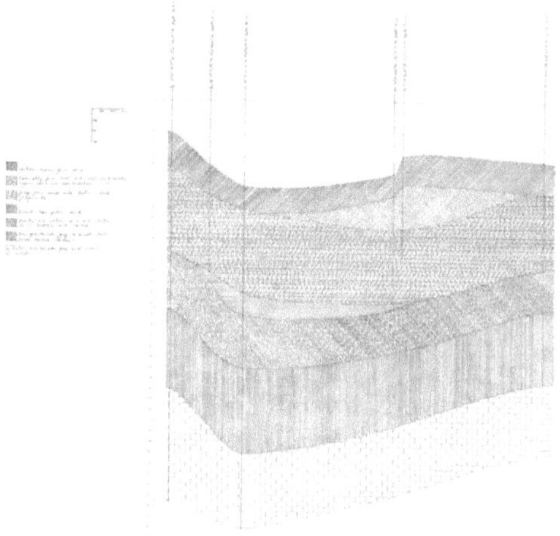

Figure 3. Drawing of a geological section of a Belgian dune landscape, 66x63cm, (2018) (Drawing by the first author).

topographic constellations that are strongly articulated and specific. It requires sections in order to fully understand the landscape's anatomy. Therefore, in our view, the construction of sections prevails over planological mapping.

> *"Pérez-Gomez suggests that the section is of utmost importance in the architect's work, as a prediction on the casting of shadows, pointing at the anatomic nature of the section that, applied by the architect 'break[s] the skin of things in order to show'"* (Pérez-Gomez in Van Den Berghe, 2012, p.69).

Architects observe topographies and draw vertical sections, probing the landscape in which they want to operate, to then intervene and cut through the earth with surgical precision. Geologists conceive the section differently. They use sections to visualize the results of their on-site investigations on earth's layers of substance. Architect and professor Elizabeth Hatz (personal communication, June 21, 2018) explained that she considers the top layers as a crust, and architectural interventions as transformations of matter in that crust. So in order to design and build, geological knowledge on that crust is essential. Architects do not have the knowledge to execute geological investigations on earth's layered composition. However, they do have the ability to interpret and analyse geological data produced by geologists. The first author of this paper is an architect and has drawn a geological section from her architectural point of view, hence the title of this chapter: drawing a geological section as an architect (Figure 3). The section has been drawn according to the base principles for drawing a geological section. One of the basic principles is that two different scales are combined in a single section; the X and Y-axis are scaled differently in order to clearly see the landscape's layers in the drawing. This requires the effort of simultaneously calculating two scales and taking countless measures, since the geological section's nature stands in contrast to the architect-as-draughtsman's own drawing conventions.

Drawing by hand allows for slowing as an indispensible condition through which an embodiment occurs. It increases the draughtsman's physical level of consciousness because the tools are connected to the hand's slowness (Flores & Prats, 2014). While hands move, the draughtsman has the time to think and observe what he or she is doing. The late Flemish writer Hugo Claus claimed that the time between lifting his pen off the paper, dipping it in the ink and bringing it back allowed him just enough time to sharpen his phrases (De Wit, 2018). He used a 'balloon'-pen, because he liked the resistance and hooking of the pen to the paper (Claus, 1956). The geological section has been drawn on thin, smooth tracing paper with the pencils B, HB and H (Figure 3). The smoothness of the paper requires steadiness of the hand, causing for a high pressure of the hand on the pencil, which results in quick wearing of the pencil's tips. Similar to Claus's writing, the time between lifting the pencil off the paper, sharpening its tip, and bringing it back, gives the draughtsman enough time to decide what lines will be drawn next and at which pressure. This is where the moment of embodiment comes into the scope. The geological data is interpreted (mental space) and than materialized (the drawing as an object) through the act of drawing. The latter constitutes a mediating moment, rather than a space, i.e. the moment of embodiment, and takes place between the material and mental space. Moreover, the pencil-tracing paper combination and the condition of having to calculate two scales substantially slow down the drawing process even more. These insights cause for the emergence of a consciousness that fragility can occur in the act of drawing, and thus in the moment of embodiment, when a form of pressure is exerted on the way a draughtsman is used to make drawings.

More than two decades after the previous geological investigations, geographers measured the topography of the dune landscape again. By transferring these new topographic results to the drawing as well, the draughtsman was able to compare them with the results of the previous geological investigations (Figure 4). Substantial differences in the topography between old and new results are visible. The top layer shows significant changes in the landscape's topography over the past 20 years. A dune landscape is constantly in motion due to conditions such as weather, vegetation, animals and people. In other words, every step taken in the landscape might mean altering its microscopic surface and can be tantamount to losing the very qualities that make it valuable (Otero-Pailos, 2016). The experiences of which the mental space consists are thus momentary and moreover, fragile.

In this case, besides making topographic sections, we consider making a geological section as essential for the architect-as-draughtsman to develop a consciousness about fragility and thus to activate empathy. Other types of analytical drawings such as isometric drawings and plans can activate empathy as well. However, the mental action that takes place when the architect-as-draughtsman uses techniques of another discipline such as geology might challenge architecture's disciplinary conventions when it comes to drawing. Here it becomes interesting because this opens up avenues for other types of drawings that enable another consciousness about fragility and the activation of empathy. From this insight one can extrapolate that adopting drawing techniques from disciplines other than geology and architecture, might further expand one's consciousness and understanding of fragility. This will be addressed in further stages of the research.

### 4.2 *Portraying the author as draughtsman: Analyzing a drawing archive*

According to Joseph Brodsky *"we make everything in our own image, because we do not have a more reliable model; the objects produced by us describe us better than any confessions of faith"* (Brodsky, 1994, p.55-56). To gain a better understanding of the first author's way of drawing and obtain insights in her mental space, her drawing archive has been investigated. We argue that analyzing and reflecting on the first author's drawing archive and creational history establishes new insights in the author's way of drawing, so as to 'unveil the secrets' of her creativity or to address failures.

The archive encapsulates drawings made as a child, a youngster, an architecture student and an architectural researcher. The collection of folders, sketchbooks and loose drawings

Figure 4. Drawing of a Belgian dune landscape's topography measured in 2016 (2018) (Drawing by the first author).

was brought together, photographed and chronologically ordered.[1] A portrait of the author-as-draughtsman was gradually laid out in doing so. The drawings have been categorized by their years of production. Laying the sequence of drawings next to one another in a chronological order indicated that drawings made two decades ago are interlinked with drawings made over the past five years in the following way(s):

By closely observing the archive, we can ascertain that crosslinks between drawings from different time periods can be gathered in different categories such as (1) techniques (e.g. etching, lino cutting, painting), (2) tools (e.g. charcoal, pencil, pastel, watercolour, ink), (3) methods (e.g. sequential drawing, storyboard methodology) and (4) subjects (e.g. still lives, animals, portraits, marine views).

In every new drawing, elements of previous drawings are embedded, some of which are clearly traceable, others are indirectly present. If one drawing had not been made, the other would not have become and would not be. E.g. a storyboard methodology, clearly present in drawings from the past 5 years, can be traced back to the earliest drawings.

Here, we can identify that the act of drawing (the body: the drawing hand) constitutes a central moment of embodiment as a mediator between material and mental space. The drawing being made is (part of) the material space. The mental space is a repository of images and experiences from the past, previously made drawings are thus part of it. A moment is situated between those spaces rather than another space, only there when drawings are being made, i.e. the moment of embodiment.

Furthermore, chronologically ordering the archive indicated that the vibrancy with which the early drawings had been made, was gradually limited and deplorably subdued through architectural education. This was exemplified by a gradual transition to fewer use of colours, an increase of conventional and 'safe' drawing techniques and stricter formats.

Furthermore, this archival investigation indicates that reverting to uninhibited ways of drawing might allow for multisensory and embedded experiences to enter the drawing process, so as to further unlock consciousness about the fragility of the material space (the drawing as an object), the moment of embodiment (the body: the drawing hand) and the mental space (the mind).

### 4.3 *(Sk)etching*

Without wanting to annihilate the most recent drawings of a topographic and geological section, and banking on the findings of the analysis of the first author's drawing archive, we argue that reverting to uninhibited ways of drawing might be an unlocker to include haptic, tactile and sensory experiences in the drawings. To do so, a series of sketches and handmade etchings have been developed since the formal start of this doctoral research.

First, sketches have been made on-site (Figure 5 and 6). Contrary to the topographical and geological section where the draughtsman imaginarily places himself in the dune landscape, the sketches have been made from a physical position in that dune landscape. By drawing on-site, external effects influence the body (the drawing hand) and its act of drawing. E.g. there are few possibilities to sit down on site and the wind and cold cause for the hand to cramp up, making it difficult to draw straight lines. Drawing on site creates a haptic contact between the draughtsman, the drawing and the subject being drawn because that what is perceived can directly be transferred from the mind through the hand onto the paper. The sketches are focused on one aspect of the draughtsman's experiences of the landscape, namely its desolation, which incited the draughtsman to make drawings of windows, closed window blinds and deserted landscapes. They have been made in sketchbooks on different site visits, which allowed the draughtsman to make adjustments to the tools and manner of drawing. A first series of sketches have been drawn with a black Artline pen on ruled paper (Figure 5). The combination of thin, smooth, ruled paper and a smoothly rolling black pen was not optimal in the

1 The archive consists of approximately 400 drawings of which a selection of 150 drawings has been analysed.

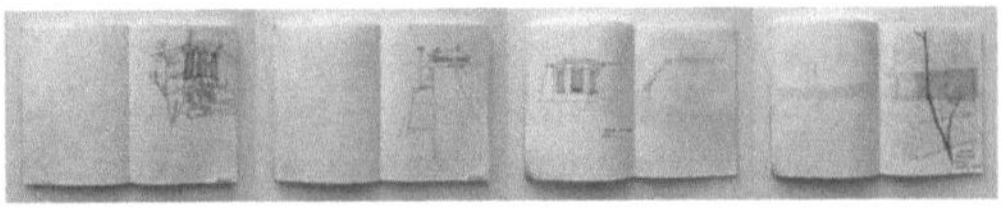

Figure 5. Selection of drawings out of the first series of sketches (2018) (figure by the first author).

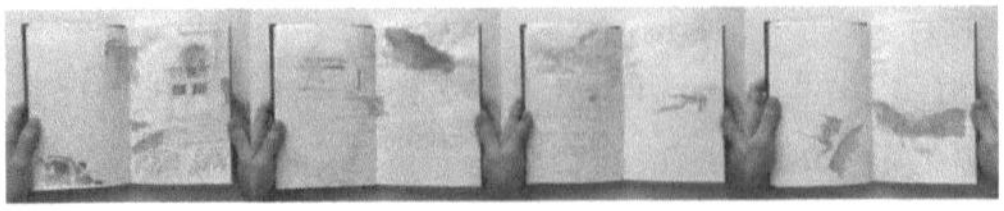

Figure 6. Selection of drawings out of the second series of sketches (2019) (figure by the first author).

Figure 7. Etchings of a Belgian dune landscape and its surroundings (2019) (Figure by the first author).

windy, rough environment of the landscape. A second series of sketches have been made with 2H, HB and 2B pencils and watercolour in a sketchbook on white sketching paper (Figure 6). The grain of the paper provided a roughness that enhanced the hooking of the pencil onto it.

Subsequently, a number of the sketches have been lifted out of the sketchbook format and have been transformed into etchings by redrawing them on plastic etching plates (Figure 7). By making etchings, the draughtsman reverted to a technique she frequently used in the past. This does not mean that the etchings are not sketches. Etching obliges the draughtsman to put a high level of detail in the drawings. The etching needle becomes the pencil that carves into the substance of the plastic. Through etching, elements of the landscape that interlock with the senses might be captured and provide the detail that moves beyond the mere visual to hapticity and tactility (Holl, 1989). The carved lines are so thin that every small scratch is transferred onto the paper in the printing process, causing for every bump in the landscape's surface to be felt through the drawing. Even though the prints are static objects, they arouse a sense of movement; it is almost possible to feel the dune grasses rustle through the wind as the finger moves over the drawing's surface. Moreover, etchings are inevitably covered under a light haze of ink, which enhances the desolate atmosphere the draughtsman has in mind (Figure 7). We argue that drawing in uninhibited ways is inevitable to develop ways of understanding fragility and thus to activate empathy. These uninhibited ways of drawing can differ, depending on the draughtsman's own way of drawing. Hence other drawing techniques can activate empathy as well; the condition is that they have to register the draughtsman's own perceptions and experiences of a site.

## 5 CONCLUSION

This paper presents the first step of an iterative process that will be continued in the following stages through new drawing processes, established through Critical Sequential Drawing (Van Den Berghe, Sanders & Luyten, 2017). *Critical* as a constant and critical (self) assessment based on meticulous (self) observations during the drawing session itself, *sequential* as a cyclical process of drawing new and more accurate versions of previous drawings. It is a method of constant improvement by turning observations into high quality observations by, in a second round of observations, turning them into an observable in their own right, generating improvement of action in doing so (Glanville 2002). This cyclical process is repeated until saturation in the interrogation of drawing with regard to fragility occurs.

Only through more cycles in the drawing process, empathy can be further and fully be activated. The activation of empathy depends on how and in what circumstances the drawings are made. Activation is needed first in order to subsequently develop ways of understanding fragility. An essential part of understanding fragility inevitably needs to go through the moment of embodiment. The written description is one form of making fragility explicit yet drawing-as -verb can make fragility perceivable and the drawing-as-noun can make fragility apparent to the readers' experience when he or she observes the drawings. Through the three phases of fieldwork, we ascertained that the material space, the moment of embodiment and the mental space possess an inherent fragility, but it only occurs when an amount of pressure is exerted on them, only then they become fragile. It thus becomes clear that fragility is in a stage of becoming. Both analytical drawings, such as topographic and geological sections, and sketches and etchings, are important and fruitful in the drawing process to activate empathy and to create a way of understanding fragility.

## REFERENCES


Bertocchi, G., & Guerzoni, A. (2011). *The Fragile Definition of State Fragility* (Vol. 16). Retrieved from https://www.researchgate.net/publication/227488253_The_Fragile_Definition_of_State_Fragility

Besse, J.-M. (2001). Cartographier, construire, inventer. Notes pour une épisemologie de la démarche de projet. *Les Carnets du paysage*, (7), 126–145.

Birnbaum, D. (2008). *The Hospitality of Presence: Problems of Otherness in Husserl's Phenomenology*. Berlin, Germany: Sternberg Press.

Brodsky, J. (1994). *Watermark*. Helsinki, Finland: Farrar, Straus and Giroux.

Carroll, A. (2010). Women in Ballet. Retrieved from The Australian Ballet website: https://australianballet.com.au/behind-ballet/women-in-ballet

Claus, H. (1956). Interview met Hugo Claus. Retrieved from http://theater.ua.ac.be/claus/html/1956-03-10_claus_interviewmethugoclaus.html

De Smet, A., Janssens, N. (2016). Probing the future by anticipative design acts. In P. Lloyd & E. Bohemia, (eds.), *Future-Focused Thinking: Proceedings of DRS2016: Design + Research + Society* (Vol. 7, pp 2761–2794). doi: 10.21606/drs.2016.79

De Wit, V. (2018, February 26). *11 jaar na zijn dood: Veerle De Wit, weduwe van Hugo Claus*. (Y. Desmet, Interviewer). Retrieved from https://www.humo.be/humo-archief/387877/10-jaar-na-zijn-dood-veerle-de-wit-weduwe-van-hugo-claus

Flores, R. and Prats, E. (2014). *Thought by Hand: the Architecture of Flores and Prats*, Mexico City, Mexico: Arquine.

Glanville, R. (1990). *The Self and The Other: The Purpose of Distinction*.

Glanville, R. (2002). Doing the Right Thing: the Problems of ... Gerard de Zeeuw, Academic Guerilla. In Glanville, R. (guest editor), *"Gerard de Zeeuw—a Festschrift", Special Issue of Systems Research and Behavioural Science* (Vol. 19 n°2).

Harley, J. B., & Woodward, D. (1987). *The History of Cartography (Vol. 1)*. Chicago, IL: University of Chicago Press.

Hill, J. (2005). Drawing Forth. In M. Belderbos & J. Verbeke, (eds.), *The Unthinkable Doctorate: Proceedings of the colloquium 'The Unthinkable Doctorate'* (pp 223–227). Brussels: Network for Theory, History and Criticism of Architecture and Hogeschool voor Wetenschap & Kunst – School of Architecture Sint-Lucas.

Holl, S. (1989). *Anchoring*. New York, NY: Princeton Architectural Press.

Lagrange, T. (2016). *Look Space! A story of Analogous Spaces*. Amsterdam, Netherlands: Idea Books.

Otero-Pailos, J. (2016). Experimental Preservation. *Places Journal*, (2016). doi: 10.22269/160913

Pallasmaa, J. (2002). Lived Space. Embodied Experience and Sensory Thought. *OASE*, (58), 13–34.

Tranquilli Leali, P., Muresu, F., Melis, A., Ruggiu, A., Zachos, A., & Doria, C. (2011). Skeletal fragility definition. *Clinical Cases in Mineral and Bone Metabolism*, 8(2), 11–13.

Van Den Berghe J., Sanders M., Luyten L. (2017). *Windows into an Architecture of Darkness and Depth.* In: Des Traces et des Hommes Imaginaires du Château de Selles. Milan, Italy: Silvana Editoriale.
Van Den Berghe, J. (2013). Architectural Drawing as Verb, not as Noun: extending the concept of architectural drawing. In J. Verbeke & B. Pak (Eds.), *Knowing by Designing: Proceedings of the 'Knowing by Designing' conference* (Vol. 1, pp. 664–673). Brussels, Belgium: LUCA, Sint-Lucas School of Architecture Ghent/Brussels; KU Leuven, Faculty of Architecture.
Van Den Berghe, J., Blythe, R., & Verbeke, J. (2012). *Theater of Operations, or: Construction Site as Architectural Design*. RMIT, Royal Melbourne Institute of Technology University, Melbourne.
van Schaik, L. (2008). *Spatial Intelligence: New Futures for Architecture*. Chichester, UK: John Wiley & Sons Ltd.
Verschaffel, B. (2012). The World of the Landscape. *CLCWeb: Comparative Literature and Culture, 14*(3). doi: 10.7771/1481-4374.2036
Vico, G. (1725). *Scienza Nuova: Principi di Scienza Nuova d'intorno alla Comune Natura delle Nazioni.*

*Research & Education in Design: People & Processes & Products & Philosophy – Almendra & Ferreira (eds)*


# Integrating visual and narrative reflective components in architecture pedagogy

K. Shoham, E. Eizenberg & I. Aravot
*Faculty of Architecture and Town Planning, Technion – Israel Institute of Technology, Israel*

ABSTRACT: Reflection, an old concept, is relevant to nowadays higher education undergraduates and graduates, architecture and design students among them. While reflection is considered as a powerful learning tool that enables students to use personal resources, the definitions and concepts associated with it are diverse and sometimes even contradictory. This paper briefly reviews the roots of reflection and the leading theories connecting reflection to learning, pedagogy and architecture. The paper also presents insights regarding the verbal and non-verbal expressions of the reflective design process, based on a case study of an architecture course. In conclusion we suggest a new reading of the different dimensions in Schön and Kolb reflection theories, through the lens of our research findings.

## 1 INTRODUCTION

The concept of reflection is as old as the documented human metacognition – thinking about thinking – and can be traced back to ancient times, in the Socratic commandment "Know Yourself" and Aristotle's concept of the virtues, yet it is relevant more than ever. The term reflection itself was defined by the philosopher, psychologist, and educational reformer John Dewey, as a unique way of thinking, which is active, constantly changing and considering the unknown (Dewey, 1933). Those dimensions are highly pertinent to design thinking (Rowe, 1987) that focuses on what is called "ill-defined problems": challenges that have multiple solutions, and moreover, that can be answered but sometimes not questioned properly.

In higher pedagogy, reflection has special importance, since it is perceived as allowing students to observe the strategies they used in the process of learning, in order to learn their advantages and disadvantages and to improve their future course of action (Schön 1983, 1987; Kolb 1984, 2005). It becomes even more significant due to the current change rate at any professional field, which indicates that what is learned today might not be relevant soon (Mazur, 2018). As a result of the information revolution, technological progress and globalization processes there is a shift in higher education, from a deductive lecture-based learning to inductive learner-centred methods (Salama, 2016). In architecture too, we do not see teaching as merely a process of passing knowledge from an "all know" professional to a novice student (Webster, 2008). It was therefore suggested that students need tools that will enable them to use personal resources and develop self-awareness to the knowledge they have acquired (Kober, 2015). Metacognition and self-evaluation strategies are considered highly effective tools for higher education graduates nowadays, expected to be lifelong learners (November, 2012). Reflection is considered as one such tool and it is used likewise in architecture education (Austerlitz, Aravot, Ben-Ze'ev, 2005).

While reflection and its philosophical and pedagogical definitions are well established, the architectural expressions in the reflective process are of a unique nature – constructed by both verbal and nonverbal expressions, which include various artefacts. By deciphering the complex and implicit integration of verbal and nonverbal expressions – the narrative and the visual integration – in the reflective design process, this paper aims at contributing to the future development of further much subtle tools for architecture education and design thinking pedagogy. Therefore, in this paper we will share our insights regarding the reflective process, integrating visual and narrative component in architecture education; Those insights were acquired by observing a case study: a new course entitled 'Architectural Editing', given as part of the curriculum at the Faculty of Architecture and Town Planning at the Technion – Israel Institute of Technology, since 2015. The objectives of the course, according to the course syllabus (2017) are: development of critical thinking, complex analysis, identification of intuitive thinking processes and dialogical observation. We will also review main theories of reflection in higher pedagogy and design thinking, and suggest rethinking Schön and Kolb's reflection models, due to our findings.

## 2 REFLECTION ROOTS, THEORIES AND MODELS

Reflection is discussed in diverse pedagogical contexts and curriculums (Ellmers Brown & Bennett, 2009; Seevinck & Lenigas, 2013; Edwards, 2017 and others). The terms "reflection" and "critical reflection" often appear in descriptions of educational programs and approaches. However, their meaning is often vaguely defined, therefore, they are attached to a wide range of concepts and strategies (Hatton & Smith, 1995). This review briefly presents the roots of the concept of reflection and its accepted definition. It also reviews leading theories that have significantly influenced the concepts of reflection in architecture pedagogy and higher education.

### 2.1 *The roots of reflection*

The definitions of reflection as an essential component of thinking and learning, characterize thought and research during the 20th century onwards. Yet the foundations to those identifications and discussion were laid by the journey to Delphi taken by Socrates' friend – Chaerophon, who asked the omniscient oracle if there was anyone wiser than Socrates, and the priestess replied that: "no one was wiser" (Plato, Apology 21a, in: West, T. G., & Platon, 1979. P.69). Socrates, perplexed by the claim, sought out and questioned Athenian men who were highly esteemed for wisdom, in order to prove the prophecy wrong. He realized they seemed to think that their expertise in one field allowed them to speak authoritatively in other issues, about which they knew nothing. Thus, he concludes, he is wiser than other men because he does not think he knows what he does not know. Hence, Socrates connects wisdom to awareness of one's knowledge and its limitations. Socrates and Plato after him, propagated the good philosophical life as a life of self-inquiry, that resemble latter day ideas of self-reflection.

Plato's prized student Aristotle described the virtue of Phronesis (φρόνησις) – as a special mental activity, which according to Birmingham (2004) is very close to the contemporary concept of reflection. In his book "Nicomachean Ethics" Aristotle distinguishes between Phronesis and epistemic knowledge, theoretical scientific knowledge and knowledge of practical skills. Aristotle's Phronesis refers to morality, the ability to distinguish good actions from bad ones and to judgment. Knowledge of Phronesis, in Aristotle's view, refers to action and relates to a fundamental change in the world (Birmingham, 2004).

The term reflection as we perceive it today was contributed by John Dewey, who drew on the ideas of earlier philosophers and educators (Houston, 1988). Dewey sees reflection as a unique way of thinking that influences learning and teaching processes. Reflective thinking according to Dewey's publication How We Think is defined as:

*"Active, persistent, and careful consideration of any belief or supposed form of knowledge in the light of the grounds that support it and the further conclusions to which it tends" (Dewey 1933, p.118).*

Dewey emphasizes the active, unknown, and constantly changing dimension of reflection, referring to two concepts: reflective thinking and reflective action; Reflective thinking is characterized by doubt, hesitation, confusion and mental distress that lead to search, investigation and persecution in order to solve the doubt and eliminate the confusion. Reflective activity is related to dealing with a current situation, from which one has to reach something that does not yet exist (Hatton & Smith, 1995). Dewey does not refer directly to design thinking – a term that was coined much later – but a proximity can be identified between his concept of reflective action and contemporary thoughts.

### 2.2 *Schön's reflective practice*

Donald Schön, who studied reflection in the context of architecture design studio, also ties it closely to thinking under uncertainty:

> *"[...] the practitioner allows himself to experience surprise, puzzlement, or confusion in a situation which he finds uncertain or unique. He reflects*[1] *on the phenomenon before him, and on the prior understandings which have been implicit in his behavior. He carries out an experiment which serves to generate both a new understanding of the phenomenon and a change in the situation" (Schön 1983, p.68).*

Schön's contribution to the study of reflection cannot be underestimated pedagogically. His ideas have become a dominant 'theory of practice' for professional and vocational education (Webster, 2008), as Barnett (1992, p.185) suggests: "we're all reflective practitioners now".

While for Dewey reflection is an individual process, devoid of interaction and dialogue (Cinnamond & Zimpher 1990), Schön separates the types or phases of reflection that he identified in his studies of the studio and expands the definition of reflection as an ongoing process. The learner reflects while working and thereafter when looking back on what he or she has done, alone but also with the help of the tutor (Schön, 1983, 1987). The entire process, he suggests, is a loop, constantly repeated to achieve professionalism. Schön conceives two modes of reflection: reflection-in-action and reflection-on-action (Figure 1). Reflection-in-action can be defined as thinking about what one is doing while one is doing it (Schön, 1987), whereas reflection-on-action is the act of reflecting on an action after it has occurred.

### 2.3 *Reflective-active axis in Kolb's learning cycle*

In the Experiential Learning Theory (ELT), David Kolb explains the role of reflection as part of the learning process. Kolb extended the ideas of John Dewey, Kurt Lewin Jean Piaget and others about adult learning, into a model representing the way learning is experienced (Kelly, 1997). Kolb views learning as a transformative process, with a strong individual dimension: the internalization of knowledge by the learner is related both to the way the information is revealed and to the way it is processed (Kolb, 1984, 2005). According to Kolb, the learning style is not fixed and is related to personal experiences.

Reflection in Kolb's model is essential for transforming concrete experiences into abstract concepts, from which one can deduce new implications for action. In Kolb's learning model, the Concrete-Abstract axis refers to the way information is perceived, while the Active-Reflective

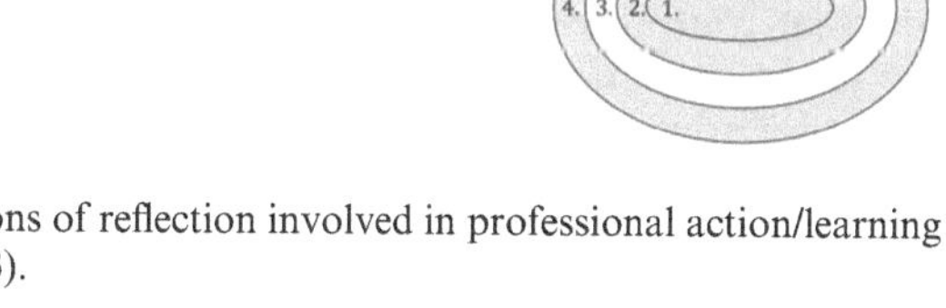


Figure 1. The dimensions of reflection involved in professional action/learning (Schön) Source: Webster, H. (2004, p.103).

[1] Here in the sense of deep thought.

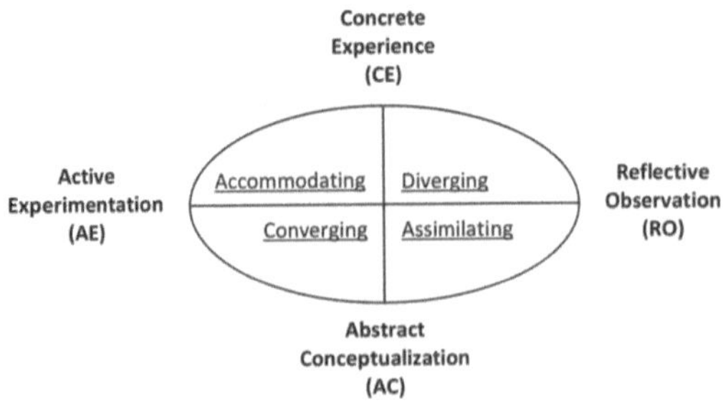


Figure 2. The experiential learning cycle and basic learning styles. Source: Kolb et al. (2001, p.229).

axis is the information processing axis. The experiential learning process is described as a non-hierarchical cycle, so that each stage is a possible entry point. Kolb's model (Figure 2), which views reflection as a fundamental element of abstract thinking, was updated by Kolb himself and others in the wake of further studies in experiential learning (Kolb, 2005).

The ambiguity and contradictory nature that characterize the desire to define and identify the reflection are already evident in the careful reading of Schön's dimensions of reflection compared to Kolb's model. While the reflections identified by Schön consist of both Reflection in Action and Reflection on Action, Kolb's model suggests an axis at the opposite ends of which is Active Experimentation and Reflective Observation.

## 3 FINDINGS

Four themes emerged from the analysis of student-tutor dialogs during course instructions. These are: contrast and hierarchy; turning the gaze from the inside out; the importance of doubts; and feedback between the narrative and the visual. The following presents these themes with a focus on the roles of verbal and non-verbal expressions in the design reflective process.

### 3.1 *Contrast and hierarchy*

Preliminary analysis revealed a perception of the relationship between the verbal and the nonverbal expressions as two different and distinct aspects of the design reflective process (Figure 3). This contrast was also reflected in the way some students testified about their processes of thought and creativity: "...my preoccupation is lacking in narrative and it is very morphological and very formal" noted a student while observing previous work.

Collected data indicated other complex connections between the verbal and nonverbal components of reflection in the pedagogical process of the course. The text written by the students is sometimes referred to as a complementary product to the visual work, as one of the tutors related to the student's verbal products: "...It will be a catalogue alongside the work." Such statement can also be considered as attesting to a stratified perception; the importance that some of the participants in the course, students and teachers, attribute to the visual product as the primary goal, as opposed to the text as a by-product: "When you go to the exhibition there is a catalogue, there is the inspiration of the artist, there are the texts, there is the history, there is the blah blah blah blah... but there is one creation." While the perception of the narrative and the visual as contrast aspects does not take a stand on their status, the hierarchical conception sets the narrative as a prior stage in the process (Figure 4).

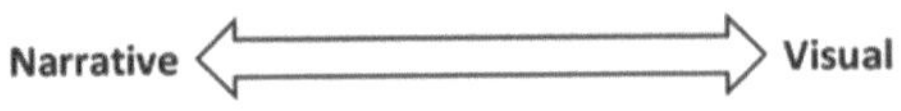


Figure 3. The narrative/verbal and visual/nonverbal axis.

Figure 4. Contrast (right) and hierarchy (left) between the verbal and nonverbal.

### 3.2 *Turning the gaze from the inside out*

The analogy to art and to the display of a museum or a gallery exhibition is a repeated theme presented in dialogues that took place during tutoring. It appears in relation to the aspects of representing the students' work. Instructors comments such as "There are many precedents in art. Yes, from where does one watch [the work]?". Or "Think you're exhibiting now like a sculpturer in a museum", were common. This analogy to 'a viewer in an exhibition' serves also to discuss the observer's point of view and the need to get out of ones' self and see what he or she had created from an external position, as a tutor instructed a student presenting a visual product: "It's your control of how I as a guest see it. How does a visitor see it?". This ability to examine their work from a different perspective is required of the students for both visual and narrative products. Here it is evident regarding a verbal product reviewed by a tutor: "Try reading this as if you have never read it before - it's hard...". While the student is asked to 'step outside' and re-acquaintance own text, the tutor acknowledges the difficulty and challenge.

Developing such a viewpoint is a one of the goals manifested in the way tutors mediated the course objectives for students: "...this course talks about ...how you get to your project. And a person who does not understand the process you have been through and is not aware of it - what does he understand? It is the mirror that you put in front of your work". Note that the point of view is the physical but also the emotional distance from one's work; it may be close or distant, as well as the point of view in the sense of an internal or external position (Figure 5). This ability to turn the gaze from the inside out, to direct the personal observation to the internal process that the students go through and to convey it for an external viewer, brings us back to the sources of reflection in the eye of the classic philosophers we reviewed: self-awareness.

### 3.3 *The importance of doubts*

Students expressed doubts concerning the complex connections between the narrative and the visual and the role of the verbal and nonverbal products in their work process. In the next dialog the tutor also expressed perplexity considering the role of the text in the design process:

> *Student: Because I took the words [key words from the students' text]... that at least represent my abstract [the students' text]... this is not the game I make [referring to the visual product], but I did feel that it also allows me to use words [text] and I do not know if this is supposed to be any product related to the process...*

In reply the tutor responded: "The intention is really umm ...(to) use the text as another kind of raw material to... work, present your idea. Umm ... okay, I think I myself, I must

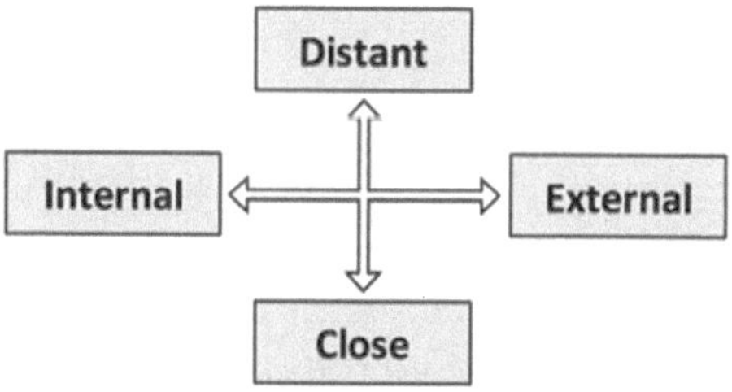


Figure 5. The point of view matrix.

say I do not always feel it...". Note the tutor being self-reflective, as a response to hesitation and uncertainty. Later in the same interaction, the tutor addressed the subject again, leading the discourse to the issue of the answer versus the journey while looking for it: "As for what you asked me before, then again, I do not feel the need to come up with an answer. It's okay you ask and flirting with these ideas in another way, in another form of representation". In this and other discussions doubts led to further thinking about the process and allowed both tutors and students to reconsider the steps along the way.

### 3.4 *Feedback between the narrative and the visual*

Uncertainty expressed by students, regarding the ability to reach their 'destination'[2], can also be related to the complexity of the connection between the verbal and the nonverbal expressions. The nature of the reflective conclusions that students can draw from the challenge of rephrasing the questions they faced during the design process was eloquently formulated in a student's reflection on preoccupation with the "philosophical" vis-a-vis the "architectural":

> *[...] there are a lot of interesting philosophical questions that I am a bit, let's say, I am going towards the philosophical sometimes and not towards the architectural. And then tying it to this architecture does not necessarily comes out of it, does not necessarily yield fruit... tying it to architecture does not necessarily yield the desired fruit.*

Different students approached the process from different starting points. While some felt at ease with the narrative as an impetus for design, others were more comfortable with the visual nonverbal domain, as observed by a tutor: "...there are people for whom the text was very instrumental, and there are people for whom the [visual] product was very instrumental." This is true to students and tutors or professionals alike, as one tutor self-observed: "Because some people, including myself, work less well with text than with visual products. Your refinements have passed through the visual products and they bring you back to the question and sharpen the question... from one question to the next...". The connections described here points to a repeated feedback between the narrative and the visual components.

## 4 CONCLUSIONS AND DISCUSSION

The complexity of the field has been revealed during the course of the research, both in the variety of definitions and concepts related to the term reflection in theory, but also in the various participants perceptions and in the different modes of expression of reflection in the case study. Those reflective expressions can be divided into two main groups/categories: verbal reflective products and non-verbal products. Both the written text and the visual product the students created during the course become mediums to convey their massage (McLuhan, 1964). Yet the distinction between them serves to enable a more subtle understanding of their roles. McLuhan (1964), refers to the verbal as "...an actual process of thought, which is in itself nonverbal" (p. 8). The pre-verbal thought can be expressed in words, but also in nonverbal actions and products, as in Schön's concept of 'knowing-in-action' (Schön, 1983). The verbal and non-verbal expressions have varied connections between them, through which these connections the specific reflection that take place in the of course 'Architectural Editing' can be characterized.

Reconsidering the role of verbal and nonverbal expressions in the reflective process of the 'Architectural Editing' course, reveals a new link between the Narrative and the Visual as a circular process. This process may, as suggested by Kolb, start from the individual based on

[2] Unlike perceptions of architecture as a pragmatic and "scientific" field (Gregory, 1966), aimed at a functional product - the 'destination', reflection is characterized by search (Dewey, 1933) and circular repetition (Schön, 1983, 1987).

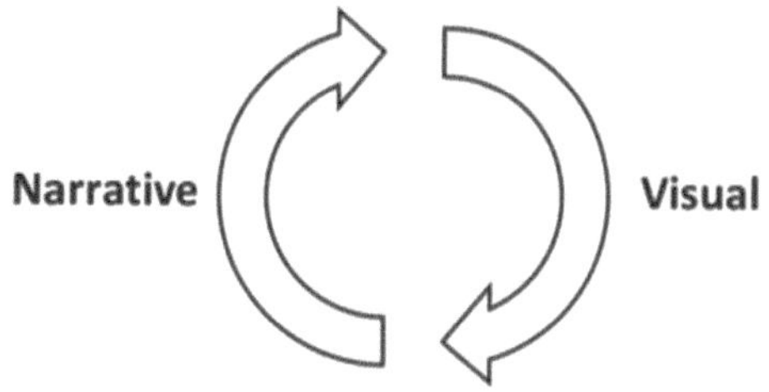


Figure 6. The narrative/verbal and the visual/nonverbal expression cycle.

his/her personal variables and preferences (Figure 6). In addition, this process is repetitive in a proximity to the reflection loop unveiled by Schön.

Moreover, the concept of the Narrative-Visual cycle is the lens through which we suggested to reread the terms that lay in the foundation of the theories of reflection, especially Schön and Kolb's reflection models. We argue for reading the verbal expression as the reflective, whereas the nonverbal expression as the active. In other words, two groups or sequences can be distinguished: one is the verbal–narrative, the reflective, and the other is the non-verbal and visual, which is the active. Such interpretation of the data offers a direction for further research that is needed in order to better ground it.

Therefore, this research should be seen as a first exploration that illuminates further directions for investigation. Most notably, further research is needed to focus on the student's internal personal process, since the presented student-tutor dialogues analysis gives a view to reflection-on-action and specifically to reflection-on-action with others (Schön, 1983; Webster, 2004), but has limitations in exploring reflection-in-action by the students alone. Also, to go beyond the specific case study, interrogations of the visual-narrative reflective components in other courses and educational programs is required; these can be for example different studio courses in architecture and other fields of design, or similar experimental courses in other institutions.

This paper, then, proposes a different view of Kolb's model: it reshapes Kolb's Active-Reflective axis as a cycle that parallels the Narrative-Visual cycle suggested above. By adding the narrative-visual dimension to Kolb's model, we allow more complex relations between the Active and Reflective expressions, akin to the different reflection modes characterized by Schön. As suggested by the case study, linear relations between the narrative and the visual expressions are only one possibility. Furthermore, the interpretation proposed in this article reconciles some differences between Schön and Kolb's reflection models, when applied to architecture pedagogy. By recognizing the Active-Reflective cycle, we instate a more diverse interaction between reflection in action (i.e., Schön's model), and Kolb's idea of action and reflection as two opposite ends of an axis. A better understanding of the narrative and the visual reflective components roles and relations have also practical implications for architecture pedagogy, in designing educational programs and future learning tools.

## ACKNOWLEDGEMENTS

Keren Shoham is grateful to the Azrieli Foundation for the award of an Azrieli Fellowship.

## REFERENCES

Austerlitz, n. Aravot, I. Ben-Ze'ev, A. (2005) *Emotions and student-instructor relationship: their educational aspects in the architectural design studio.* Haifa: Thesis (PHD), IIT.

Bennett, R. E. (2010). *Cognitively based assessment of, for, and as learning (CBAL): A preliminary theory of action for summative and formative assessment.* Measurement, 8(2-3), 70–91.

Birmingham, C. (2004). Phronesis: A Model for Pedagogical Reflection, *Journal of Teacher Education*; 55; pp. 313–324.
Cinnamond, J. and Zimpher, N. (1990). Reflectivity as a function of community. In R. Clift, W. R. Houston, and M. Pugach (Eds.), *Encouraging reflective practice in education* (pp. 57–72). New York: Teachers College Press.
Dewey, J. (1933). H*ow we think: A restatement of the relation of reflective thinking to the educative process*. New York: D.C. Heath and Company.
Edwards, S. (2017). Reflecting differently. New dimensions: reflection-before-action and reflection-beyond-action. *International Practice Development Journal, 7*(1).
Ellmers, G., Brown, I., & Bennett, S. (2009). Graphic design pedagogy: Employing reflection to support the articulation of knowledge and learning from the design experience. In *Proceedings of the Conference on Experiential Method, Knowledge and Methodology*, 66–81.
Gregory, S. A. (1966). Design science. In *The design method* (pp. 323–330). Springer, Boston, MA.
Guba, E.G. & Lincoln, Y.S. (1989). *Fourth Generation Evaluation*. Newbury Park, CA: Sage Publication.
Hatton, N., & Smith, D. (1995). Reflection in teacher education: Towards definition and implementation. *Teaching and teacher education, 11*(1), 33–49.
Houston, W. (1988). Reflecting on reflection in teacher education. *Images of reflection in teacher education, 7*.
Kelly, C. (1997). David Kolb, the theory of experiential learning and ESL. *The Internet TESL Journal, 3*(9), 1–5.
Kober, N. (2015). *Reaching students: What research says about effective instruction in undergraduate science and engineering*. National Academies Press.
Kolb, D. A. (1984). *Experimental learning: Experience as the source of learning and development*. Englewood Cliffs, NJ: Prentice Hall.
Kolb, D. A., Boyatzis, R. E., & Mainemelis, C. (2001). Experiential learning theory: Previous research and new directions. *Perspectives on thinking, learning, and cognitive styles*, 1(8), 227–247.
Kolb, D. A. (2005). *The Kolb learning style inventory* version 3.1 London, UK: Hay Group.
Mazur, E. (2018, February 7). *Flat space, deep learning* [Web lecture]. Retrieved from https://www.youtube.com/watch?v=2Jxt0-dCN6U.
McLuhan, M. (1964). *Understanding Media: The Extensions of Man*. Cambridge. New York: McGraw-Hill.
November, A. (2012). *Who owns the learning? Preparing students for success in the digital age*. Solution Tree Press.
Rowe, P. (1998). *Design Thinking*, 1987, MIT Press.
Schön, D. (1983). *The reflective practitioner: How professionals think in action*. New York: Basic Books.
Schön, D. (1987). *Educating the reflective practitioner: Toward a new design for teaching and learning the professions*. San Francisco: Jossey-Bass.
Seevinck, J., & Lenigas, T. (2013, April). Rock, paper, scissors: reflective practice for design process in the novice landscape architect. In *Proceedings of the 2nd International Conference for Design Education Researchers* (Volume 3) (Vol. 4, pp. 2145–2159). ABM-media.
Shkedi, A. (2010) *Words of Meaning: Qualitative Research–Theory and Practice*. Tel-Aviv: Tel-Aviv university Ramot.
The Technion – Israel Institute of Technology. (2017). *205010: Architectural editing* [Course syllabus]. Available from The Technion UGPortal website: ugportal.technion.ac.il.
Tzabar ben-Yehoshua, N. (2006). *Traditions and Trends in Qualitative Research*. Lod: Dvir.
Webster, H. (2004). Facilitating critically reflective learning: excavating the role of the design tutor in architectural education. *Art, Design & Communication in Higher Education, 2*(3), 101–111.
Webster, H. (2008). Architectural education after Schön: Cracks, blurs, boundaries and beyond. *Journal for Education in the Built Environment, 3*(2), 63–74.
West, T. G., & Platon. (1979). *Plato's Apology of Socrates: an interpretation, with a new translation* (p. 66). Ithaca, NY: Cornell University Press.

*Research in Design*

*Research & Education in Design: People & Processes & Products & Philosophy – Almendra & Ferreira (eds)*
*© 2020 Taylor & Francis Group, London, ISBN 978-0-367-49296-0*

# Mapping the territories around Design Research: A four-layer analysis

V. Clemente
*School of Design, Management and Production Technologies Northern Aveiro and ID+ Research Institute for Design, Media and Culture ID+, University of Aveiro*

K. Tschimmel
*Economy Faculty of Porto University, Porto Business School and ID+ Research Institute for Design, Media and Culture, University of Aveiro*

F. Pombo
*Department of Communication and Art and ID+ Research Institute for Design, Media and Culture, University of Aveiro*

ABSTRACT: Despite great progress in the last five decades, Design Research still reveals fragilities in comparison with other academic fields. To avoid stagnation and lack of impact, it needs to strengthen its theoretical and methodological foundations. Following previous work aiming to contribute to Design Research consolidation, we propose in this paper a Map where four categories of Design Research are positioned in relation to territories of Design Research, Education and Practice. The Map also supports the examination of those four Design Research categories based on a four-layer analysis resulting from the conference title keywords: Processes, Philosophy, People and Products. The Map intends to help design researchers, especially inexperienced ones, like PhD students, to visualise where their own research is located within the Design universe and, by that, understand the ontological, epistemological and methodological implications.

## 1 INTRODUCTION

Design Research was defined by Archer (1981) as a "systematic inquiry whose goal is knowledge of, or in, the embodiment of configurations, composition, structure, purpose, value, and meaning in man-made things" (p. 30). Since the first steps of Design Research in the 60' and 70's, the ambition to provide a strong and coherent basis for Design Research has been pursued. However, as is recognised within its own community, Design Research still remains scattered and confused with some well-known weaknesses (Margolin, 2010; Dorst, 2016).

Within universities, Design Research faces theoretical, methodological, and scientific challenges with consequences on its impact and relevance. According to Cash (2018, p. 97), "lack of methodological development, validation and standardisation limits design researchers' ability to provide convincing evidence to researchers in related fields where such standards are common". The result is, that while Design draws extensively on related fields, "the reverse does not occur" and the more pessimistic believe that "Design risks being superseded by other fields eager to include Design Science in their portfolios".

At the level of PhD Design Courses, which are the origin of academic Design Research, as they educate future professional researchers, the immaturity of Design Research is commonly revealed in poor research orientation, sometimes provided by educators who "are indifferent, if not antipathetic to research, some of them some resentful of their colleagues who involve themselves in research and publishing" (Er & Bayazit, 1999, p. 41).

At the same time, Design Research seems to be disconnected from the day-to-day reality of designers, not only due to weak communication between universities and practitioners, but also because decisions about what to investigate are not always directed at improving design practice (Dorst, 2016).

In previous works (Clemente, Tschimmel & Pombo, 2017), we intended to contribute to the field of Design Research theory, with a special focus on doctoral research, by examining the boundaries between Design Research and Design Practice. We started by synthesising the contributions from authors such as Frayling (1994), Cross (2007), Friedman (2008), Findeli, Brouillet, Martin, Moineau & Tarrago (2008) on a three-category Design Research taxonomy. Following that, we conducted an empirical analysis from which a fourth category emerged, resulting on a four-category Design Research Classification Model that includes research ABOUT, THROUGH, FROM and FOR Design. In the resultant work, we moved to the paradigm level, explaining the ontological, epistemological and methodological differences between the four previously presented categories (Clemente, Tschimmel & Pombo, 2018). REDES 19 conference provided us with the opportunity to go further with the discussion and extend our reflection to Design Education.

Through the Map presented in the next section, we clarify the relative positions between the different places and agents around Design Research, aiming to find a consensual, common-ground language to include all kinds of research around design and its relations, at the same time providing the opportunity to find occasions to bring different parties together and enhance fruitful connections between them.

## 2 A 4-LAYER MAP OF DESIGN RESEARCH, EDUCATION AND PRACTICE TERRITORIES

The Map (Figure 1) is organised around the four design research categories and their positions in relation with Design Research, Practice and Education. Our reflection was guided by the keywords that compose the conference's thematic: Processes, Philosophy, Products and People, which constitute four different layers of analysis, as described next.

### 2.1 *Processes*

The Map presents three main processes within the Design field, identified at the bottom, in the darker area: Design Academic Research, Design Higher Education and Design Professional Practice (Figure 2).

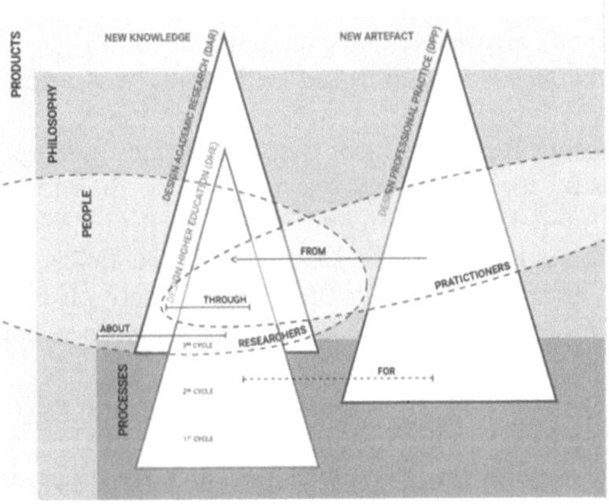


Figure 1. Design research processes, people, philosophy and products map (paper authors).

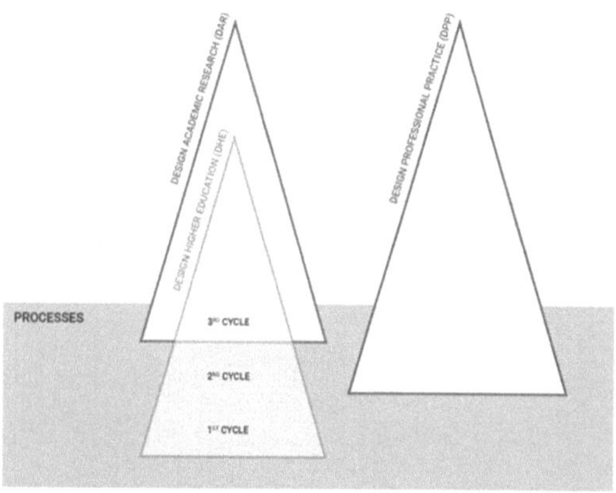


Figure 2. Design research processes partial map.

These processes can also be seen as personal trajectories that an individual undergoes as design student, researcher and practitioner. Although we recognise the three processes are interrelated, we chose to represent them by three different pyramids to clearly identify the singularities of each of them and the interconnection between them.

At the centre of the Map, the Design Academic Research (DAC) pyramid represents acceptable research in the academic design field – which is the core of our discussion. The pyramid below, Design Higher Education (DHE), refers to the typical hierarchical organisation of academic design education starting with the $1^{st}$ cycle level (Bachelor) at the base, followed by $2^{nd}$ (Master) and $3^{rd}$ cycles (PhD) above.

Since the PhD is recognised worldwide as the pinnacle qualification for scholarly endeavour, we obviously locate it near the top of the DHE pyramid. At the same time, "a PhD is awarded on the basis of good practice in research" (Pedgley & Wormald, 2007, p. 71). Er & Bayazit (1999, p. 35 – 36, 39-40) describe a PhD in Design as a certificate or licence that states that "this person has successfully demonstrated the ability to undertake independent research that contributed to knowledge" and not that he or she is "able to design a better product". In accordance with those authors view, PhD, $3^{rd}$ cycle Design Education, is simultaneously located at the top of the DHE pyramid and at the bottom of the DAR.

PhD research can't be seen as just another, bigger, longer and more complex, design project similar to those conducted at $1^{st}$ or even $2^{nd}$ cycle studies. This position is not acceptable within the vision of doctoral research as producing new knowledge and contributing to educate future independent design researchers (Findeli & Coste, 2007). To that end, $3^{rd}$ cycle education must provide structured and explicit "training in research skills, such as literature review", research proposal writing and theoretical basis grounding of knowledge about different research paradigms and methodology. PhD students must learn about a variety of research methods, adequate to address different kinds of research questions. Even if they are not going to use all the learned methods, the exposure to different approaches will help students to develop research skills and promote research methodological reflexivity, avoiding "mistakes in methodology that are seen in the design area" (Melles, 2009, p. 256, 262).

At the right side of the Map, and outside 'university walls', is the Design Professional Practice (DPP) pyramid, representing the design professional activity. The pyramid shape was also chosen, in this case to represent the personal growth of a design practitioner.

## 2.2 *Philosophy*

Going up to the Map's Philosophy area (Figure 3), we find the four categories of Design Research presented and discussed previously (Clemente, Tschimmel & Pombo 2017, 2018): Research ABOUT Design, Research THROUGH Design, Research FROM Design, and

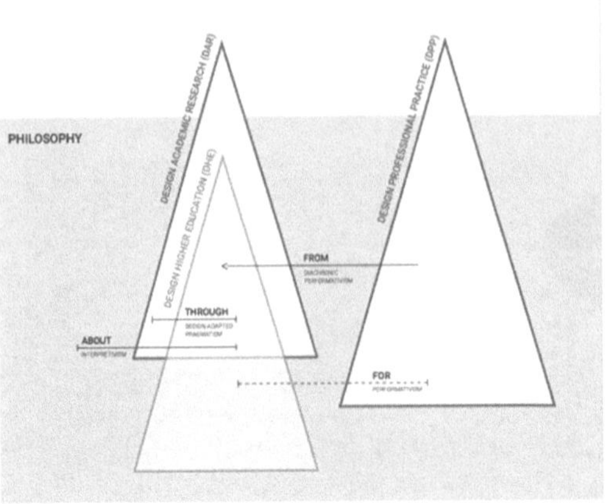


Figure 3. Design research philosophy partial map.

Research FOR Design. The horizontal lines are positioned to show the relations between each category and the pyramids described above.

Research ABOUT Design is usually performed by disciplines outside the design field, following scientific standards already well established in the academic community. The issue about Research ABOUT Design is on its relevance for the design field. Frequently conducted by other disciplines' scientists, its main goal is to contribute to the advancement of such disciplines, and not necessarily to Design. It should be the design community which decides if such knowledge is relevant for designers and, if such is the case, how the new knowledge may be implemented in their respective practices (Findeli et al., 2008).

On the extreme bottom right of the Map, and clearly out of the range of the academic realm, is Research FOR Design which is the same as project research and is mainly associated with "information-gathering activities" required by design projects (Pedgley & Wormald, 2007, p. 74). The main outcome of Research FOR Design is a product, service or process, and even producing some new tacit knowledge, it does not necessarily create new communicable and explicit knowledge, and it does not follow rigorous scientific standards. Frayling (1994), Friedman (2008) and Findeli et. al. (2008) all agree that this kind of research is not considered scientifically acceptable.

However, it is recognised that design practice produces tacit knowledge that, if made explicit and communicable, contributes to the advancement of the design field. As stated by Cross (2007), for practice work to qualify as research, "there must be a reflection by the practitioner on the work, and the communication of some re-usable results from that reflection" (p. 126). That leads to Research THROUGH Design and Research FROM Design. The difference between these two categories lies in the time and context in which that reflection takes place. Table 1 summarises the relation between Design Project and Research ABOUT, THROUGH, FROM and FOR Design.

Research THROUGH Design, which according to Pedgley & Wormald (2007) would be more appropriately termed "Research through Designing", explicitly refers to "research with a practical design element" or "research incorporating a design project". That means that "selected periods of a research study are occupied by a design project carried out by the researcher" since "integration of design activity must be a means to an end, and not and end in itself" (p. 72-73).

The Research FROM Design category was introduced by Clemente et. al. (2017, 2018) and refers to research that results from the diachronic study of one's own relevant and professionally validated design activity. Therefore, in both categories, design projects assume a central role. However, they differ on the place where the design project is developed and when the author's reflection and analysis occur. Research THROUGH Design involves design projects developed inside universities. In this kind of research, author's reflection, research project and

Table 1. Characterisation of the four design research categories.

| Acceptable academic design research | | | Non-acceptable academic design research |
|---|---|---|---|
| Research ABOUT Design | Research THROUGH Design | Research FROM design | Research FOR design |
| No researcher own design projects involved | Design project developed ***inside*** the academy | Design project developed ***outside*** the academy | |
| | Theory produced ***inside*** the academy | | No ***explicit*** theory production |
| | Researcher reflection and analysis as design author occur ***in parallel*** with the design project | Researcher reflection and analysis as design author occur after market's validation of design project(s) outputs | No ***structured*** reflection and analysis from the designer as author |
| | Theory precedes practice (practice being an application, illustration or validation of a previously developed theoretical intentionality) | Practice precedes theory (theory resulting from the translation of implicit knowledge embodied in the products and process) | Theory embodied in the process and final products but not made explicit or communicable |

design project, all occur in parallel, at the same place and within the same period of time. In Research FROM Design, on the contrary, the studied design project(s) belong to the researcher's past professional activity, developed outside the academy. The author's reflection and analysis is diachronic because it only happens after the output of the studied project(s) have been validated by the market. Research THROUGH and Research FROM Design also differ in the way theory and practice are related. In Research THROUGH Design, theory precedes practice, practice being an application, an illustration or a validation of a previously developed theoretical intention. In Research FROM Design, practice precedes theory, theory resulting from the translation of implicit knowledge embodied in the design products and processes. That is the reason why the solid line representing Research THROUGH Design is fully contained inside the DAR pyramid, while the line representing Research FROM Design is positioned between DAR and DPP and oriented from knowledge origin towards knowledge theory production.

This Philosophy layer also provides the opportunity to reinforce our argument that a coherent theory for Design Research needs to be supported by a consensual and widely spread discourse about Research Paradigms. The fact that a great part of published design research misrepresents paradigmatic assumptions, reveals that researchers are frequently unconscious of those "silent, implicit or even hidden, but fundamental" philosophical assumptions underlying their own research and their consequences and implications (Lukka, 2010). In spite of some voices claiming "aparadigmatic" approaches (Shannon-Baker, 2016, p. 320), we argue that any research is always conducted under a certain system of beliefs about how the research problem should be addressed, including what is to be studied, what kind of research questions are supposed to be asked and how they should be formulated, with which methods these studies should be conducted, and how their results should be interpreted. That means "aparadigmatic" research doesn't really exist. It is probably just a "shortcut" to avoid the paradigmatic question. However, usually shortcuts come with pitfalls. It is not uncommon to see inexperienced researchers, especially, PhD students, already at an advanced stage of the research, rambling and struggling with methodological questions which should have been addressed earlier. A clear establishment of the research paradigm is not a waste of time but, on the contrary, a strong basis to guide research. Even when unforeseen obstacles emerge during the investigation, it becomes easier to find an alternative way to address the research problem respecting the same belief system. Beyond this utilitarian perspective, the explicit identification of the paradigm that has been followed is a requirement of honest research, informing the audience about the values underlying and influencing the investigation, also contributing to the legitimation of Design Research by other academic disciplines.

Because, as was explained above, Research ABOUT Design can be performed by disciplines outside of the design field, it is natural that it follows research paradigms inherited by those well-established disciplines. When Research ABOUT Design consists of descriptive, historical and phenomenological studies, it is considered to follow an Interpretative (or Constructive) paradigm. Interpretive research methods are usually qualitative, including Case Studies, Phenomenology, Hermeneutics and Ethnography. Interpretive theory is usually grounded (inductive). The applied techniques include, for example, Open-ended Interviews, Focus Groups or Think Aloud Protocols. When Research ABOUT Design involves researchers' values, their critical position, their intention to change, their agenda, it can also be conducted under the Critical paradigm. Socio-critical methods include, for example, Critical Ethnography and Action Research. Applied techniques can include Open-ended Interviews, Focus Groups, Open-ended Questionnaires, resulting usually in qualitative data, similar to Interpretivism, but the data analysis is influenced by the researcher's explicit intention to change reality, instead of just describing it (Guba, 1994, Scotland, 2012).

Design Research resulting from design projects, however, seems to "not easily fit within existing paradigms" (Isley & Rider, 2018, p. 359), with some arguing the value of pre-existing paradigms and others claiming the need of a totally new, specially fitted, paradigm. Melles (2008), Morgan (2007, 2014) and Rylander (2012) are among those defending the virtues of enlarged or adapted versions of classical Pragmatism, because it accepts both objective and subjective ontological orientation, moving back and forth between inductive and deductive epistemological approaches. For these same reasons, we agree that Pragmatism is appropriated to Research THROUGH Design, where the 'rigorous' research project is punctuated by periods of design project, where creativity, intuition and nonconformist thinking takes place. We go further in suggesting Design Thinking as a research method that perfectly fits into the methodological pluralism that characterises Pragmatism. Under this paradigm, that we would designate as Design-Adapted Pragmatism, Design Thinking techniques must follow academic standards as much as possible (for example when applying Surveys, Interview or Focus Groups). At the same time, space for intuition and imagination is also permitted and material and visual elements of design such as Sketching and Prototyping are accepted.

Among the arguments of those defending a disruptive paradigm to accommodate Design Research is "The manifesto for the Performative Paradigm", by Haseman (2006). The first peculiarity of this paradigm is the fact that research is not led by one problem or research question, but instead by practice itself. The author argues that while conventional problem-led research flows from a central research question, practice-led research does "not commence with a sense of a problem" but, instead, with an "enthusiasm of practice" from which the problem emerges. This description is aligned with Rosenberg's (2000) concept of Poetic Research that also isn't "channeled by a research problem" because "the focal territory is found through the process", emerging "from a questioning of practice" (p. 2). A second peculiarity of the Performative Paradigm is related with research outputs. It is stated that embodied knowledge, resulting from practice, doesn't need to be translated into numbers and words as in traditional research paradigms, because performativity is not primarily about artefacts' meaning, but rather about their effect on the world (Bolt, 2009).

Following that, and because academic "good research" needs to be purposeful, based on the identification of an issue or problem worthy and capable of investigation, and communicable, generating and reporting results which are testable and accessible by others (Cross, 2007), we clearly oppose the idea that academic research can be addressed by the Performative Paradigm. However, we accept that Performative Paradigm is suited for non-academic design research, conducted in a professional context, of which the main outcome is a product, service or process. Even though it may produce some new tacit knowledge, it does not necessarily create new communicable and explicit knowledge. It doesn't follow rigorous scientific standards (and it doesn't have to) and that's why it is not scientifically acceptable (Clemente et. al, 2018).

However, because Research FROM Design arises from tacit knowledge resulting from professional practice once it is made explicit and communicable, we accept it could be framed by a modified version of the Performative Paradigm, that we would name Diachronic-Performative Paradigm. In this modified version, ontology remains the same (knowledge and the research question itself are embedded in practical results), however, epistemology and methodology are modified because it is recognised that this knowledge, to be academically acceptable, needs to be translated and transferred by its author (epistemology), through a diachronic and idiosyncratic reflection process (methodology).

## 2.3 *People*

Although people moving around the three universes of Design Education, Research and Practice include students, professors, users, among others, the core of this discussion is Design Research, so we focused our discussion on those who conduct research. In the middle of the Map (Figure 4) we find Research FROM Design and Research THROUGH Design both developed by those who are, at the same time, design researchers and practitioners and their research involve their own design project(s).

Research FROM Design occurs inside the academy but deals with data coming from a designer's own projects developed previously, as design practitioner, and already validated by the clients and the market. For that reason, the line representing Research FROM Design is placed between the DAR and the DPP pyramid, because it lies on this connection between Design Practice and Research. It is also intentionally positioned at the top of the DPP pyramid because research FROM Design should be conducted only by experienced professional designers with a relevant history of already validated design projects.

On the contrary, Research THROUGH Design is well centred within the DAR pyramid because even when the design project (which is only part of a bigger project of research) is not just an academic exercise but, instead, a "real" project asked for by the market, the main goal of the researcher is to provide an answer to the research question. For that reason, project methodology, procedures and decisions are governed by academy research rules. Some had argued that to became an academic design researcher, which involves being familiar with academic research rules, and formulating and approaching problems according to the rules of a well-established scientific discipline, a designer almost needs to "forget" what is to be a designer (Findeli & Coste, 2007). In fact, until recently, people with this dualistic profile of researcher and practitioner were mostly professional designers or graduates seeking an academic career, for which holding a PhD is a precondition. However, as it is described by Dorst (2016), "with more and more design researchers working in companies, design research has already found multiple homes. A good deal of the best academic design takes place in

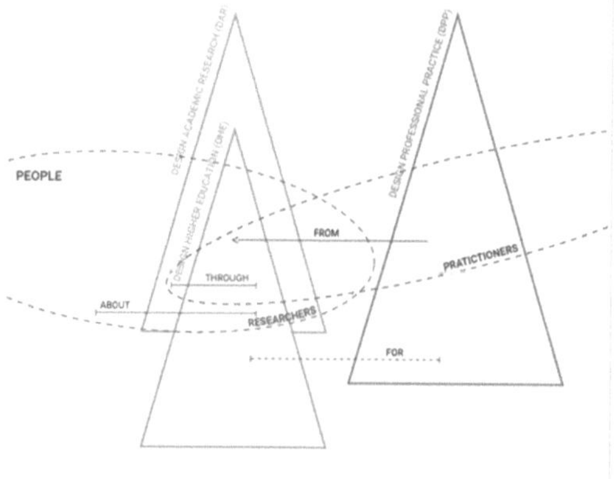


Figure 4. Design research people partial map.

companies like IDEO", (p. 7). He claims the distance between academic design research and professional day-to-day design reality can be reduced by increasing people with a "nomadic" profile, with a "foot in practice as well as in universities".

Moving to the left, we find Research ABOUT Design which is done by academic researchers but not necessarily designers since it does not include researcher's personal design activity. It can be, for example, about "other people's designing, artefacts", or about "people who use artefacts" (Pedgley & Wormald, 2007, p. 71). This means that academic researchers without a design background can conduct Research ABOUT Design. That's why the line representing Research ABOUT Design extends to the left of the DAR pyramid. Because Research ABOUT Design welcomes research done by people from other academic fields such as psychology, education, sociology, engineering, with the proviso that the produced knowledge contributes to the advancement of design knowledge.

On the right side, we find Research FOR Design which is carried by design practitioners who are not necessarily design researchers, in the academic sense of research, because design projects usually don't have to (and should not) follow academic rules. However, it is possible to connect People from ABOUT and FOR territories through research. Although a designer "practicing activities when creating work (...) cannot be considered research, it is possible for an external observer to do research into how", a designer "is working on his or her work (...) to make a contribution to common knowledge" (Bayazit, 2004, p. 16). Although it is also possible to connect Design students at 1st and 2nd cycle levels with professional designers/researchers, for example, by including "real market" problems within academic courses, that possibility is not represented on the Map because it is not directly related with the discussion core which is academic research.

### 2.4 *Products*

In the Map, products from Design Research and Design Practice are positioned at the top of the DAR and DPP pyramids, respectively (Figure 5).

The main products of Design Practice, and for that reason, of Research FOR Design are artefacts (products, services, spaces, images, etc.) including outcome such as design registrations, patents, sales, global recognition, between others. On the other hand, new knowledge, and only in a communicable form, is the main output of academic Design Research, including Research ABOUT, THROUGH and FROM Design (Pedgley & Wormald, 2007).

New knowledge resulting from Design Research may be focused on the designer/design team, design outputs, design processe(s), design management, creativity, cognition, innovation, users/customers, cultural issues, emotional responses and there are infinite possibilities to add to this list (McMahon, 2012).

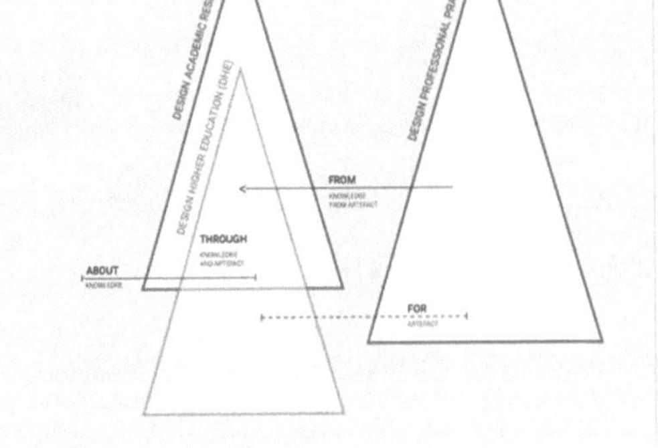


Figure 5. Design research products partial map.

In Research ABOUT Design, new knowledge is the only expected output. In Research THROUGH Design, (although it is not its main objective), designed outputs coexist with knowledge as research outputs. In Research FROM Design, the research product is the explicit and communicable translation of knowledge embodied in designed outputs of previous practice projects, of which the legitimacy and appropriateness was already appreciated and demonstrated in the professional universe.

## 3 CONCLUSIONS

With this paper, we intend to provide a visual conceptualisation about Design Research and its adjacent and sometimes intersecting areas. The analysis is based upon the four sources of design knowledge which constitute the REDES 19 Conference themes: Processes, Philosophy, People and Products. The resulting Map, and its partial versions, allows a better understanding of the four categories of Design Research.

The Map visually shows where Design Research is situated in relation with design professional activity, design doctoral education and academic research outside design. Additionally, it clarifies the characteristics of each Design Research category by indicating their distinct outputs and the different profiles of involved researchers, enabling a deeper understanding of the underlying philosophical assumptions. Ultimately, it contributes to the epistemological basis required to academically legitimate design knowledge by providing a common shared discourse facilitated by a common visualisation of all the places and agents around Design Research.

## REFERENCES


Archer, B. (1981). A view of the nature of design research. *Design: science: method, 1*, 30–47.

Bayazit, N. (2004). Investigating design: A review of forty years of design research. *Design issues, 20*(1), 16–29.

Bolt, B. (2009). Performative Paradigm for the Creative Arts, Working Papers in Art and Design, *5*, University of Melbourne, Melbourne.

Cash, P. J. (2018). Developing theory-driven design research. *Design Studies, 56*, 84–119.

Clemente, V., Tschimmel, K., & Pombo, F. (2017). A Future Scenario for a Methodological Approach applied to PhD Design Research, Development of an Analytical Canvas. *The Design Journal, 20*(sup1): S792–S802.

Clemente, V., Tschimmel, K., & Pombo, F. (2018). Methodologies in doctoral research in design: the role of research paradigms. In O. Moret (Ed.), *Back to the Future The Future in the Past. ICDHS 10th+1 Barcelona 2018 Conference Proceeddings Book* (pp. 694–698).

Cross, N. (2007). From a design science to a design discipline: Understanding designerly ways of knowing and thinking. *Design Research Now*, 41–54.

Dorst, C. H. (2016). *Design practice and design research: finally together?* Paper presented at 2016 Design Research Society 50th Anniversary Conference, Brighton.

Er, H. A., & Bayazit, N. (1999). Redefining the Ph. D. in Design in the Periphery: Doctoral Education in Industrial Design in Turkey. Design Issues, *15*(3), 34–44.

Findeli, A., & Coste, A. (2007). De la recherche-création à la recherche-projet: un cadre théorique et méthodologique pour la recherche architecturale. *Lieux communs, 10*, 139–161.

Findeli, A., Brouillet, D., Martin, S., Moineau, C., & Tarrago, R. (2008). Research through design and transdisciplinarity: a tentative contribution to the methodology of design research. In *Swiss Design Network Symposium* (p. 67).

Frayling, C. (1994). Research in Art and Design. *Royal College of Art Research Papers 1*(1).

Friedman, K. (2008). Research into, by and for design. *Journal of Visual Art Practice*, *7*(2), 153–160.

Guba, E. G., & Lincoln, Y. S. (1994). Competing Paradigms in Qualitative Research. In N. K. Denzin & Y. S. Lincoln (Eds.), *Handbook of Qualitative Research*, (pp. 105–117). London: Sage.

Haseman, B. (2006). A manifesto for performative research. *Media International Australia incorporating Culture and Policy, 118*(1), 98–106.

Isley C., Rider, T. (2018) Research Through-Design: Exploring a design-based research paradigm through its ontology, epistemology, and methodology. *Proceedings of DRS2018*, (357). Limerick.

Lukka, K. (2010). The roles and effects of paradigms in accounting research. *Management Accounting Research, 21* (2), 110–115.

Margolin, V. (2010). Doctoral education in design: problems and prospects. *Design Issues, 26*(3), 70–78.

McMahon, C. A. (2012). Reflections on diversity in design research. *Journal of Engineering Design, 23*(8), 563–576.

Melles, G. (2008). An enlarged pragmatist inquiry paradigm for methodological pluralism in academic design research. *Artifact: Journal of Design Practice, 2*(1), 3–13.

Melles, G. (2009). Global perspectives on structured research training in doctorates of design–what do we value? *Design Studies, 30*(3), 255–271.

Morgan, D. L. (2007). Paradigms lost and pragmatism regained: Methodological implications of combining qualitative and quantitative methods. *Journal of mixed methods research, 1*(1), 48–76.

Morgan, D. L. (2014). Pragmatism as a paradigm for social research. *Qualitative Inquiry, 20*(8), 1045–1053.

Pedgley, O., & Wormald, P. (2007). Integration of Design Projects within a Ph. D. *Design Issues, 23*(3), 70–85.

Rosenberg, T. E. (2000). The reservoir: towards a poetic model of research in design. *Working papers in art and design, 1*, Unpag.

Rylander, A. (2012). Pragmatism and Design Research. *Ingår i Designfakultetens serie kunskapssammanställningar.*, 42.

Scotland, J. (2012). Exploring the Philosophical Underpinnings of Research: Relating Ontology and Epistemology to the Methodology and Methods of the Scientific, Interpretive, and Critical Research Paradigms. *English Language Teaching, 5*(9), 9–16.

Shannon-Baker, P. (2016). Making paradigms meaningful in mixed methods research. *Journal of Mixed Methods Research, 10*(4), 319–334.

*Research & Education in Design: People & Processes & Products & Philosophy – Almendra & Ferreira (eds)*
*© 2020 Taylor & Francis Group, London, ISBN 978-0-367-49296-0*

# A framework to analyse PhD theses in design

R. Almendra & J. Ferreira
*CIAUD, Lisbon School of Architecture, Universidade de Lisboa, Portugal*

ABSTRACT: This paper presents the analytical framework used to conduct a survey of PhD theses in Design completed between 2008 and 2019 (n=106) at the Lisbon School of Architecture. The framework is described in how it relates to the broader research project started in the REDES research group. Each PhD thesis is detailed according to four main levels of analysis: research identification, territory, processes, and results. The framework produces detailed results concerning the research reported in each thesis manuscript and the aggregate of the individual analysis allows the production of meta-data about the research conducted in the PhD course.

## 1 INTRODUCTION

This paper presents an analytical framework applied to the analysis of PhD theses in Design from the Lisbon School of Architecture. The work presented here is the result of 10 years of teaching by one of the authors in the Lisbon School of Architecture's PhD Design course. Through the years, the author's close engagement with dozens of PhD students led to a reflection on the PhD course at several levels, these issues are presented in detail in Author, et al. (2019a).

In particular, the work done in the *Object Critique* module of the PhD course was crucial to uncover particular difficulties that the students repeatedly encounter when first structuring their PhD research. During the module, the students are required to develop their PhD research proposal, and the crucial point is that they must apply critical thinking skills at each level of research design. This means every decision is carefully examined to determine if the proposal is developed with a consistent inner logic.

A PhD proposal should include the definition of a problem, clearly articulated research questions, often the formulation of a hypothesis, the scope and aims of the research, planning of tasks, an overall research schedule, and the methods to be used. Working closely with the PhD students, the author observed that a pattern of specific difficulties emerged when dealing with the challenges of developing a research proposal (Author, 2012) in short, it was clear that the students struggled with the challenge of critically examining information in order to expand design knowledge.

From these observations and reflection, the author developed a methodology (Author, 2019b) to help PhD students produce a relevant, rigorous, and innovative PhD proposal that establishes a solid foundation for meaningful research that has a positive impact and contributes to human flourishing. However, the author's reflection also indicated that a systematic review of the PhD in Design course was needed; as such, the first step was to look into the output of the PhD course, i.e. all the completed thesis. To achieve this, the author launched a research project and developed an analytical framework to examine the final manuscript of every PhD thesis whose details are presented here.

These concerns are part of the ongoing discussion in the design academic community about the role of design research and its connection with design education and practice. Reflection on these issues can be traced back to the turn of the century when many schools of design began their doctorate programs (in the Lisbon School of Architecture, for instance, the PhD course in design began in 2006), an illustrative example being the discussions compiled in the book *Doctoral Education in Design: Foundations for the Future* (Durling & Friedman, 2000) which presents a selection of articles presented at a conference held to discuss and advance issues of doctorate in design. It is worth noting a similar analysis to the one presented in this paper was conducted by Horvath (2008) in the context of the PhD in design of the TU Delft in The Netherlands. In his study, Horvath delved into 117 PhD theses with the aim of describing a pattern in what he termed 'design inclusive research' — meaning research that *includes* design as a research method within a methodology of scientific inquiry as opposed to a purely scientific investigation *into* design or a purely practice-based approach.

Horvath's work is particularly relevant because it presents an analysis of the state of design research at doctorate level based on empirical data, and empirically-based studies are still lacking in design research. The framework we present here also seeks, in part, to address this gap by detailing our methods and encouraging other institutions to conduct similar analysis as a way to construct a meta-analysis of design research at the doctoral level.

Empirical approaches add to a discussion that has often concentrated on theoretical proposals of how and why design research and education should change. These reflections are by definition abstract and therefore removed from the world of practice (be it professional design practice or the educational and research practice that goes on in institutions of higher education); further, a theory should be tested or at least compared with the empirical reality of the actual educational and research practices it describes.

For instance, in the introduction to the paper entitled "Rethinking Design Education for the 21st Century: Theoretical, Methodological, and Ethical Discussion" (Findeli, 2001)the author admits that "It is, therefore, not really original to claim that we are in a period of necessary change, be it in design education, practice, or research." (p.5) The author then argues for a paradigm change for design education that should focus on developing a new theoretical model of design, an appropriate epistemology of design practice, and how to problematise the issue of a design ethics. In a similarly far reaching article, Friedman (2012) reflects on the issue of the future of design education and proposes a taxonomy of four design knowledge domains, as well as a skill-set the author considers relevant for designers. In another article, the same author (Friedman, 2003) presents a wide-ranging reflection on university-level design education in the context of the emerging knowledge economy.

Discussing how doctoral education in design can move forward, Margolin (2010) reinforces the need to conduct meta-studies of design research to work towards a "core curricula for all doctoral programs in design," that is to say, to draw from what has been done to insure that any doctor in design will be familiar with the same core literature and methodologies; and Buchanan (2001) reflecting on the Future Directions for Doctoral Education in Design also reinforced the goal of working towards a common corpus of design knowledge based in doctoral research when he states "a community of inquiry has formed in the field of design and is moving ahead to consolidate what is known about the field in its most sophisticated and well-grounded form and to prepare researchers and educators who will expand that knowledge through original inquiry." (p.23)

It seems there is an emerging consensus that the time has come for the design academic community to analyse and consolidate its knowledge foundations, determine what works and what needs to change, and move forward. We should look back and examine what have been the results of doctoral programs in design and work towards the necessary changes in design courses (at every level) with a solid foundation on empirical data. It is crucial that design consistently develops the ability to self-analyse; the design research community should look into the impact of its research projects, critically examine methodologies, and understand if there is an effective connection between the education, research, and practice of design.

This paper contributes to this aim by presenting and detailing an analytical framework and research tool that was designed to examine in detail the content of PhD theses in Design. By applying the tool systematically, we can generate meta-data and analysis of design research in general. In the next section, we will present the framework, detail its analytical elements, and describe the procedures on how to apply it.

## 2 THE ANALYTICAL FRAMEWORK

### 2.1 *Data sources*

The main goal of the analytical framework we will describe in this paper is to analyse the content of PhD theses in design; as of writing, the framework has been applied to the theses completed at the Lisbon School of Architecture, but the tool is not context-dependent and so can be applied to any PhD thesis in Design. The framework focuses on the analysis of research scope, research processes, and research results and recommendations described in a final thesis manuscript. This means that the final document of each PhD thesis is the only source of information for the study, in other words, documents such as annexes, appendices, or separate volumes, are not collected nor considered for analysis.

This limitation guarantees consistency of collected data which facilitates meta-analysis. Furthermore, time management is a critical factor for the success of any research study, and the collection of data in documents other than the thesis manuscript is incompatible with the completion of the study in a timely manner. It is important to note that this restriction limits the type of information we can collect since, as an example, data about the author's biography is not required in the Lisbon School of Architecture's regulations.

### 2.2 *Four key areas*

The framework is designed to analyse a PhD in Design thesis' manuscript; its main objective is to obtain a broad and detailed perspective of the research presented in the manuscript, with that in mind, the framework looks into four key areas for data collection and analysis:

1. Identification
2. Territory
3. Processes
4. Results

#### 2.2.1 *Research identification*

At this level, the analysis centres on data concerning author, supervisors, and thesis identification. For author and supervisor(s) we record first and last name, gender, and affiliation; regarding the thesis, the entries are more detailed and include: Title, subtitle, year of completion, word count, page count, typology (theoretical or practice-based), degree (PhD or Master thesis[1]), and Language (Portuguese, English, or Spanish).

Notice that while the data gathered at this level is quite generic, it is already possible to compare some data. For instance, it is possible to compare theoretical and practice-based thesis in terms of length by looking into word and page count.

#### 2.2.2 *Research territory*

Research territory concerns the analysis of the theoretical territories explored during the doctoral investigation; this is achieved by detailing the following five elements:

1. Keywords
2. Discipline
3. Scope
4. Research topic
5. Taxonomic classification

Points 1 through 4 can be determined within the thesis' introductory chapters. A list of *keywords* is a mandatory requirement for a PhD thesis in Design in the Lisbon School of Architecture; the *discipline* can be any of the following: Product, Communication, Ambient,

[1] In subsequent stages, we will expand the study to cover the master thesis in design as well.

Fashion, or General Design; the *Scope* concerns a deeper level of detail, for example, a thesis in Communication Design may explore the scope of Typography; and finally, the *research topic* is a short sentence that expresses the subject matter of the investigation, for instance: "Movement applied to digital typography."

Point 5 requires the categorisation of the research according to a taxonomic classification; this classification is based on the work of Nigel Cross (2007) in which the author noted that design knowledge can be found in people, processes, and products. To these categories Cross corresponded a type of research: epistemological, praxiological, and phenomenological. Cross' categories are a useful framework to characterise the territories of design knowledge, but to the author's three types of research we added an Ontological dimension, we added this category in order to cover investigations that delve into philosophical reflection on design as a discipline, since a mature discipline should self-reflect on its nature, on its epistemological and ontological basis, on the ethics of its practice and so on. This decision emerged from the observation that a researcher of the REDES research group had dedicated his doctoral research to a study on the nature of design (Monteiro, 2014); from this insight, a cursory survey of the PhD thesis database suggested that a category of *design philosophy* was needed to be able to cover the full spectrum of research in design.

Let us take a closer look into each category of the taxonomy:

**People**

Design knowledge can be found mostly in people. Firstly, in designers themselves, i.e. the professionals that through years of study and professional embody the knowledge inherent to design. From the perspective of professional design practice, a good part of design knowledge is tacit, that is, it consists of personal action expressed through behavioural patterns that depend – to some extent – on information stored in long-term memory (see Polanyi [2009]) for a detailed view on Tacit Knowledge and Friedman (2008) for a discussion on its role for design practice and research).

In other words, as is the case with all expert professional practice (Schön, 1983) designing draws on spontaneous behaviours and activates implicit information without the practitioner being aware of doing so. Tacit design knowledge is embedded in the individuals that mastered the skills and thinking necessary to develop new and useful artefacts. It is then the aim of research to carefully collect this knowledge and systematically develop theories that build a corpus of explicit design knowledge, thus rendering the implicit, hard to grasp, tacit knowledge into clearly articulated principles that can be successfully transmitted to others.

On the other hand, we begin with people also because people benefit or suffer from the (material or immaterial) artefacts that make up the artificial world in which we live; here we expand the notion of design knowledge to include users. The knowledge of how to use, appropriate, transform, and combine artefacts emerges from the unfolding interactions of people with things and their surrounding context, and is as relevant to understand *Design* as the professional skills and thinking required *to design*. In short, design research should look into how people design, how people learn to design, and how people use design.

**Processes**

Design knowledge can also be found in the tactics and strategies of design, or in the study of tools to facilitate designing, in other words, in design processes. At a practical level, the study of design processes may focus on the traditional tools and skills of designing such as drawing, computer assisted design, parametric design, modelling, or on a strategic level to explore techniques that enhance creativity, foster collaborative and co-design practices, or improve communication with stakeholders.

Also, the study of the design process itself has been the concern of design researchers for decades (see for example Jones [1970], Lawson [2005], or Rozenburg & Enkels [1995]); so, the discovery and detailed description of a general model of the design process (a high-level abstraction model able to describe the working process of any design discipline) has been pursued diligently and is therefore an integral part of design research.

Thus, the study of design processes aims to describe the inner mechanisms and logic of designing, to identify common aspects in the workings of different design disciplines, to map out and report best practices as well as to identify failings and mistakes. Understanding the design process can have an impact on professional design practice but also on design education, since understanding how designing unfolds in practice may lead to better ways to teach it.

**Products**

Design knowledge can also be found in products themselves, that is, in the material embodiment of artefacts. We take artefact in a general sense to mean any object that is the product of human skill and ingenuity(Erlhoff & Marshall, 2008), which means that any designed entity can fall into the category of artefact, in other words, not only material objects but also designed spaces, images, software, systems, clothes, or environments can be considered a designed entity — an artefact.

The study of artefacts may reveal the embodied information about the processes that guided their creation. Also, the study of artefacts is naturally related with how people use them, which means that the study of objects can not only reveal the processes that lead to their creation but also their impact on people's everyday lives. Notice that while the development and production of artefacts is the end result of a design process, the end of the story is found on how people use it, how long it lasts, or how well it functions, in short, the complete story of design implies the understanding of the interconnectedness of people with processes and products.

Further to this point, an artefact does not exist in a vacuum. In fact, most products imbody past references, that is, previous incarnations of solutions that attempted to solve similar problems. This historical thread leaves an evolutionary mark, in the same manner that human beings still carry a genetic inheritance from the species that preceded them. As such, the study of products can reveal, for instance, the connections between craftwork and design, as well as issues of the implicit knowledge carried within human-made objects (like clues from earlier species in the genetic code), issues of copying, and the study of prototypicality and archetypes.

Finally, the study of products may include research on morphology (see Christopher Alexander's Pattern Language (1964) for a classic investigation – in this case in the field of Architecture – of this sort) that detail issues of form semantic and syntactic (see Wim Muller's Order and Meaning in Design [2001]); and also applied research concerning economic or ergonomic efficiency, as well as the increasingly urgent study of product sustainability in its broadest sense: which means to work towards full environmental, social, and economic balance.

**Philosophy**

As was stated above, to Cross' three categories we added Philosophy of Design. It is clear that knowledge about design can also emerge from philosophical exploration of what the nature of design is; thus, studies of Design philosophy can include reflection on the ethics of design practice, the aesthetic dimension of design, and also enquiry on the nature of design ability itself (for instance: to what extent is design ability learned or innate? Is there such a thing as *natural* design talent? Is design thinking a unique form of human intelligence?). Studies of this nature can help us position Design in relation to other areas of human endeavour such as the arts and humanities or the natural and social sciences.

To conclude, to each of the four domains corresponds a type of research which is summarised as follows:

Epistemology of design – study of ways of knowing in design.

Praxiology of design – study of design practices and processes.

Phenomenology of design – study of artefacts.

Ontology of design – study of the nature of design.

It is clear from the description of each domain that, even though each area of design knowledge is discrete, they are nonetheless connected. Therefore, it is useful for a research project to have a foothold on a particular domain of design knowledge because it clarifies the type of research to be conducted (epistemological, praxiological, phenomenological, or ontological), but one must keep in mind the inherent interconnectedness of these domains to have a full understanding of design.

The definition of the research territory – and with it the positioning of the research – allow us to assess the relevance of the research, that is, how the project fits with what is already known, how it relates to other knowledge areas besides Design, and in which aspects of human-life it proposes to have an impact.

Having described how the framework tool analyses the research territories, we will now explain how it addresses research processes.

2.2.3 *Research processes*

At this level, the analysis focuses on the specific methods used during the research studies. The goal is to detail how the studies were conducted, in which research phase, the type of information that the study generated, the object of study, and the role of the researcher.

The analysis of research processes is based on the research design framework of Martin and Hanington (2012) entitled *Universal Methods of Design*. Martin and Hanington's framework has been presented to the PhD candidates in the Lisbon School of Architecture's PhD in Design course since 2012 by one of the authors of this paper. In Universal Methods of Design, the author's describe distinct dimensions of research methods in which some dimensions rule the others. The different dimensions provide a detailed picture and characterisation of design research methods.

Thus, we drew on the dimensions described in Martin and Hanington's framework to analyse the research processes of the PhD thesis according to the following criteria:

1. Study sequence: the architecture design of the methods being used; the sequence of studies/methods used is not arbitrary, it establishes a roadmap that guides the research process.
2. Method designation: a short name describing the study (e.g. 'Literature Review') used consistently throughout the analysis of all the thesis (to facilitate meta-analysis).
3. Research phase: exploratory, generative, or evaluative (this is the crucial category and is elaborated below).
4. Type of information: this item applies to methods that study people, the types of information are attitudinal (concerning the self-reporting of feelings, emotions, opinions, disposition, positions, and so on) or behavioural (the reporting by a researcher of the actual actions, behaviours, reactions, etc, of the persons observed); attitudinal information is a self-expression of internal disposition (as captured in an interview or questionnaire for instance) whereas behavioural information are patterns of actual behaviour observed by a researcher.
5. Type of method: if the method is traditional of disciplines other than design; if it is an adaptation, of existing methods to design, or if it is an original design research method.
6. Quantitative or qualitative approach: here the terms qualitative and quantitative research refer to the methods used, and not to the data being studied or generated, which means that a qualitative method may use quantitative data by exploring interpretations and ramifications of such data, and likewise, quantitative methods may use qualitative data as well.
7. Researcher role: when the method had a direct intervention of the researcher their role is described as participative, observational, design process (in the case of practice-based research), or self-reporting.
8. Object of study: here we describe the study's focus. Depending on the nature of the research, the object of study may focus on people, products, processes, design itself, or a combination of these. The object of study may concern the whole research problem or a part of it, it may be the context or key actors within the context that frames the research; whatever the case, each method has a specific focus point which is captured and briefly described in this category.

Within these criteria, the dominant dimension is the *research phase*; this is the case because it is possible to use the aim of the method (to explore, to generate or to evaluate information) to categorise the research process phases into three stages: (1) *exploratory* – the initial phase is typically a divergent stage in which the researcher probes the territory defined by the literature review; (2) *generative* – a phase whereupon after positioning the research one needs to generate information about the object of study or its context – and finally (3) *evaluative*, the research phase in which the data gathered is examined according to its value in terms of knowledge contribution. While discrete categories are useful to guide the researcher's process and to support the conceptualisation of how the research is put together, a researcher should, nonetheless, consider possible overlap between these stages. In fact, the three categories above are not a cage but rather stages along a continuous path.

The structure and sequence of methods are the engine of the research, that is, they propel the investigative process forward; when we take the engine apart, we can see how the different parts fit together and work (or fall short) as a coherent whole. Also, the methods are employed to answer and explore the problematic that was established a priori. Therefore, we can see how the sequence and connections between methods are the core of a PhD investigation, that being the case, when applying the analytical framework we described, a researcher should be prepared to spend considerable time in taking the research methods apart, analysing its components, and understanding the logic behind the sequence of studies.

It is after completing the analysis of the research processes that we can connect the problematic, research territories and aims described in the introductory chapters of a thesis, i.e. the research starting point, with the results presented in the concluding chapters and bits of discussion and reflection that may be spread throughout the manuscript. Reaching this stage, we are prepared to delve into the research results.

### 2.2.4 *Research results*

Finally, in *the research results*, the analytic framework reports on two areas: (a) the research questions and hypotheses, and (b) the potential for expansion.

The analysis focusses on how the thesis connects the results with the problematic described in its introduction. An important part of this effort concerns answering the research questions and (when applicable) verifying the hypotheses. It is this work that completes the research and closes the investigative circle; it is essential to conclude the thesis with a fully developed discussion that connects the findings with the theoretical framework and thus interprets any results from a broader perspective.

To achieve this, it is necessary to identify if the author clearly and directly attempted to answer the research questions and hypotheses raised. Normally, the *Discussion* and *Conclusions* chapters of a thesis revisit the problem statement in light of the results generated during the investigation; however, note that the structure of a PhD thesis may vary and a general discussion of the results could be embedded in previous chapters in which case the thesis will not have an explicit *Discussion* chapter. With that in mind, one should make a thorough read-through of the thesis manuscript to uncover any insights that may be missing from the concluding chapters.

On another level, the analysis also looks to the thesis' potential for expansion by identifying insights, results, suggestions, or new research questions that could lead to further research projects. These expansion items are also described in terms of the discipline to which they are related (for instance 'Product Design') and if it applies to research, education, or practice (society).

By looking into these concluding aspects, it is possible to disclose what are the thesis main contributions. Furthermore, by examining if the investigation reached answers to its research questions, we can also reflect on the appropriateness of the research design, that is, if the research process was adequately put together.

Below we present a summarised view of the elements of the framework.

1. Identification: Name and affiliation of author and supervisor; title, wordcount, page count, typology, and language.
2. Territory: Taxonomy; discipline; scope; topic; keywords,
3. Processes: Sequence; designation; research phase; type of information; type of method; quantitative or qualitative; researcher role; object of study.
4. Results: Research questions; research hypotheses; expansion.

## 3 LIMITATIONS

It is important to acknowledge that the analytical framework generates data that is dependent on and *relative* to the researcher's interpretation, in other words, it is the result of a qualitative research approach. There are a few exceptions in the case of the factual data reported in the first level of analysis – *identification* (entries such as "author name", "thesis title", or "word count" are factual pieces of information and thus not the product of interpretation). Notice that the same limitation can be identified in the aforementioned Horvath's work (2008) which followed similar procedures. This limitation is inherent to any research that deals with qualitative methods; to examine the epistemological limitations of qualitative research falls substantially beyond the scope of this paper, nevertheless, we are looking into ways of mitigating the relative aspect of the data generated in order to have as an objective analytical tool as possible.

For instance, we plan to have the data examined by a different researcher and conduct an intercoder reliability analysis (Krippendorff, 2011) to reflect on the results obtained. Also, we plan to provide the analytical framework to the actual authors of the thesis and ask them to self-analyse their work; this would have the additional benefit of presenting an internal perspective which could in turn be compared with the external perspective generated by our analysis.

Furthermore, from a methodological perspective, we adopted some of the Grounded Theory principles suggested by Glaser and Strauss (2006) for qualitative research in which the authors argue that qualitative results should be thoroughly reported in terms of procedure, criteria, and rationale so that an independent researcher/reader may fully examine the reliability and validity of the data being presented. As such, this paper also contributes to that goal by detailing the analytical framework used to analyse the PhD thesis.

Finally, we should note that the framework is effective in guiding the examination of *levels 1 identification, 2 territory*, and *3 processes*, but level *4 results* is considerably harder to describe with the same precision. The main problem is that the discussion of research results is founded on the interpretation of the findings of the author of the thesis; as such, our examination is fundamentally an *interpretation* of an *interpretation*. To mitigate this ambiguity, we limited the extent of the analysis to uncovering if the author of the thesis reasonably and clearly tried to present an answer to the research questions and hypotheses, that is, we restrained from examining how thoroughly the author answered the research questions and instead focussed on identifying if the research questions were acknowledged and dealt with.

## 4 CONCLUSIONS

To conclude, we would like to emphasise the ways in which the analytical framework we presented here may contribute to design research and education. It is our aim, with this paper, to disseminate this research tool and encourage researchers to adapt and use it in their own studies. This way, the design research community can collectively reflect on the outcomes, processes, and the underlying epistemologies of its investigations.

It is clear that the analytical framework we present is an effective research tool. The results are useful to examine each thesis separately and – crucially – to aggregate those results and build an expanding meta-analysis of design research. This represents a contribution to the current state of design research and design education since we can compare strategies, research designs, and patterns that produce solid PhD thesis in design with the ones that fall short; examining this will necessarily have consequences to the structure and content of PhD courses in design.

The tool can be adapted and applied in other academic disciplines since the principles that underly the analytical framework are general enough to be applied in other disciplines (for instance in Architecture or Urban Planning) and in other cultural research contexts as well. Furthermore, the framework is versatile which means it could be used in parallel with other research tools to support necessary in-depth studies on design research, education, and practices.

As such, we gather that this framework could be useful as a diagnostic tool for the people involved in the development of PhD courses, on the other hand, it may also be helpful to prospect PhD candidates in providing an overall perspective on the territories of design research, its methods and processes, as well as the research themes and areas that have been heavily studied or alternatively still lack systematic inquiry.

## REFERENCES


Author, (2012).

Author, (2019a).

Author, (2019b).

Alexander, C. (1964). Notes on the Synthesis of Form. Cambridge (MA): Harvard University Press.

Buchanan, R. (2001). Design Research and the new learning. Design Issues, 17(4), 3–23.

Cross, N. (2007). Designerly Ways of Knowing. Basel: Birkhäuser Verlag AG.

Durling, D., & Friedman, K. (Eds.). (2000). Doctoral Education in Design: Foundations for the Future. La Clusaz: Staffordshire University Press.

Erlhoff, M., & Marshall, T. (2008). Design Dictionary: Perspectives on Design Terminology. Basel: Walter de Gruyter.

Findeli, A. (2001). Rethinking design education for the 21st Century: theoretical, methodological, and ethical discussion. Design Issues, 17(1), 5–17.

Friedman, K. (2003). Design Education in the University: A Philosophical and Socio-Economic Inquiry (Hot Debate) Design Philosophy Papers, 1(5), 243–253.

Friedman, K. (2008). Research into, by and for design. Journal of Visual Arts Practice, 7(2), 153–160.

Friedman, K. (2012). Models of Design: Envisioning a future design education. Visible Language, 46(1/2), 132–153.

Glaser, B., & Strauss, A. (2006). The Discovery of Grounded Theory: Strategies for Qualitative Research. London: Transaction Publishers.

Horvath, I. (2008). Differences between "research in design context" and "design inclusive research" in the domain of industrial design engineering. Journal of Design Research, 7(1), 61–83.

Jones, J. (1970). Design Methods: seeds of human futures. New York: John Wiley & Sons.

Krippendorff, K. (2011). Agreement and information in the reliability of coding. Communication Methods and Measures, 5(2), 93–112.

Lawson, B. (2005). How Designers Think: The Design Process Demystified (4th ed.). Amsterdam: Elsevier.

Margolin, V. (2010). Doctoral Education in Design: Problems and Prospects. Design Issues, 26(3), 70–78.

Martin, B., & Hanington, B. (2012). Universal Methods of Design. Beverly (MA): Rockport Publishers.

Monteiro, P., (2014). A Questão do Design: (in)definições e equívocos entre forma, fama e função (Doctoral dissertation). Faculdade de Arquitetura da Universidade de Lisboa, Lisboa.

Muller, W. (2001). Order and Meaning in Design. Utrecht: LEMMA Publishers.

Polanyi, M. (2009). The Tacit Dimension. Chicago: The University of Chicago Press.

Roozenburg, N., & Eekels, J. (1995). Product design: fundamentals and methods. Chichester: Wiley.

Schön, D. (1983). *The Reflective Practitioner: How professionals think in action.* New York: Basic books.

*Research & Education in Design: People & Processes & Products & Philosophy – Almendra & Ferreira (eds)*
*© 2020 Taylor & Francis Group, London, ISBN 978-0-367-49296-0*

# Tools, methods or theories in design research?

R. Herriott & C. Akoglu
*Design School Kolding, Denmark*

ABSTRACT: The use of terms in design research is not clear. A significant number of papers discuss how to design with reference to "tools for" without making clear what constitutes a tool (or method). This paper reviews work on the concept of tools; it examines the use of terms in selected papers and reviews three possible ways to define and categorise tools. It recommends the term "tool" be used as little as possible and only with reference to tangible objects. This can help distinguish between methods as ways to carry out an action and tools as the physical means to do so.

## 1 INTRODUCTION

Work on this paper was prompted by the general sense the authors had of a proliferation of *tools* and *methods* in the design literature. A preliminary literature review using design research databases was conducted. It became apparent that the initial question would have to wait until the terms "tool" and "method" could be defined.

This paper should be considered a preliminary inquiry into how common, important terms are used in design research. It is, if you will, a rough prototype. Each step of the literature review could have been expanded much more than space allows here.

The literature showed that the two terms (and others) were being used interchangeably and that there existed a lack of ontological order such that the primacy of terms and their relation to each other was confusing. For example, Sanders and Stappers′ much-cited 'New Landscapes' paper of 2008 has six instances of the term "tools and methods"; it refers to co-creation as a tool (p.8), has three reference to "tools and techniques" and has two references to "co-designing tools" but at no point explains their relationship to each other.

This finding is in line with the observations of Love (2000) that a) "there exists a substantial amount of confusion with respect to the basis of many theories, concepts and methods" (with "tool" being a concept and a thing); b) "there exists an unnecessary multiplicity of design theories and concepts"; and c) "the terminology of design research has become unnecessarily and unhelpfully confused by dint of the above points" (Love, 2000, p.295)

From the above, it became clear to the authors that a definition of terms would be not a start point for this paper. Instead, we begin with be a presentation of the use of the terms. The *result* is some form a comment and proposal regarding the use of the terms in research.

The value of this inquiry rests on this assumption: researchers must have either a common set of terms or else be clear what they mean if design research is to lead to an understanding of design practice. To use Dalsgaard′s (2017, p.21) words, a clear understanding of the concept "tools" could "provide a foundation for structured critical analysis of existing tools employed in design" and "create a better basis for selecting and employing tools in a given design practice". It could "offer a starting point in planning and developing new tools that better support design" (*ibid*).

The alternative is to abandon attempts to generalize. That means not being able to group like-concepts under broad terms. This is akin to giving up the useful transformation of data into knowledge. In design education it is necessary to be able to transmit ideas clearly and this work will permit researchers who teach to be better able to convey complex ideas relating to the interplay of tools, methods and theories.

## 2 A REVIEW OF DESIGN RESEARCH TEXTS

We first look at some work on the concept of tools in design research. Dalsgaard (2017) usefully provides an overview of work in the nature and role of tools in design. Dalsgaard argues that there "are few frameworks for understanding how and why they work" and states that "the use of tools pervades most forms of interaction design practice" (p.21). More generally, it could be said the use of tools pervades most forms of *all* design practice. But what is a tool?

Tools, according to Dalsgaard, are used by designers to "create future products and services" and "to understand the design situation and the problem at hand". Dalsgaard's premise is that tools not only achieve some end (e.g. "augmenting designers' capabilities for carrying out intended actions") but also "guide perception and understanding of design problems" (ibid, p.22)

Dalsgaard does not define what tools *are* but discusses the role of tools. Citing examples, the term is used both as reference to tangible objects e.g. sketching tools (in Buxton, 2007), mock-ups (in Ehn & King, 1991) and prototypes (in Lim *et al.* 2008). Less clearly tangible, cases of "tools" include ones for "co- ordinating and managing collaborative design" (in Bannon, 1993) and "card-based design techniques" (in Wolf & Merritt, 2013). Notice how the word "technique" has been deployed, a sign of ontological disorder.

Marchand (2017) also treats tools as a "means to understand the design situation and problem at hand" (echoing Dalsgaard, 2017 and Schoen 1984) in a book on craftwork as problem solving. In craftwork "tools are used to think" and indeed craftwork, which uses tangible tools, was the basis of modern western scientific inquiry (Marchand citing Pamela-Smith, 2006) and in accord with Buchanan (2001). The tools referred to in Marchand (2017) are explicitly and implicitly the physical devices used in the crafts of woodwork, textile, bike repair, ceramics but also horse-training.

### 2.1 *Analysis of the use of terms*

For this article a sample of 30 papers, books and articles were chosen. Key terms were abstracted with the intention of showing what they referred to. To narrow the search, we have focused in this part on papers from participatory design and related areas. We have attempted to find the main terms and try to make clear, from the text, what the categorical relationship is between the terms.

We have attempted to identify what the author thinks "tools" are by the instances of the word's use. An example of this process is as follows. Lee *et al* 2018 write "... an increasing number of design and innovation projects apply co-creation as a process, agenda or tool". This can be taken to mean that the authors think co-creation is a tool and co-creation is a process and co-creation is an agenda. This implies the authors think co-creation = process = agenda = tool.

There now follow three expanded examples.

Andrews and Revolo (2004) write about participatory design in the development of digital products. They describe a firm's programme called Understanding Digital Experience (UDE) which is presented as "paradigm" which can be used to identify "design principles" (p.53). The "paradigm" is also described as "a theoretical framework" and assists or serves to develop "strategies and tools" (p.54). A plausible interpretation of this is that the term "strategy" corresponds to "method" and that strategies are not tools. But the UDE theory has helped designers devise four "tools". One is the user-experience model. Here a model equals a tool.

That model or tool can be interpreted as a statement describing how customers should interact with the products. Thus "a model = a tool = a statement". The second tool is the "experience qualities matrix" which is a graphical device structuring the visualization of two sets of concepts of varying degrees of abstraction e.g. "environments" and "care". The third tool is "personas", which is a way of representing data in an empathic way. The fourth is "co-creation" which is described as a checklist and as a structuring of information.

Hussain, et al (2012) deals with implementing participatory design with economically-weak people in developing countries. The article uses the terms tools, techniques, approaches, methods. The role of prototyping is mentioned but not assigned a category. By implication it seems to come under the category workshop which is implied to be a "technique". The term participatory design was used; it was reported that workshops were undertaken. This activity was termed a technique and described in detail: two workshops are described. This workshop was used to prime the participants for ideas. In the second, participants used a variety of tools [actual, physical tools] and materials to make prototypes. Refinement of the prototypes took place in Norway and these were shown to disabled children in Cambodia to get feedback. The authors used audio and video recordings, field notes and photographs during the participatory design practices. In this paper there is no clear assignment of a hierarchy of terms for the concepts of participatory design, workshops or the technological items used such as audio and video equipment, which could have been called tools.

Mortati and Villari (2013) discuss giving people networking, participation and networking skills to people taking on the role of social innovators in urban communities. The article uses the term tools, methodologies, processes and approaches. The imparting of the skills required the following: a) "mapping and collecting interesting cases and people" which involved "interviews, questionnaires and a workshop" (p. 130); b) conducting an educational project which generated "service ideas" by the "use/creation of collaborative tools and processes aimed at co-designing" which "in particular" required "contextual analysis". c) testing "using a design research project" to "define new urban services". It was not explained which entities/activities were tools.

### 2.2 *Summarised selected papers*

There now follows some shorter summaries of a selection of papers.

Al-Kodmany (2001) discusses participation design using the terms "tools" and "methods". The "tools" refer to visualization software (p.111) and geographic information systems, slide projectors and maps. The author implies "tools and methods" (p.113) and "visualization methods" are equivalent. "visualization tools" are inadequate (p.114) and the remedy is a "visualization method". The term "workshop" was used in which "techniques" were deployed and the techniques consisted of "printouts of maps and models". The term "method" is used (p.113) and the case discussed the failure of these. The examples of "methods" included slide projectors.

Lee & Lee (2009) mention *methods* in regard to "focus group interviews and design workshops" so here, focus group is a *method.* The paper described "tools" (p.17) to help focus groups and the "tools" were tested in a focus-group interview. The *tools* proposed are 1) "pre-question cards" (p.23) which are described as a "technique for "context mapping"; 2) "mini-me dolls" which are small toy figures 3) a random-turn selection procedure which determines who can speak up at the focus group 4) a form of role-playing themed around television.

Bjögvinsson et al. (2012, p.104-105) discuss the reporting of results arising from design research. Bjögvinsson et al. (2012) don't directly use the term methods or tools, but there is an interesting emphasis by the general listing of concepts: "Involved practices being reportable", "representing in different forms", "controversies among participants" are gathered together. This type of a perspective suggests that 'methods, tools and techniques' are somewhat undifferentiated.

Yang and Sung (2016) report on service design tools to facilitate multidisciplinary ideation and co- creative social innovation. It is difficult to establish a hierarchy here since the high-level concept "service design" provides lower-level tools which are used to achieve co-creation. That is not a process here but more understood as an outcome: it is via service design tools that co-creation is attained. Examples of tools "provided by" service design are: "stakeholder maps, service blue-prints and customer journey maps".

### 2.3 *And methods?*

The point could be made that this search process has only looked at articles and papers selected for the term "tools". On the one hand, it is reasonably evident that the search has shown a common interchangeability of the terms "tools" and "methods" merely by looking for usage of the term "tool". However, we conducted a preliminary search for articles using the term "methods" in order to see if there was a similar equivalence.

Hanington (2003) uses the term methods in his reviews of the state of human-centred design. The items on this list all fall under Hanington's methods category: focus groups, surveys questionnaires, interviews, archival methods, trace measures, experiments, design workshops, collage, card sorting, cognitive mapping, velcro modelling, visual diaries, camera studies, document annotations. For some other authors those are all tools.

Hendriks et al (2015) discuss co-designing with people with cognitive disorders. They use the term methods and techniques interchangeably e.g. "Codesign with people living with cognitive or sensory impairments poses challenges for researchers and designers, due to differences in their mutual experiences and due to the fact that many well-established codesign methods and techniques may not be appropriate and need adjustment" (p.70). A section entitled "Searching for a dedicated codesign approach" (p.71) could be reduced as follows: codesign is an approach; the approach makes use of techniques and a study of the use of these could provide tools and guidelines. The tools and guidelines could be for the selection of an action or could refer to the actions and decision-making used directly in designing.

Later in the paper the authors refer to "a methodological approach for co-design" (p.75). If we recall that on p.71 of Hendriks et al. co-design is itself an "approach" then the authors seem to mean a methodological approach to an approach (co-design) which is a thing consisting of methods and techniques. Clearly there is a circularity problem in dealing with levels of abstraction in the article.

Halskov and Brodersen-Hansen (2015) investigate the diversity of participatory design (PD) research practice by examining the papers from a conference series in PD. Under the heading of participatory design methods probes, personas, whiteboard usage, and experience clip techniques are identified explicitly (ibid. p.83). As a matter of interest, the authors group "(theoretical) approaches" with models and concepts: "Theoretical contributions to participatory design: this category includes papers that explicitly introduce or establish theoretical approaches, models, or concepts." (ibid. p.84) This understanding of "approach" is incompatible with the use by Hendriks et al (2015). The term "tool" only makes one appearance, a description of Buur's (2005) work involving the use of videotape.

## 3 ATTEMPTS TO CLASSIFY TERMS

We having reviewed (albeit in a preliminary fashion) the situation of how the terms in question are not defined and not always consistently related to each other hierarchically. Now it is time to look at some possible ways to handle the terms so as to mitigate the situation, the work Dalsgaard (2017) and Love (2000).

### 3.1 *Dalsgaard's (2017) pragmatism*

We return to Dalsgaard's (2017) attempt to present a framework consisting of five qualities of "instruments of inquiry", his term for tools. The description of the five qualities is a form of definition. The five qualities are presented from a Deweyan pragmatist perspective since Dewey's key concepts "including inquiry and technology are particularly apt for understanding creative design and the role of tools in it" (p.22) It is perhaps worth noting that Dalsgaard considers the strong division between theory and practice untenable since they are intertwined. Theories are a means for understanding the world and at the same time the various means at our disposal (these indeterminate "tools", "methods" and "techniques" etc) have the effect of not only changing the world but also our understanding ("theories") of it.

Dalsgaard uses the broad terms "technology" and "instruments" where technology is "the use of instruments" and an instrument is an artefact used as a means to transform the situation. Dalsgaard deploys the term instrument as defined by Hickman (2001) who argues "they serve as tools to carry out specific actions" and "also help us understand the problem at hand" (p.24). The attempt to define instrument yields this sentence "Instruments can be defined as instruments that scaffold the process of inquiry" (ibid) in which the word *instrument* is used to define itself.

Having settled on this understanding of tools as "instruments of inquiry" Dalsgaard explains their five qualities without defining necessity or sufficiency. The qualities of the instruments are that they enable and 1) support perception, 2) support conception 3) externalization (meaning representation) 4) knowing-through-action and 5) mediation between actors and artefacts.

The desired goal of an understanding of a "tool" is to be able to place it hierarchically in relation to theory and methods so that terms are not interchangeable. It is the interchangeability demonstrated in the sample research above that is problematic. So long as the understanding of the term "tool" or "instrument of inquiry" permits tangible and abstract usage, then it is plausibly demonstrable that both "tool" and "method" can be explained in terms of the five qualities. This happens because the term "tool" is understood as a thing and a metaphor and so tools and "methods" which are abstractions become indistinguishable.

If we look at each quality in turn, we can see the extent of this overlap. A method could "*enable and support perception, revealing facets of a design situation that would otherwise be hidden*". Consider an interview, which is a method of asking questions, or a mind-map which is a process of organizing words in two- dimensional space. Drawing is a method for placing marks on paper to show things which do not exist. It "reveals facets of the design situation" (in Dalsgaard's terms) e.g. making evident a problem of geometry or construction. This is not perceiving a thing in the way binoculars enable and support perception of, say, distant gazelles.

A method "*could enable and support conception*" and this applies to an interview, mind-map and possibly drawing in that by trying to draw something realistically we alter our conception of how it could be. Statistical analysis is a method of processing numbers and the resultant, new numbers form our conception of a condition such as say processed data yielding an average value comparable to other average values.

Could a method enable and support externalization "*by which designers can make imagined design solutions part of the world and allow them to manipulate, evaluate and develop them*"? Dalsgaard suggests as examples sketches models, mock-ups and prototypes and this is the most plausible category of tool and they do have this quality of externalizing a mental state which is a form of representation. Plausibly abstract models and theories are also forms of externalization. Like a sketch of an imagined car, a theory is an externalisation of how one imagines the world to be. And if we connect together the method of car design which encompasses sketching, clay modelling and CAD modelling we have plausibly shown how a method can support externalization.

Finally, we turn to how Instruments of inquiry enable and support "knowing through action". It is argued that by using the tools available, designers find out about the state of the world. It is true that tools have this character. The screwdriver tool helps the user learn that the screw has been turned too tight and cannot be removed from the metal plate. However, like Simon's definition of design (1969) it would appear that "knowing through action" allows in too many trivial instances. Almost all design actions at all levels of tangibility and abstraction yield knowledge of the world. The application of a method for X will help us know something we did not before. The method for painting the front door will help me know about the behavior of emulsion paint on corduroy, for example.

To summarise the previous five characters, we can say that all things we commonly call a method (Hanington and Martin, 2012, list a hundred) have the same characteristics of the things we commonly call tools. Since design as research and practice involves a high number of methods, we have learned that design has these qualities too. However, we have not been able to distinguish design tools from higher or lower-level design methods.

### 3.2 *Love's (2000) analysis*

Is there another plausible way to distinguish between tools and methods? Love (2000) proposed a meta- theoretical structure1 for design theory. It aimed to divide the compound concepts of "design" into its constituent levels of complexity so as to "assist with the establishment of coherence and compatibility between concepts in disparate theories" (p.293). To repeat the problem, if we can't agree about the basic terms of design activity it will be hard to discuss rival theories of design process. This would make work on the nature of design activities and reflection on the fundamental concepts of design (Cross's 1984 categories) tricky.

Love (2000) uses critical analysis to clarify design theory "because its purpose is the clarification of individual relationships between individual concepts and theories" (p.299). Love outlines the problems of applying critical analysis being to do with a term having a variety of meanings (see above). We could argue that since the terms "tool" and "method" have been used in so many different ways critical analysis is of no use in solving the problem. We will present the first six classes (of ten) in Love's design meta-theory. Having explained them briefly, we will then see where "tool" and "method" could fit in. What we will be proposing is that tools should be understood in such a way as to put them in a different category level from methods.

The first five categories are:

1. Direct perception of realities e.g. the pen makes a mark on the page.
2. Description of objects. "This level encompasses simple descriptions of objects, processes and systems" (ibid. p.305).
3. Behaviour of elements. "The level at which the behaviour of elements which may be incorporated into objects ... e.g. the hammer is made up of two parts" (ibid. p.305).
4. Mechanisms of choice. "The level of descriptions about the choices are made between different objects, processes and systems and how solutions are evaluated e.g. choosing one method over another" (ibid. p.305).
5. Design Methods. "The level in which theories and proposals for design methods and techniques are described" (ibid. p.306).

The later categories do not apply to tools or methods. Category 10 is the ontology of design which is what this paper is about.

What we learn about tools and methods in relation to categories 1 to 5 is that category 2 will handle descriptions of *both* tangible and abstract entities meaning a pen *and* interview method. Category three will be about how tools and methods operate. Four is about how we choose a tool or a method and finally, five is about methods but since tools are conflated with methods, we get no further down the road.

It would appear there is something elusive about tools and methods which two well thought-out analytical schemes fail to identify and help disentangle. Part of the problem may rest on the way in which tools are both actual objects e.g. a clay modelling blade and metaphorical e.g. a user-experience model (e.g. Andrews and Bevolo, 2004).

### 3.3 *Thagard's (1988) pragmatism*

Leaving behind the science of design (Love) and Deweyan pragmatism (Dalsgaard) we might turn to the theory of science. Thagard's (1988) work on the demarcation of science might be the way forward. Thagard contends that we must make do with a "general profile" of science. There is no precise definition that includes only all that we want and excludes only all that we don't want from a description of science. Thagard lists some attributes of science. None of them are necessary or sufficient but the more of them are present, the more science-like an activity is, argues Thagard. This makes it a pragmatic approach.

By analogy, we might consider a range of abstraction from tangible to metaphorical, with tools at one end and methods at the other. At the tangible end are physical objects such as knives, pens and recording devices. At the metaphorical, abstract end are mind-maps, personas and user-journey diagrams. Attractive as this may seem, it doesn't reduces the amount of uncertainty, only redistributes it. And it does not avoid the need for definitions either, but merely uses more of them for more concepts.

## 4 A SUGESTION

Can this terminology problem be solved? One approach is stringency, to avoid as much as possible the use of the word tool entirely and adopt the term "method" for ways of doing something. This puts the emphasis on the verbal aspect of the activity and not the object involved. This approach would likely better retain the hierarchical relations of object, tool (if used), method and theory. It would be especially helpful in the intangible design arenas of service design, strategic design and social design. In these fields, the dominant means to attack a problem are largely ways of communication. Using the term tool distracts from that condition and makes it harder to cross-compare reports of projects.

Work is needed on a more comprehensive review of design research literature. This would require a more rigorous method to determine authors′ intent regarding the hierarchy of the terms such as *tools, methods* and *theories*. It would also be useful to examine further what tools are and how they work in design having arrived at a definition that excludes methods and can distinguish between tools for changing things e.g. a blade and objects for measuring e.g. 3D scanners, thermometers and the like.

## 5 CONCLUSION

Clarity about terms is essential in teaching undergraduates who are coming to terms with the conceptual complexity of design research. Design research is a bridge to design education and absent consistent terminology, the complexity is only harder to manage. This paper has shown there is a lack of hierarchical order in the component terms of the design discourse.

Further work might usefully be done to see if this blurring of concepts applies in languages other than English. It is unfortunate that from a stylistic point of view, authors seem to have used a word other than tool simply to avoid sounding repetitive. Editors and peer reviewers need to take a long and hard look at the use of terms. In many instances cited above, the problem of terms would not exist if there had been more scrutiny of language.

## REFERENCES


Andrews, A., Revolo, M (2004) Understanding digital futures. Design Management Review. Winter 2004; 15, 1.

Al-Kodmany, K (2001) Bridging the gap between technical and local knowledge: tools for promoting community-based planning and design. Journal of Architectural & Planning Research; Summer 2001, Vol. 18 Issue 2, p110–130, 21p.

Bjögvinsson, E., Ehn, P., and Hillgren, P-A (2012) Design Things and Design Thinking: Contemporary Participatory Design Challenges. Design Issues 2012 28:3, 101–116.

Buchanan, R (2001) Design research and the new learning Design Issues Vol 17, Number 4 Autumn 2001.

Yang, C., Sung, T., (2016) Service Design for Social Innovation through Participatory Action Research. International Journal of Design; Apr2016, Vol. 10 Issue 1, p21–36, 16p.

Dalsgaard, P (2017) Instruments of Inquiry: Understanding the Nature and Role of Tools in Design. International Journal of Design; Apr2017, Vol. 11 Issue 1, p21–33, 13p.

Hanington, B (2003) Methods in the Making: A Perspective on the State of Human Research in Design. Design Issues Volume 19 | Issue 4 Autumn 2003. p.9–18. doi/10.1162/074793603322545019.

Hanington, B., Martin, B. (2012) Universal Methods of Design. Rockport, MA: Rockport.

Halskov, K., Brodersen-Andersen, N (2014)The diversity of participatory design research practice at PDC 2002–2012 International Journal of Human-Computer Studies. Volume 74, February 2015, Pages 81–92.

Hendriks, P., Sleghers, K, Duysburgh, P (2015) Codesign with people living with cognitive or sensory impairments: a case for method stories and uniqueness. CoDesign, 11:1, 70–82, DOI: 10.1080/15710882.2015.1020316

Hickman, L.A (2001) Philosophical tools for technological culture; Putting pragmatism to work. Bloomington, IN: Indiana University Press.

Hussain, S,. Sander E, B., Steinert, (2012) Participatory Design with Marginalized People in Developing Countries: Challenges and Opportunities Experienced in a Field Study in Cambodia. International Journal of Design; Aug2012, Vol. 6 Issue 2, p91–109, 19p.

Lee, J-J., Jaatinen, and six others (2018) Design Choices Framework For Co-Creation Projects. International Journal of Design, Vol 12., No 2 pp 15–31.

Lee, J.J., and Lee. K.P. (2009) Facilitating Dynamics of Focus Group Interviews in East Asia: Evidence and Tools by Cross-Cultural Study. International Journal of Design; Apr2009, Vol. 3 Issue 1, p17–28, 12p.

Marchand, T.H.J (ed.)(2016) *Craftwork as problem solving – ethnographic studies of design in the making.* Oxford: Routledge,7.

Mortati, M and Villari, B. (2013) Crafting social innovators: Designing collaborative, participative, networked solutions in urban contexts. Craft Plus Design Enquiry; 2013, Vol. 5, p126-140, 16p.

Sanders, E. B-N, Stappers, P.J. (2008) Co-creation and the new landscapes of design, CoDesign, 4:1, 5–18, DOI: 10.1080/15710880701875068

Simon, H (1969) The science of the artificial. Cambridge, MA: MIT Press.

Taylor, A (2) Developing a Design Brief for a Virtual Hospice Using Design Tools and Methods: a preliminary exploration. Visible Language; Apr2015, Vol. 49 Issue 1/2, p96–111, 16p.

Thagard, P. (1988) Computational Philosophy of Science. Cambridge, MA: MIT Press).

*Research & Education in Design: People & Processes & Products & Philosophy – Almendra & Ferreira (eds)*
*© 2020 Taylor & Francis Group, London, ISBN 978-0-367-49296-0*

# A roadmap for a hot air balloon journey? A grounded design research approach

I. Veiga, P.C. Monteiro & J. Ferreira
*CIAUD, Lisbon School of Architecture, Universidade de Lisboa, Portugal*

ABSTRACT: Looking over the Design theses from the Lisbon School of Architecture's Ph.D. in Design course we noticed that the research design diagrams were similar regardless of the investigations were theoretical or practiced- based. The common template prescribes a sequence of steps that places the research process along a predetermined path in which the design project takes a secondary role. This observation raises questions concerning the research through design paradigm. Using a roadmap is a contradiction to the kinds of research that unfold through the practice of design, which are open-ended and exploratory in nature. The implications of having a template for all kinds of design research needs examination. In this paper, we build on this analysis and explore an alternative we termed Grounded Research Design.

## 1 INTRODUCTION

This paper emerged from an interrogation: having gone through 100 design theses from the Lisbon School of Architecture's PhD in Design course, we grew intrigued by how similar the research designs presented in the final manuscripts were. Every PhD is required to include a research design in the introductory chapter; the research design is a visual summary of the research project that accompanies the articulation of the research problem, research questions, objectives, hypotheses, methodology, and so on. Such a diagram is established in a very early stage of the research process, thus working — as it should — as roadmap for the work ahead.

Design is a young discipline with something unique to offer academia: being the happy encounter of creativity and rationality, pragmatism and utopia, problems and solutions, the discipline has its distinct things to know, ways of knowing them, and ways of finding out about them (Cross, 2007). Taking this into account, we expected to find a diverse array of approaches to design research (see Frayling [1993] and Stappers & Giaccardi [2017]) reflected in the manifold PhD thesis.

However, this was not the case, which was particularly intriguing when we looked into practice-based PhD thesis. Surely, we should find differences between the research design of a theoretical thesis and a practice-based PhD one?

Our perception is that some problems that arise along the research are the result of the fact that everyone is using a similar map, one that does not necessarily suit each particular journey, one that might not take into account every starting point, and that guides each research(er) along known routes to foreseeable destinations.

## 2 WHAT DOES A PRACTICE-BASED PHD LOOK LIKE?

Our initial drive was centred on comparing the research design of practice-based thesis with theoretical or empirically based ones. We expected to find fundamental differences between these two types of research in the respective research designs. Also, we expected the research design of practice-based theses to be different (perhaps even unique) between themselves, considering that there is still no consensus on what a PhD practice-based research in Design should be.

Yet, while differences may exist in the way the investigations actually unfolded, there is a repeating pattern and a common research approach that permeates most of thesis we analysed. As it stands, we found little difference in the research design of a theoretical thesis or a practice-based one, since the research approach that supports the different types of PhD investigation roughly follows a fairly similar road-map with these sequential steps: (1) define a research problem, (2) conduct a literature review, (3) establish a state of the art, (4) derive clearly defined research questions and hypotheses, (5) test the hypotheses, (6) draw conclusions. Sometimes, even, point 4 precedes points 2 and 3, operating as a goal to be achieved by design — to put it another way, the outcome of a design project is used as an instrument to answer a question.

This underlying logic is illustrated in research designs that adopt the following template:

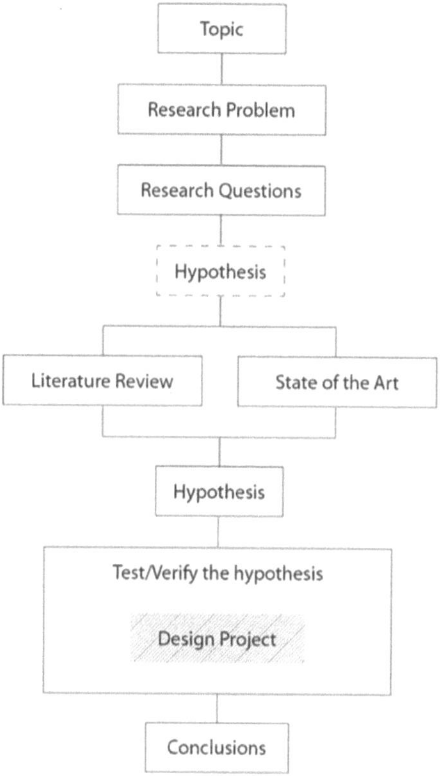


Figure 1. Research design template.

The adoption of a common template may or may not be the most adequate way to conduct a PhD investigation; that is, if the template is versatile it can accommodate different types of research projects and therefore be a suitable way to structure the work of researchers who have limited experience in academic investigations.

However, there should be a clear difference between the research design of a practice-based thesis and a conventional one: both have the aim of generating new and useful knowledge, but one proposes to do so through the practice of design while the other is a systematic and careful investigation about Design.

As it stands, we could not distinguish the research design of a theoretical thesis from a practice-based one. This situation is clear when we consider the place of the design project within the research design: more often than not, the design project is given equal treatment to every other method used during research; that is, the design project seems to be just another research method alongside a questionnaire, an interview, an expert panel, a case-study, and so on.

Furthermore, the majority of thesis places the design project towards the final stages of the research process. In some cases, the aim of the design project is to make a "pilot study" or "action research" to implement some concrete design product or process in reality. The design project is taken for a one-off experiment or prototype that operates as a validation tool to prove a hypothesis. In these cases, research results have already been anticipated and enclosed before the research process even began.

Consequently, we observed in most practice-based thesis a lack of open-endedness in terms of 'what' the research might generate since design practice is oriented towards convergence to a predefined design product. As it stands, in the design process developed during research there is no room for surprise and failure is simply not an option, since the outcome of the design project is to validate a hypothesis.

As such, the role of the design project within the research process is paradoxical. Notice how, on the one hand it is expected to be the *rosetta stone* of the whole research, that is, the designed product/prototype must answer a research question and verify a hypothesis, and failing to do so means the failure of the research endeavour; on the other hand, the project is but a side character in the cast of tasks, methods, and activities determined by the hierarchy of the research design.

While we agree that design research still needs to find a stable and to some degree consensual format to apply in practice-based PhDs, surely, whatever the format may be, the design project should occupy a more central place within the overall research design. If not, it will be difficult to argue that we are in the presence of a practice-based research in design.

A logico-deductive research model presupposes that we can identify a problem that is possible to define, detail, and delimit from the start, in other words, a problem whose boundaries are as well-known and unambiguous as a crossword puzzle. Further, it establishes and defines a linear sequence that follows the stages of problem-statement, literature review, state of the art, development of hypotheses, testing, validation, and conclusions.

Design problems, however, do not lend themselves to linear problem-solving and are seldom well-defined, thus, they contrast with goal-oriented problems where we come across a limited set of finite solutions. On the contrary, in design you must wrestle with a wide range of solutions that are good enough. Like putting together a lego-set in which the image on the box has faded and the instruction manual was lost, you build it up by making educated guesses about which pieces fit where, disassembling and assembling it again, all the while trying to discover what the finished product's final form might be; and so the designer advances by exploring local combinations and taking steps back to check if the big-picture makes sense.

If the design project is to have a more central role in practice-based research, we need to confront the issue that the design process does not follow a linear, sequential path with well-defined problems on one end and designed solutions on the opposite end[1]. How is this compatible with the kind of scientific requirements that prescribe a well-defined problem at the start from which questions and hypotheses are derived and tested?

## 3 WHAT SHOULD A PRACTICE-BASED PHD IN DESIGN LOOK LIKE?

When we talk about design problems, we are referring to the problems of design practice, which raises the question: to what extent is a design research problem similar to a typical

[1] In fact, design problems are often described as wicked (Buchanan 1992) or ill-defined (Rittel & Webber 1973), meaning problems that have no definitive formulation and require solutions that are neither completely true or false but rather better or worse. When designers explore a solution, they start to develop a different idea of what the problem actually is, which means that the solution starts to change as well and so on as the design process unfolds; in other words, in Design, problem and solution co-evolve (Dorst & Cross, 2001).

design problem encountered in (professional) practice? In our view, there should be a parallel between research through design and designing in general, that is, the act of designing should be the engine that propels a practice-based research project. In research through design, there should be no clear separation between designing and scholarly analysis; in other words, the act and experience of designing should include an analytical intention and the explicit aim of producing new and useful knowledge from the act of designing and its consequences.

Therefore, even if the problems of design research and the problems of a professional design project are not exactly the same, the mindset and actions inherent to designing could be carried into design research with benefit. So, to place designing at the centre of research through design carries a unique potential. Designers are trained to be comfortable with uncertainty; with no clear route between problem and solution, the designer explores alternatives without knowing exactly where the process is heading. Here we find a fortunate resemblance with an explorer setting out into unknown territory without a pre-established idea of where the exploration ends and what to find. In fact, the purpose of an exploration is precisely to discover something new, not to reach a predetermined place/conclusion which, what is more, is an impossible mission since the terrain is unknowable to begin with. An explorer may have the goal of drawing a map that is useful for others who wish to venture in the same terrain, but it is misguided and premature to determine, for instance, what kind of fauna and flora to gather before moving into the field.

To answer this question, we have to make a distinction between two things: a regular every day, professional design project and a research design project(s) developed for a PhD. Clearly, it is not the case that any design project (as is usually conducted in the professional domains) merits a PhD — it would be strange if it were so, since a design project does not start off with the goal of adding to the bulk of human knowledge since its goals are more likely to be found on the needs of clients and other stakeholders. A regular design project is a sequential process that unfolds towards the achievement of some concrete design product that materializes or responds to these goals. On the contrary, the focus of a design project developed for a PhD should not be located in an artefact that validates a hypothesis, but rather in a practice of designing that is conducted with the deliberate purpose of developing new knowledge that is useful for the whole community.

The cornerstone of this kind of approach to research (research through design) is found on the process first, with the knowledge-value of any resulting artefacts to be interpreted in light of the process that generated them[2]. When design research is not only about design but aims to be performed through the practice of design, the regular everyday professional design approach is insufficient.

A problem now emerges: How does one (candidate, supervisor, institution) convey what type of design project is admissible as a PhD research? This problem is crucial since research depends on the institutional support afforded, for instance, by grants; also, a PhD candidate goes through several rounds of evaluation in which juries determine if a research proposal should go through or not. Therefore, for a research through design methodology to be accepted as a valid form of research it must argue convincingly of its rigour and merits as a research proposal.

## 4 GROUNDED DESIGN RESEARCH

Although we suggest an *ungrounded* approach (in the sense of being open-ended and exploratory) we are aware of the contradiction of adapting a research approach that purports to be *grounded* — we do so because we are building on the Grounded Theory methodology. Be that as it may, the crucial point is that the notion of grounded we propose means that the design research is founded on design practice which is open-ended by nature; which means that we

[2] Notice how this is the opposite of what is usually observed in professional practice in which the artefacts are *everything*, and the design process left for scholars to uncover or for publications of anthologies in which it usually describes a mythologised version of what actually unfolded.

suggest to ground the research in the journey thus acknowledging the pivotal point of serendipity, accidents, even failure, and incorporating these into the research process.

Considering the distinctions made so far between the design project and the design research project, and the fact that it seems clear that design practice comes up rather late in the process/diagrams, it seemed only logical that the discussion of possible frameworks for design research might be ignited by turning the pattern upside down (literally). Thus, what would it mean for a design research process to start with the *conclusions* and the *design project* and everything else following from there until we reached a better comprehension of the research problem and topic in the end? What would happen if, instead of a validation tool, or just another research method among others, the design project is present not only from the outset but at the centre of the research? That is, what if the unfolding of the design process propels the research process itself?

Of course, to place the conclusions in the beginning is problematic, if not entirely bizarre; but, if we interpret *conclusions* not in a linear sense but instead as a broader understanding of a research problem, then flipping the research diagram 180 degrees becomes more viable and insightful. Usually, the conclusions to a PhD thesis include a general discussion of what the insights and findings mean in relation to a broader context that includes previous studies done by others, the ongoing discussion(s) within the discipline, and how we can generalise from the specific results generated by the research[3].

So, taking *conclusions* here to mean a wide-perspective view of the research setting that is placed at the beginning of the research process we find a parallel to the approaches usually adopted by some qualitative research approaches, we are referring in particular to Grounded theory[4].

Grounded Theory is a qualitative research approach founded on the construction of theories through the gathering and analysis of empirical data. It is different from the traditional model of research, where the researcher chooses or constructs a theoretical framework as well as hypotheses to test first, and only after that collects data to validate them. On the contrary, the grounded theory (GT) paradigm of research proposes a move from data to theory (and not the other way around) so that new understandings can emerge and be articulated as theories that fit the context that sustain it. The contextual grounding of GT makes it potentially interesting for design research since design deals with specific situations with particular stakeholders and other variables that render each design situation, to some degree, unique.

The basic underlying mechanism of GT is that a study may begin with a question, or sometimes with the collection of qualitative data that refers to an area of concern or a topic of interest; in fact, the original proponents of GT (Glaser & Strauss, 2006) report that their first study was set off by a common interest in studying hospital wards, their study later evolved into a theory that describes and explains how the *awareness of dying* impacts hospital staff, patient, and family members involved. The Grounded Theory Method was born from the authors' reflection on how this study unfolded. As researchers examine the data repeated ideas, concepts or other elements emerge as patterns that are analysed, interpreted, and categorised. These patterns then suggest further data to be collected and so the studies proceed. From the joint data gathering and analysis the patterns fall into categories that describe similar phenomena, then categories become concepts which are connected and form the basis for a new theory.

In short, the researcher approaches a central issue, problem, or specific context with an open mind guided only by broad questions. Then, through examination of data, in whatever form, the aim is to develop pertinent abstractions that describe and explain significant aspects of the object of study.

Such an overarching research approach could serve as an umbrella concept to frame design research, both practice-based and the more traditionally theoretical kind. A paradigm

[3] Granted, the conclusions also include recommendations for future studies or other applications deriving from the thesis, for the sake of argument, we will leave these out of the broader interpretation of *conclusions* we are conducting here.

[4] Grounded Theory (GT) was first introduced by Glaser and Strauss (2006), Stauss and Corbyn later developed variations of the original method (2008), and Bryant (2017) recently presented a review on the different approaches to GT and elaborated on a pragmatic version of this research approach.

endorsing that design research should be *grounded* in the observation and analysis of specific contexts and empirical data. In other words, a *Grounded Design Research* approach as a paradigm for scholarly investigation.

Notice how both theoretical and practice-based studies can share the procedure of moving from empirical data to higher-level abstraction; in the case of research through design this is accomplished with design experiments, while in theoretical investigations the same procedure is set in motion with empirical studies and observations. The table below captures and summarises what we mean.

Table 1. Grounded Design Research.

| Grounded Design Theory | Grounded Design Practice |
|---|---|
| Design studies that follow a grounded theory approach (adapted to design) to generate new theory from empirical data. | A research through design methodology that applies design experiments framed by a design program with the aim of contributing to the development of design knowledge. |

So, on one side we have Grounded Design Theory, which would correspond to a more traditional research effort, that is, a sequence of studies that investigate a topic and generate new theory; while on the other side we find Grounded Design Practice, which is a specific research through design methodology we will detail below. Both types of research have the common aim of producing new design knowledge, but one follows a progressive move from observation and analysis of empirical data towards higher-level abstraction (new theory) while the other builds on design experiments to explore a design program that answers a research setting.

### 4.1 *The dynamic of program and experiments*

In this section we will elaborate and describe what we mean with Grounded Design Practice. What we propose is a research approach that can take advantage of designerly ways of knowing, while being compatible with both *designing* as usually found in professional design practice as well as with other types of scholarly practice.

At its core, the approach establishes that 'what to research' is different from 'how to research'. The 'what to research' is not a normative set of items to be accomplished, instead it serves as a guideline of possibilities for investigation that may grow differently through 'how' these possibilities are explored; therefore, through design practice becoming more elaborate, unique, and concrete.

The Grounded Design Practice research approach is a research through design methodology anchored in repeated iterative movements between a research program and several design experiments in which design is not a practice that results in a single coherent design product (the program is not a design brief for action). Instead, designing takes the form of short design experiments that explore the space opened by the research program with the aim of discovering insights that contribute to the development of design knowledge and the clarification of the research setting.

The model adopts concepts such as *experiment* and *laboratory* from the sciences to develop a vocabulary for design research methodology that captures the open-mindedness of design inquiry, yet maintains an ambition of arguable rigor and reasoning. Further, the term *program* is also drawn from the traditional practices of design and architecture (where it means a formal and concrete set of constraints that the project must answer); Brandt, Redström, Eriksen, and Binder (2011)[5] argue about the dependency on framing a *program* that forms the

[5] This model builds on the proposals of the XLAB design research laboratory from the Design School of the Royal Danish Academy, see (Brandt et al, 2011) for more details on the work developed there.

basis of research to explore what can be achieved through design practice, thereby expressing the possibilities of the *program* i.e. topic/field or challenge to work as a departure point for design *experiments* and evolved/grow through them as a thesis in construction.

The concepts of *setting*, *program*, and *experiments* determine a framework to develop and propose research challenges, and guide the journey of a design research. In what follows we will describe each element and how they work together.

### 4.1.1 *Research setting*

The setting clarifies what the contribution of the research project is. Drawing inspiration from framing the *conclusions* at the beginning of research, the setting can include a broad research question, broad aims and objectives, but these are not normative, that is, they will change and evolve as the research through design unfolds. Nevertheless, the setting defines the research theme, topic, area of concern, or problem, and presents a tentative state of art (including a review of the ongoing discourse of the discipline, a review of what is known and previous studies on the same topic).

All the elements enumerated above, configure the research setting as a wide-perspective view of why the research is relevant; however they do not determine how the research process will unfold, instead, they set the stage for the concrete design experiments that will function as the laboratorial scenario where the research unfolds in practice.

The main thrust of the research process is found in the dynamic between program and experiments, but the setting connects the research results with a wider perspective. So, the research process requires a constant return to the research setting to relate the results and insights of the experiments with a broader context. This is achieved by posing questions such as: *what does it mean? Why is this important?* Or *so what?* Such an effort cannot be left to the end of the research process, disconnected from the iterative dialectic of program and experiments, but rather defined into the methodology from the start, so that with each experiment we clarify both the program and the research setting itself. To put it another way, the setting ensures that research results are not fragmented and unfocused efforts but integrated into the wider aim of producing new knowledge that is relevant.

### 4.1.2 *Program*

It is important to clarify the distinction between the program in design research compared to the program in professional design work. Fundamentally, both the aims and outcomes are different. In design work the intended outcome is the finished product (artefact) that fulfils a brief, while in design research the intended outcome is knowledge. Therefore, the crucial difference is that in design practice the aim is to create a finished product (an artefact of any kind) that answers a brief. Whereas, in design research the outcome is knowledge, that is, the goal is to contribute to the advancement of human-knowledge.

The program should include a critical question or a statement about the present (how things are) and a suggestion(s) about alternative ways of doing things (how things could be). For instance:

Explore a new set of perspectives and possibilities on energy consumption in everyday life, by adopting a concept of energy that expands its technological domain to include its aesthetical, functional, material, interactive or other dimensions. (Adapted from Brandt et al [2011])

A design research program should also contain a set of concise statements that open up a research space for exploration, interpretation, analysis, reflection, or in a more practical sense, a frame for how the design experiments may unfold, for instance, here is an example of a program: The aesthetics of energy as material in design: working with energy not only from a technical but also from an aesthetic point of view; reflective use by systematically reinterpreting designed things not only in terms of utility and ease of use but in terms of critical reflection through the things at hand.

As we can see, a program establishes contextual and provisional frames of reference. That is, the program is not unquestionably established beforehand, but rather functions as a hypothetical perspective that frames a situation and opens the possibility of exploration and experimentation. Like all design framing (Schön, 1984; Dorst, 2011) it opens up the space to explore how things could be, which incidentally is the ethos of Design. As the research process unfolds (through cyclical experimentation within the frame set up by the program) the results will substantiate or challenge the perspective set up by the program.

4.1.3 *Design experiments*
Brandt and her colleagues define experiment as the "exploratory probes into what the program may entail and how it can be expanded and sharpened to account for how the experiments unfold." (2011, p.19)

A design experiment is not a short design project with a clear beginning middle and an artefact in the end, but rather a concise designing experiment framed by the research program. The program frames the boundaries of the experiment and in turn the results of the experiment feedback into the program, sharpening it. Thus, the design experiments do not refer to a single activity but several practical, concise and specific possible materializations of the design program, thus, each one enacts not a complete but different or fragmented parts of a whole design project. In other words, while one experiment may entail the making of a concrete prototype, another might – for instance – consist of several outputs from a short co-design experiment that explores the same topic.

Design experiments are then short laboratorial experiences, open-ended inquiries into particular aspects suggested by the program and made relevant by the framing of the research setting.

4.1.4 *The setting–program–experiments triad in action*
The diagram below illustrates how the different elements work together.

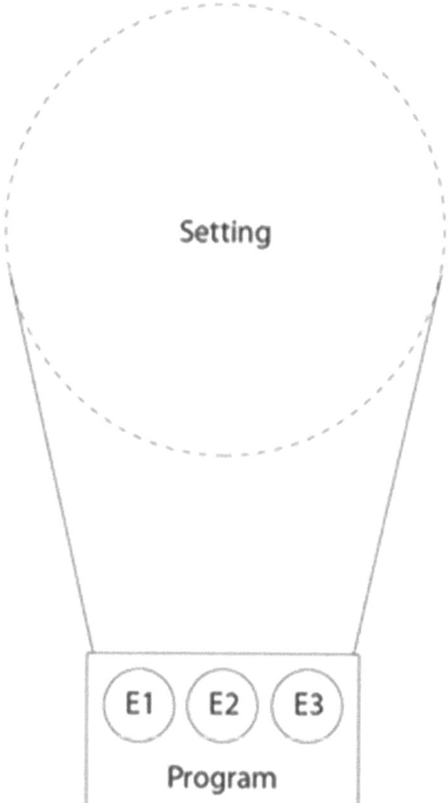


Figure 2. The dynamic of setting, program, and experiments.

Program and experiments do not exist in a hierarchical format, rather, it is the combination of both that explores the problematic raised by the research setting. In other words, the experiments help to make sense of the program, since with each design experiment the researcher revises and sharpens the program's statements, and a similar iteration then occurs between the synthesis of the program/experiments dynamic and the research setting.

In short, each experiment clarifies the research program, which in turn allows for further, more focused, cumulative or growing experimentation; at the same time the researcher relates whatever insights emerge from the experiments with the research setting, so that the exploration of the program through concrete design experimentation is related to higher-level abstraction and knowledge generation.

## 5 CONCLUSION

Before we conclude it is important to acknowledge a few important limitations concerning this paper. First, the *grounded design research* paradigm may not fit every object of study or research problem. Some areas of design research have been more thoroughly studied than others, and in some cases a deductive approach may be entirely adequate. Further, it is possible that in some circumstances, a single completed design project may indeed explore a research question and contribute to advance design knowledge in general. Finally, the research approach we propose should be read as a broad proposal emerging from a specific study and intended to encourage reflection and dialogue, therefore, it is not a tried and tested methodology but guidelines to discuss.

To conclude, this paper drew on the identification of a particular pattern for design research which needs to evolve. In our view, a prescriptive model of research is inadequate as a template for practice-based design research. To approach design research in a deterministic and mechanistic way is a contradiction with the open-ended nature of designing.

From our reflection, we outlined a common research approach for theoretical and practical design research called *grounded design research* which purports to build investigations from empirical, observational, and experimental actions grounded on concrete experience towards higher-level abstractions and knowledge production.

Knowledge grows in the confrontation with the unknown, specifically in Design it unfolds in the tension between what things are and what things could be.

## ACKNOWLEDGMENTS

The authors would like to thank Prof. Rita Almendra (coordinator of the REDES – Research & Education in Design research group) for developing, promoting, and coordinating the research project that allowed us to conduct the analysis that supports this paper.

## REFERENCES

Brandt, E., Redström, J., Eriksen, M., & Binder, T. (2011). *XLAB*. Copenhagem: The Danish Design School Press.

Bryant, A. (2017). *Grounded Theory and Grounded Theorizing: Pragmatism in Research Practice*. New York: Oxford University Press.

Buchanan, R. (1992). Wicked problems in design thinking. *Design Issues*, *8*(2), 5–21.

Corbin, J., & Strauss, A. (2008). *Basics of Qualitative Research: Techniques and Procedures for Developing Grounded Theory* (3rd ed.). Thousand Oaks (CA): Sage Publications Ltd.

Cross, N. (2007). *Designerly Ways of Knowing*. Basel: Birkhäuser Verlag AG.

Dorst, K. (2011). The core of "design thinking" and its application. *Design Studies*, *32*(6), 521–532.

Dorst, K., & Cross, N. (2001). Creativity in the design process: co-evolution of problem–solution. *Design Studies*, *22*(5), 425–437.

Frayling, C. (1993). Research in art and design. In *Royal College of Art Research Papers (Vol. 1) Research Papers (Vol. 1)*. London: Royal College of Art.
Glaser, B., & Strauss, A. (2006). *The Discovery of Grounded Theory: Strategies for Qualitative Research*. London: Transaction Publishers.
Rittel, H., & Webber, M. (1973). Dilemmas in a general theory of planning. *Policy Sciences*, *4*(2), 155–169.
Schön, D. (1984). Problems, frames and perspectives on designing. *Design Studies*, *5*(3), 132–136.
Stappers, P., & Giaccardi, E. (2017). *Research through Design*. Delft: Delft University of Technology.

*Research & Education in Design: People & Processes & Products & Philosophy – Almendra & Ferreira (eds)*
*© 2020 Taylor & Francis Group, London, ISBN 978-0-367-49296-0*

# On the brink of dissipation: The reactivation of narrative heritage and material craftsmanship through design research

H. Alvelos, S. Barreto, A. Chatterjee & E. Penedos-Santiago
*ID+/Unexpected Media Lab, Universidade do Porto, Portugal*

ABSTRACT: This paper introduces two ongoing research projects based at the Research Institute for Design, Media and Culture (ID+), pertaining to the preservation of specialised knowledge on the brink of disappearance:

Anti-Amnesia focuses on the recovery and documentation of human, social, and material narratives surrounding four dissipating traditional industrial contexts in Portugal;

Wisdom Transfer establishes a groundwork for acknowledging, communicating, and activating the contributions to knowledge, culture, and society made by retired Portuguese art and design academics.

Both approaches are human-centric, relying on ethnography to gather the essence of the respective realms — foremost from the point of view of the original masters and makers — to arrive at a position from where it is possible to duly appraise the involved historical and cultural legacies. The convergence of learnings forms the basis for substantiating the forming of a European-level consortium of academic institutions dedicated to legitimizing empirical and practice-led knowledge in design and media research.

## 1 INTRODUCTION

The following paper discusses the strategic convergence of two ongoing research projects based at ID+, Research Institute for Design, Media and Culture/Unexpected Media Lab (LUME) that seek to sustain and inscribe specific streams of specialised traditional knowledge:

Anti-Amnesia (AA): Design research as an agent for narrative and material regeneration and reinvention of vanishing Portuguese manufacturing cultures and techniques. The project focuses on the recovery and documentation of human, social, and material narratives surrounding four dissipating traditional industrial contexts in Portugal. It seeks to utilize the retrieved knowledge towards building a scholarly infrastructure that can aid in the articulation of design-based reanimation strategies.

Wisdom Transfer (WT): Towards the scientific inscription of individual legacies in contexts of retirement from art and design higher education and research. The project aims to establish the groundwork for acknowledging, communicating, and activating the contributions to knowledge, culture, and the social fabric made by retired Portuguese art and design academics. It seeks to recuperate elements of associated wisdom that may not have been scientifically inscribed during the academicians' time of active service, and explores further contexts of resonance and re-applicability.

Despite their individual vocations and research objectives, a series of connecting threads provide both a broader view of a shared mission between the two interventions, and strengthen LUME's approach to attaining its overarching objectives. A common ground of argument sustaining both projects pertains to the preservation of specialised knowledge at risk of disappearance. With this in mind, it is worth noting that:

Both projects are based on the observance and interpretation of knowledge connected to particular communities, as well as the potential recovery and development of further community related contexts;

Recovery, sustenance, and reapplication of traditional learning notably in terms of tacit knowledge and expertise may be largely regarded as the driving force in both instances;

Both projects contribute to the discourse on contemporaneity vs. convention by addressing the notion of value persisting in analogue heritage and practices in relation to modern-day digital alternatives.

Both projects are in the process of establishing collaborative ties with national and international higher education institutions. The corresponding intention is to develop research related reciprocally advantageous pedagogical content for students of art and design studies at undergraduate, postgraduate and doctoral levels.

The paper accordingly presents an individual summary of each project's premises, followed by a report of research actions, both concluded and in progress. Further on, the methodological framework is outlined, followed by an overall conclusion which includes a set of recommendations for further implementation, within the scope of the respective projects, and as reference material for related parallel/future interventions.

The resulting inscriptions and outputs are meant to showcase the persevering agency and relevance to contemporaneity that is encapsulated within these antiquated knowledge systems, and to encourage further discourse and action towards safeguarding the underlying structures and embedded forms of cultural, artistic, and intellectual wealth. Accordingly, both projects argue that active pedagogy is the primary vehicle for a functional perpetuation of the legacies in question, and it can be further aided by ensuring a sustained dialogue with creative entrepreneurship. A crucial aspect of the intervention is its location within a business incubator: The Creative Industries Incubator of the University of Porto's Science and Technology Park (UPTEC). The nature of either project and the expected outcomes are consistent with UPTEC's mission of sharing multidisciplinary knowledge and expertise towards generating value in and through creative industries. This placement, thereby, provides unique opportunities for synergy between project partners, higher education students and institutions, and start-up culture, ensuring compatibility with contemporary challenges and fostering prospects of mutually advantageous knowledge and technology transfer.

### 1.1 *Anti-Amnesia*

The project 'Anti-Amnesia' is a design research and mediation process dedicated to the recovery and sustenance of traditional knowledge systems that are embedded in four typical industrial practices of Portugal — handloom weaving, shoemaking, Azulejos tilework, and letterpress typography. The research focuses on recovering and interpreting elements of the related identities, customs, knowledge and material culture, towards developing a scholarly infrastructure that can help endure their tangible and intangible value to society and culture. Actions accordingly comprise of ethnographic documentation, archiving and interpretation in art and design, alongside pedagogical interventions that are intended to address tactical objectives connected with industrial and cultural heritage preservation.

Project 'Anti-Amnesia', in its purview as a design research and mediation initiative, is dedicated to the recovery and restoration of unique forms of traditional knowledge associated with craft and small-scale industrial practices. The project is based on growing concerns over the dissipation of traditional craft and manufacturing practices in North and Central Portugal (Gomes, 2016; de Almeida and Chatterjee, 2016; Albino, 2017) due to new global commercial and industrial realities that are rendering older techniques and technologies obsolete. The project takes into account four such typical industries from the said regions, namely, handloom weaving, shoemaking, Azulejos tilework, and letterpress typography, all of which were prevalent in past eras but whose present-day viability is increasingly being called into question. Anti-Amnesia's research and intervention prerogative in this regard is to observe, retrieve, and interpret elements and aspects related to these making cultures by conducting ethnographic studies, documenting and archiving testimonies, material culture and processes, and exploring further application avenues through design and design pedagogy.

Figure 1. Tipografia Damasceno.

1. The project's directive towards the revalorisation of endogenous making is its constructive response to the uniformity of present-day globalization. The issue is timely and pressing, and among the cultural concerns being discussed at the forefront of European and international research projects, design philosophies and practices (Clarke, 2011; Manzini, 2015; Verganti, 2015; Morrison, 2017), and finds its own vivid examples in contemporary Portugal through instances such as the practices presented above.
2. The research comprehends that factors such as the emergence of radical new paradigms, dictated obsolescence, and accelerated shifts in consumer preferences are presenting untenable terms of trade to traditional makers, as a consequence of which the associated practice communities are facing severe disadvantages in market/maker retention. The ensuing decline is marked with depopulation in the traditional bastions, and many such crafts are striving to endure their practice beyond the prevailing generation of active practitioners.
3. A connected aspect of the decline is that the human, social, and material narratives surrounding the crafts are yet to be comprehensively inscribed in public and academic discourses. Such inscription can become a heritage asset of particular importance during the imminent era of knowledge economy, in contrast to its relative objectivity in recent decades. Thus, an urgent and structured recovery of material, processual, human, and social narratives, thus gains special significance, since, in the event of dissipation, the specialised knowledge that is embedded in these narratives, and which may have evolved over centuries, is otherwise perceivably at a risk of permanent loss.
4. Anti-Amnesia has accordingly identified key representative entities as partners within each of the four case studies towards gaining emic perspectives and to collaboratively articulating strategies for recovery and reanimation. These key partners comprise:
5. A community of weavers pertaining to a unique and historically significant handloom technique from the region of Coimbra: Almalaguês.
6. A typography related initiative based in a traditional letterpress studio that encourages original perspectives on the age-old art form with a view to derive unconventional forms and charter new territories: Clube dos Tipos/Tipografia Damasceno.
7. A family-run footwear factory based in the industrial town of São João de Madeira that is presently relatively stable but is witnessing an increasing rate of disinterest from younger age-groups towards shoemaking as a career option: Netos.
8. A collective that seeks to recover, sustain and promote the legacy of traditional Azulejos tilework which, despite its significance to Portugal's architectural heritage, is facing an onslaught of theft, insensitivity and invalidation: Azulejos do Porto.

Figure 2. Azulejos do Porto.

These entities individually exhibit varying degrees of technological obsolescence; however, they also represent entities that foster deep-rooted interest in mitigating the prevailing circumstances of economic and cultural volatility. By securing their collaboration, the project ensures ideal conditions for studying and documenting elements of the connected material culture and recovering the embedded human and social narratives through ethnographic means.

The research's adopted methodological framework is underpinned by well-established ethnographic methods from the social sciences (Pink, 2007; Rose, 2011) that prove highly effective when applied in articulation with intuitive methods of design (de Almeida, 2012). In this regard, the research team works closely with its partnering community of makers to uncover and document elements that can potentially extend their avenues of value generation. These elements include:

personal testimonies, stories, and anecdotes;
audio-visual documentation of related folkloric customs and traditions;
audio-visual documentation of materials and processes;
visual documentation of products, designs, and patterns;
specimens of products, swatches, and samples;
specimens of contextual externalities, including residual material.

The obtained source materials are correspondingly archived, and form the evidential base for further analysis into different forms of embedded value, and correspondingly, certain aspects are interpreted in art and design contexts as exemplars for future endeavours.

With respect to exercising active pedagogy as a strategic means, the project ensures an ongoing reversion of its outcomes into multiple contexts of related socio-cultural appropriation through a "build-measure-learn" loop, a significant extent of which is attained by the means of curricular participation from graphic and multimedia design students. The student participants engage directly with the project's objectives through tactical workshops and project-focused coursework, including specific theme-oriented modules and dissertations. The research fronts that are engaged with such curricular interpolation include:

- ethnographic engagement with the practice communities;
- conceptualization of new products, material applications, and packaging design approaches;
- interpreting the contextual *leitmotif* through art.

This aspect of pedagogical interpolation has allowed the project to establish a methodological approach that effectively extends the scope of design education through introducing and integrating a synchronisation of contemporary and conventional knowledge, tools, and practices towards real-life application scenarios. Here, the project takes scholastic pre-requirements and the prospects of reciprocity into careful consideration before engaging student participants. Product prototypes (footwear, packaging material, tiles), public and academic dissemination, workshops and exhibitions, and online communication have constituted the project's primary outputs.

Figure 3. Leftover from Netos shoe production.

Correspondingly, the following section on the parallel research intervention Wisdom Transfer shares much of the ethos of Anti-Amnesia, while maintaining its own scientific path and objectives. An aspect to be emphasised here is that certain redundancies exist in the respective teams of researchers associated with the projects, wherein there are members who represent both, albeit in different capacities. This facilitates the channelling of relevant information, exchange of ideas pertaining to research approaches, and ensures a sustained dialogue between either intervention towards extending individual scope.

### 1.2 *Wisdom transfer*

The design research project Wisdom Transfer bases its intervention strategy on addressing two questions regarding the preservation and perpetuation of traditional knowledge:

- How to scientifically legitimize, activate and communicate the empirical and narrative legacies of knowledge, skills and experience of an older and retiring generation of art and design researchers and academics?

  And accordingly:

- How can the incorporation of such disciplinary wisdom inform and apply to other scientific disciplines with analogous challenges and opportunities?

Wisdom Transfer acknowledges the insufficient inscription and reapplication of individual knowledge and experience of ageing and retired art and design professors and researchers in Portugal. It hypothesizes that the legitimization of practice-led wisdom in art and design research can greatly contribute to its scope and depth as a discipline, and inform its role as a multi-disciplinary interface of mediation in other intervention scenarios.

The research includes an analysis of the life and work of a group of artists who graduated from the School of Fine Arts (ESBAP) in Porto, Portugal, between 1957 to 1986, a time span containing the revolution of 25 April, 1974 as its epicentre. The project has also established collaborative connections internationally with various past and present academics affiliated with art and design related institutions, towards testing and validating research actions, gathering multidisciplinary and multicultural inputs, and procuring opportunities for further contextualisation of the research outcomes.

Considering that art and design research have been validated as scientific disciplines on a relatively recent time span, it can be argued that the available scientific heritage precedes the formalization of the disciplines themselves - and consequently resides in an older generation of protagonists who have pioneered the transition of creative activity from professional practice to academia.

However, given its fundamentally empirical nature, this heritage has remained abstracted from curricular content in present art and design higher education. This may be attributed to the lack of a reference framework that endorses and potentiates the value of earlier-era professorial experience and testimonies in the development of curricular and project-based content.

Thus, Wisdom Transfer is working towards establishing a groundwork for bringing a paradigm shift in the acknowledgment, communication and activation of relevant contributions to knowledge, culture, and the social fabric that art and design academics have made, including the tacit forms of knowledge and impromptu application scenarios that have not been previously recorded or reverted into scientific knowledge during their careers. In gathering such instances through personal testimonies, and inscribing them in public and academic discourse and thought, Wisdom Transfer seeks to further contexts of resonance and applicability.

According to a United Nations world census (2015), people over the age of 65 are the fastest growing demographic section and these numbers are expected to grow exponentially in the foreseeable future. There are a number of initiatives that seek to identify and address challenges associated with the designated "silver economy", operating in transgenerational inheritance. However, in Art and Design Academia, the production and dissemination of the inherited knowledge is paradoxically less evident.

Concerns regarding the silver economy can be prorated into two main sets of consideration. On the one hand, directives such as promoting healthy living, and ensuring productivity and meaning in order for the aging population to remain engaged in the community and maintain social well-being (White House Conference on Ageing, 2015) help orient interventions pertaining to active aging. On the other hand, safeguarding the knowledge and experience that older generations have created during their lifetime, towards developing pathways for future effectuation is also a prerogative of the present, however, the activities under the former are more visible due to their more direct economic impact.

The above concerns, in the context of this project, are appurtenant, but auxiliary, since the retired artists mostly remain active and persevere with exercising their creative drive despite health issues and sporadic contact with their respective art institutions.

In case of the latter concern, the evidence is less visible. In this regard, Leonard et al (2014), give an example of a noted business enterprise which reported that their next anticipated wave of almost 700 retirements would mean the loss of over 27,000 years of experience, and that there would be critical losses in four areas in particular — relationships, reputation, re-work, and regeneration. The interview sessions offer the project an opportunity to collect individual testimonies and anecdotes that are rich in insights pertaining to these areas, and in collating such information, the project is seeking to sustain not only specialised knowledge, but also valuable ethnographic information with respect to localised creative cultures from a preceding era and their consociation with concurrent happenings globally.

It is essential to acknowledge here that in the Portuguese context, the Foundation of Science and Technology (FCT) has itself recognised the issue of such non-inscription, as the Archive of Science and Technology (ACT) attests. FCT has been consolidating ACT through public contributions since 2011. The underlying objective, shared by Wisdom Transfer albeit on a distinct scale and degree of specialisation, is to preserve and provide a greater degree of access to intellectual heritage for researchers and the general public.

However, contributions from Art and Design research, in consequence of the fields' late legitimation as scientific disciplines, are yet relatively modest. It is therefore imperative for the project to create and sustain a contributive relationship with FCT's ACT. This purview is also reflected upon by Hartman (2009), who suggests that:

> *"Universities should do everything they possibly can to retain the intellectual capital of their retired faculty, maintain the products of their faculty's research careers (possibly in institutional repositories), and obtain as much new knowledge as they can from the continued professional activities of emeriti faculty." (Hartman, 2009)*

The project also recognizes the urgency of certain time related factors. In consideration of the advanced age of most of the retired academicians, whose personal testimonies are critically relevant to the project, the present time is highly pertinent to set the deliberated actions into motion, in order to be able to document and legitimise the gathered inputs towards creating a scholarly infrastructure in recognition of their contributions.

For developing such infrastructure, the project is undertaking a number of complementary actions:

- Aggregating individual creative knowledge and experience that is reproducible and replicable as generative resources;
- Developing projects that validate a diverse set of contributively scientific and socio-cultural reapplication scenarios;
- Developing communication templates for best practices in art and design education that can aid in perpetuating the associated learnings and provide reference of contributions to present and future generations;
- Developing means to raise a multidisciplinary awareness of "emeritus wisdom" beyond the proposed scope of art and design.

An additional objective of Wisdom Transfer is to develop a valuation system that can help appraise such scholarly wealth in conjunction with the evolving requirements of the academic and scientific community. In this respect, the project is seeking to consolidate upon Hartman's (2009) observations:

> *"The intellectual capital gained from professional activities in an academic discipline is crucial knowledge for junior faculty in that same discipline to acquire, information that more experienced—if sometimes officially retired—faculty are in a unique position to share in formal or informal mentoring relationships." (Hartman, 2009)*

Direct and indirect methods of engagement and observation, such as interviews with open-ended questions (Quivy & Campnhoudt 2008, p.164) were used as methodological tools.

In relation to the interview sessions that have been conducted, the initial set of interviewees was identified by the research team upon discussion, however, subsequent snowball sampling lead to an extension of the number of participants. Interestingly, contemporary social media platforms also played a pivotal role in raising the sample, wherein, the project found a suitable premise for promotion and validation, and which lead to a greater number of potential interviewees being more open to collaborate since they were already in the know regarding the project's outreach with their peers and contemporaries online.

Figure 4. Interview with António Quadros Ferreira.

Figure 5. Interview with Carlos Barreira.

Consequently, a diversified methodology has been applied via square reasoning, incorporating a study of oral histories, in-situ observation, analysis of records and documents, and visual analysis of images and artefacts. The project deems visual anthropology, in particular, as central to the collection, analysis and creation of knowledge, in conjunction with Collier's (2001) perspective of visual analysis being both an art and a science, a view that is also shared by Kemp (1990). Both references find resemblances in their respective epistemological expectations that are placed upon visual lexica and semantic relevance.

The applicability of this research is tied to specificities of art and design research, as well as the understanding that these may open up avenues to other disciplines:

As recent scientific fields, art and design research are in the process of stabilising their own mission and repertoire; dynamics exist between creative practice and research in the hands of a generation who led this transition, and these dynamics may have developed research before it was taken as such.

This transitional generation of art and design practitioners-academics are reaching the end of their professional and academic paths; their knowledge and testimonies need to be gathered and incorporated in the present times, or the discipline risks irreversibly losing this heritage.

The research also recognises that similar processes may occur outside of the radar of other scientific disciplines; as such, it disseminates the content, methods and outcomes in multidisciplinary contexts that may find them suitable to the furthering of other disciplines.

With this in mind, the project has adopted a convergence between ethnography, taxonomic systems, and communication of competences. It began by surveying the proposed scope of action and respective actors, and subsequently proceeded to mapping the gathered knowledge into coherent readings, and is presently testing these outlines with the respective original actors. From this joint validation, a series of communication templates, developed on the outcome of the interviews conducted, are being produced, with the aim of integrating into scientific and pedagogical contexts of art and design research.

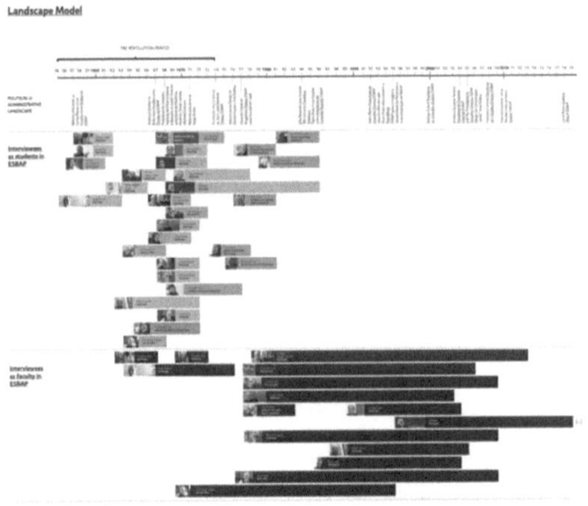


Figure 6. Timeline with our interviewees as students and lecturers at the School of Fine Artes in Porto.

A connected set of research activities has been conducted in order to ensure further content and structural resonance through pedagogic engagement and multi-disciplinary knowledge transfer.

Several workshops have been developed with undergraduate level students of art and design from different universities, wherein the students engaged directly with the interviewees. These ongoing pedagogical interventions cover different approaches: portrait illustration, graphic novels, typography/calligraphy posters and procedural programming, towards pursuing the graphic, typographic, and algorithmic translation of intellectual heritage into visual artefacts.

Figure 7. Workshop developed under the supervision of Prof. Rui Vitorino Santos at FBAUP.

Transdisciplinarity and intergenerational dialogue are axes of extreme relevance, curricularly legitimizing these initiatives as a key point to our research. The corresponding outcomes have been disseminated on a regular basis through conference and journal publications, events and exhibitions, and on social media. One of the most successful examples of this curricular reverse results from the curatorship of the exhibition "You Look Familiar", launched on October 25, 2019, in which retired artists and young students engaged as a consequence of a workshop carried out in the context of illustration.

Figure 8. Exhibition "You Look Familiar" – opening day.

Figure 9. Exhibition "You Look Familiar", Ana Campos and Elvira Leite – opening day.

A set of five seminars has also been conducted where in the project invited artists to share their insights and experiences with the general public. The seminars brought the artists in conversation with each other, thereby providing new themes and narratives.

Figure 10. Mais do Que a Soma Seminar nº4 with Elvira Leite, José Paiva, Isabel Cabral + Rodrigo Cabral at Galeria Serpente.

Figure 11. Mais do Que a Soma Seminar nº5 with Ana Campos, João Nunes e Paula Soares at Reitoria da Universidade do Porto.

The project also highlights the following results and outcomes that are in course of development under its ambit:

The development of a dedicated online platform, an audio-visual documentary archive, towards preserving personal testimonies for future recontextualization.

Publishing generated results, approaches, and events on social network platforms, in consideration of their critical value towards the project's communication, with special focus on the immediate reach and intersection between the different communities involved.

A book publication that compiles the project´s research field work: photographs, interviews excerpts, published papers, scientific communications, activities and outcomes.

## 2 CONCLUSION

Despite their specificities in scope and outreach, both projects share fundamental congruence in terms of the recovery and regeneration of specialised knowledge on the brink of dissipation. It is pertinent to note that both approaches are human-centric, relying primarily on ethnographic practices to gather the essence of the respective realms — foremost from the point of view of the original masters and makers — to arrive at a position from where it is possible to truly understand their legacy and acknowledge the associated sensibilities before introducing any design or pedagogy related aspects that may be otherwise considered insensitive, assumed or invasive.

The consequent inscriptions and outputs under the respective circumstances speak as much for the persisting agency within these knowledge systems as they do for encouraging and opening further discourse and action towards safeguarding the underlying structures. With this in mind, we argue that the primary vehicle for a functional perpetuation of the legacies in question is through the channels offered by higher education and business incubation.

The projects additionally benefit from the allocation framework of central coordination, whereby the respective Principal Investigators (PI) also represent the other project as co-PIs, while parallelly maintaining individual teams of investigation. As a result, a fertile cross-referencing of approaches, methods and progress is made possible on a consistent basis. An additional aspect is the aggregation of a diverse range of skills and disciplines among the associated members, representing five higher education institutions.

Furthermore, the projects are attending, integrating and fostering international collaborations that share aspects of the identified issues and proposed missions - as the identified dynamics between heritage and modernity point towards a broader pattern. Straightforward replication of concepts and contexts is evidently neither ensured nor desired, as the issues are inherently contextual; however, the overall underlying premises of both projects have garnered sufficient interest from an international perspective. Thus, the research team, through its approach to potential partners, is scaling the concerns on to a European level towards constituting a network of art and design institutions. The Wisdom Transfer Network will be aimed at the legitimisation of empirical and practice-led knowledge and heritage in design and media research.

Figure 12. Almalaguês weaving.


## ACKNOWLEDGEMENTS

The project *Anti-Amnesia* (POCI-01-0145-FEDER-029022) is supported by Competitiveness and Internationalisation Operational Programme (POCI), under the PORTUGAL 2020 Partnership Agreement, through the European Regional Development Fund (ERDF) and through national funds by the FCT – Fundação para a Ciência e a Tecnologia.

The project *Wisdom Transfer* (POCI-01-0145-FEDER-029038) is supported by Competitiveness and Internationalisation Operational Programme (POCI), under the PORTUGAL 2020 Partnership Agreement, through the European Regional Development Fund (ERDF) and through national funds by the FCT – Fundação para a Ciência e a Tecnologia.

This joint research intervention includes contributions from the following researchers:

- (Cláudia Lima/ID+: Universidade Lusófona do Porto)
- (Jorge Pereira/ID+: IPCA)
- (Nuno Martins/ID+: IPCA)


## REFERENCES


Albino, C. (2017). *À procura de práticas sábias – Design e Artesanato na significação dos territórios.* CEARTE and Universidade de Aveiro.

Clarke, A. (Ed.) (2011). *Design anthropology. Object culture in the 21st Century*, Edition Angewandte. Vienna: SpringerVerlag.

Collier Jr, J. (1995) Photography and Visual Anthropology. *Principles of Visual Anthropology*. Berlin: Mouton De Gruyter.

de Almeida, P. C. (2012). *Brand archives: the rescuing of locally specific brand imagery as a graphic design response to the globalization of visual identity*, PhD thesis, Exhibition catalogue (Scholarly Edition). London: Central Saint Martins College of Art and Design, University of the Arts London. URL: http://ualresearchonline.arts.ac.uk/id/eprint/5689

de Almeida, P. C., and Chatterjee, A. (2016). *Sapatilhas: cultural significance and industrial legacy*, Poster Presentation, Ciência 2016, Lisbon.

Gomes, A. J. (2016). *Design, Tradition and Craft: The Case of Almalaguês*, Presentation at PhD Design Forum, PhD Design International Doctoral Program, University of Porto, Portugal.

Hartman, K. (2009). Retaining Intellectual Capital Retired Faculty and Academic Libraries. Retrieved from https://journals.ala.org/index.php/rusq/article/viewFile/3489/3774. volume 48, issue 4 p. 385.

Kemp, M. (1997) *Behind the Picture: art and evidence in the Italian Renaissance*. London: Yale University Press.

Leonard, D.; Swap, W. and Barton, G. (2014). *What's Lost When Experts Retire*.Retrieved from https://hbr.org/2014/12/whats-lost-when-experts-retire.

Manzini, E. (Ed.) (2015). *Design, when everybody designs: An introduction to design for social innovation*. Massachusetts: MIT Press.

Morrison, J. (2017). *The hard life, Zurich*. Lars Müller Publishers.

Pink, S. (2007). *Doing visual ethnography*. London: SAGE Publications.

Quivy, R.; Campenhoudt, L. V. (2008). *Manual de Investigação em Ciências Sociais*. Lisboa: Gradiva.

Rose, G. (2011). *Visual Methodologies: An introduction to the interpretation of visual material*, 3rd edition. London: SAGE Publications.

United Nations, Department of Economic and Social Affairs Population Division. (n.d.) (2015). *World Population ageing Report*. Retrieved from http://www.un.org/en/development/desa/population/publications/pdf/ageing/WPA2015_Report.pdf.

Verganti, R. (2015), Scientific yield from collaboration with industry: The relevance of researchers' strategic approaches, *Research Policy* (44)4.

White House Conference on Aging (n.d.) (2015). https://whitehouseconferenceonaging.gov/2015-whcoa-final-report.pdf

## Related Links

Anti-Amnesia
https://endlessend.up.pt/antiamnesia/
https://www.facebook.com/antiamnesia/

Wisdom Transfer
https://endlessend.up.pt/wisdomtransfer/
https://www.facebook.com/wisdomtransfer/

Wisdom Transfer Network
https://endlessend.up.pt/wtn/

Wisdom Transfer Seminars
https://www.fba.up.pt/2019/10/31/mais-do-que-a-soma-carlos-carreiro-helena-abreu-e-lima-e-sobral-centeno/
https://www.fba.up.pt/2019/11/11/mais-do-que-a-soma-antonio-quadros-ferreira-mario-americo-e-manuela-bronze/
https://www.fba.up.pt/2019/11/15/mais-do-que-a-soma-armando-alves-carlos-marques-e-zulmiro-de-carvalho/
https://www.fba.up.pt/2019/11/29/mais-do-que-a-soma-elvira-leite-jose-paiva-isabel-cabral-rodrigo-cabral/
https://www.fba.up.pt/2019/12/09/mais-do-que-a-soma-ana-campos-joao-nunes-e-paula-soares/

*Research & Education in Design: People & Processes & Products & Philosophy – Almendra & Ferreira (eds)*
*© 2020 Taylor & Francis Group, London, ISBN 978-0-367-49296-0*

# REDES – The vision for a research group on research & education in design

R. Almendra & J. Ferreira
*CIAUD, Lisbon School of Architecture, Universidade de Lisboa, Portugal*

ABSTRACT: This paper presents REDES, a research group on Research & Education in Design that integrates CIAUD – the research centre of Architecture, Urbanism and Design – from the Lisbon School of Architecture – University of Lisbon. The vision, the process and planning of activities of this research group are here explained as well as its relationship with the ongoing discussion on design research and education. By detailing the genesis and ethos of REDES we aim to discuss how design research and education can look inward, reflect, and ultimately uncover paths to advance.

## 1 INTRODUCTION

This paper should be understood as a foundational article that frames the creation and development of a Design Research group. In this case, it refers to REDes (Research and Education in Design) and the paper outlines the research group's vision, details its creation, and presents its ongoing research projects.

REDes was created in 2017 as the result of the research conducted by coordinator of the group (a designer, professor, and researcher) since 2007. This process included the PhD research that led to the author's thesis (Author, 2010) but also several experiments and investigations about education and research in design that were conducted in a continuous reflective reasoning process about what was being taught at the Lisbon School of Architecture's design courses (across the three cycles of higher education).

The apparently disperse reflections and research previous to the creation of REDes was presented in papers published between 2008 and 2017 and resembled the labour of an archaeologist that takes samples (probes) of diverse nature in a delimited territory in order to plan and determine the approach to be done in the excavation itself.

It was the case. In fact, the proto history of REDes includes: a) work in the area of design cognition and design processes – at the level of graduate and master students; b) reflections based on empirical data regarding emergent design areas such as service design, social design, inclusive design and sustainable design and its links; c) didactics at the PhD level aiming to develop and expand critical thought; d) development of guidelines to establish a network among the three design education levels and among these and the external world – companies and society in general (see, for instance Author [2009, 2012, 2014, 2016]).

Furthermore, in terms of research areas we can identify four main ones: a) Sustainability through memory values and affection; Sustainable and inclusive public equipment; Sustainable learning environments; b) Education: Studies on didactics improvement; Studies on curricula and its impact in design practice performance; Studies on the relationship between teacher/student; Studies related with methodologies and methods to be used in the course; c) Social Design: Sustainability of social design practices; Social design entrepreneurship; Social design principles and practices; Social design education; and finally d) Design Management: Metrics and measurements of design value in companies; Better performance= better design usage; Design policies; and Design management across design education.

From these studies emerged a broader theme: the need to investigate how Design Research and Design Education are connected; this issue is central to establish Design as a mature discipline within academia. Any practice-based discipline goes through this stage of development, notice for instance, the move within Medicine from a practice-based work to an evidence-based, research intensive activity (Best & Neuhauser, 2004). While currently no one disputes the importance of scientific research to support medical practice there was a time when many physicians resisted the intrusion of researchers into their conventional practices[1]. Of course, medicine deals directly with life-or-death and the physical and psychological well-being of people; the concerns of Design are different, but the comparison holds in that design also has a tradition of craftsmanship and learning-by-doing which is naturally suspicious of research.

A mature academic discipline should establish its epistemological and ontological footing in robust, systematic, insightful, and above all useful research efforts that expand its knowledge base and constantly questions its practices. At the same time, we should be mindful not to grow too enamoured with a mechanistic view of design that is blind to the particularities of professional design practice; in fact, an effective connection between research, education, and practice must acknowledge that "[t]he designer has a specific territory of action that should be investigated using methods and tools adapted to the specificity of this domain of knowledge"[2]. (Almendra, 2013, p.189)

Noticing that there was no research programme at the Lisbon School of Architecture that focused on these concerns the next step was clear: to create a research group specifically concerned with the study of Research and Education in Design.

## 1.1 *Connect research and education in design*

One way to approach the connections between design research and education is to understand how the three cycles of design education are linked. Notice how design education grows from a teaching and learning environment centred on the practice of design (a studio-based learning context anchored on design projects) which is prevalent in the first years of undergraduate studies, and moves to a progressively more research-intense context during the Masters and finally a research-focused PhD course.

In other words, the three cycles of higher-education develop as a move towards higher levels of research skills that culminate in an original contribution to human knowledge – the PhD thesis; this is the case in any academic discipline (the three cycles are a result of the Bologna accord that defined learning objectives and length of each level of higher education within the European Union).

There is a clear difference between the kind of work that a first-year undergraduate and a PhD candidate do, but we should expect to find a common thread uniting their efforts, something approximating the *DNA* of design, that is, the fundamental and inherent characteristics or qualities of Design as a discipline.

The metaphor of 'DNA' is apt in the sense that REDes aims to identify the connections between a way of doing (designing,) a way of thinking (design thinking,) and a way of researching. At least at a molecular level there should be an identity, a trace, so that an investigation conducted by a designer produces different insights from, say, what a sociologist or an expert in classical Greek literature might do. Researchers working within different disciplines concentrate on different objects of study in different ways; for instance, let us consider an expert in English literature and a theoretical physicist, not only are their objects of study worlds apart, but both arrive to research with a coherent way of looking at the world that is manifest in specific ways of investigating. It is clearly different to apply mathematical

1 To illustrate this point, notice how physicians in the 19th century resisted the call to include hand-washing in their daily practice as the results of Ignaz Semmelweis' research in the field of infection control and germ theory (Best & Neuhauser, 2004).

2 Translation from Portuguese: "o designer tem um território particular de actuação que deve ser investigado com métodos e ferramentas adequados à especificidade deste domínio do conhecimento".

reasoning to describe how a black hole functions from interpreting the role and meaning of Odysseus in Homer's *Odyssey*. Of course, there are common aspects uniting academic investigation in any discipline; scholarly research must be held to the highest levels of rigour concerning the systematic study of materials (whatever form these may take) and sources to establish facts and reach conclusions that contribute to advance human knowledge and wellbeing.

As such, in order to stand alongside the Sciences and the Humanities, Design research must adhere to the general standards of what constitutes scholarly research; however, as it would be unproductive to analyse *The Odyssey* with the same tools and methods used in Physics, Design also has – like Nigel Cross (2007) put it – its distinct things to know, ways of knowing them, and ways of finding out about them. This is not to say that design should ignore the contributions of relevant disciplines (such as Psychology and Sociology, for instance), since complex human problems are likely to require multidisciplinary approaches; but multidisciplinarity implies many disciplines working together with each one contributing its own expertise.

So, Design has its particular way of studying the world, a way that, we argue, is unique to human endeavour. Designers are trained to consider problems and solutions to complex, ill-defined (Rittel & Webber, 1973) or wicked problems (Buchanan, 1992) as a pair, which means that there is no unique solution to a design problem only better or worse ones, and furthermore, a wicked problem rarely lends itself to a linear, or logical step-by-step scientific deduction to find a solution. Considering that most human problems are wicked (complex, ill-defined, involving multiple stakeholders with often conflicting needs, with hard to determine boundaries) and only occasionally similar to a chess problem, then Design's way of doing, thinking, working, and researching has much to offer academia and society in general. Of course, it is the responsibility of the design research community to systematise this way of looking into consistent methods of research than can be explicitly taught and communicated to others.

Also, designers must master research methods, considering that part of their work consists in observing, questioning, information gathering, and testing of ideas as a way to solve human problems. Therefore, it makes no sense to have three cycles that are hermetically closed, i.e., a practice-based intensive undergraduate course, a middle stage Masters which at the moment is more a continuation of the undergraduate than an effective pivot between design *practice* and design *research*, and a completely research-intensive PhD. Of course, a PhD candidate has to practice and eventually master research skills at a much higher level of depth than an undergraduate since a PhD is also a certificate that ensures the candidate is able to conduct research independently; however, a first step towards connecting education and research could be to introduce basic research skills that undergraduates can put to use in the research stage of their design projects, also, it would be a benefit to slowly introduce research outcomes that may be useful for their projects. This way, undergraduates can progressively come to know and be familiar with the relevant literature of design research and design theory.

Here we find an opportunity for REDes to make an impact since most members of the group are both teachers and researchers. It is the aim of the founder and current coordinator of the group to take full advantage of this situation in order to create concrete links between research and educational practice. The goal is for the research projects to feedback into their teaching practice, this objective has high potential but also brings challenges and difficulties. Most professors struggle with heavy schedules and duties of teaching and other academic responsibilities (exams, tutoring, lecturing, grading, supervision of graduate students, department coordination, bureaucratic tasks, grant applications, publishing aims, and so on) which means time-management is a critical factor. This means that the integration of research outcomes and education has to be systematised in order to be effective.

REDes is uncovering new terrain, there is no other group with similar goals and aims at the Lisbon School of Architecture. In the next sections we will detail the group in further detail in order to clarify how REDes is aiming to make a decisive contribution to advance design research and education.

## 2 REDES' AIMS

The research aims direct the group's activities towards a broader understanding of research and education in design. The aims reflect the need to adopt an exploratory and empirically based approach to research; in our view, to reach insightful results that benefit the development of design as a discipline, it is necessary to confront theoretical models and ideas with the specific contexts in which education and research take place.

To articulate research done in the areas of Design education and research (across different design disciplines of Product, Communication, Interior, and Fashion);

To promote a continuous questioning of design education and research in terms of its contents, methodologies and didactics;

To test and experiment an enlarged pool of different methods to assess the teaching-learning process in these areas;

To produce rigorous knowledge about the design education in Portugal mainly based on empirical studies;

To test and conclude about adequate methodologies and didactics not only regarding design studio but mostly about interdisciplinarity and the role of other scientific areas on a designer's formation;

To create innovative contents (to be tested and validated) in the design education field namely in the areas of sustainability, social design, and design management.

These aims describe a broad territory of intervention for the group's activities; in the next section, we will briefly describe the group's team of researchers and projects that actualise these objectives.

### 2.1 *REDES' team and projects*

REDes research group accommodates the study interests and supports the investigations of its members while simultaneously establishing a design research agenda for the whole group. In this topic we will briefly expose who are the members of REDes and the type of studies they are involved with, followed by an explanation of the REDes research agenda.

A broad picture of the members of the group reveals a team of researchers (currently 12 members) with different backgrounds that work in distinct design areas. For instance, some researchers are architects that have been working in both design and architecture for the last 30 years and teach in the design area. There are full professors with decades of teaching and research experience as well as PhD candidates starting their academic careers. We value this combination of novices and experts; the younger members can grow from the experience of the more experienced members and the group benefits from having fresh perspectives and new research interests.

The specific research projects of the members reveal a diversity of topics: there are studies on design as a discipline (in ontological terms); empirical studies regarding the development of critical and creative competencies at the graduate level; communication designers exploring multidisciplinary approaches to research and education, user experience (UX), and user interface (UI); there are also product designers (that simultaneously teach, research and some also practice) interested in the creation of methods and strategies to promote the knowledge transfer from academia to society and the growth of entrepreneurship competencies among design students either related with business or with social sector; researchers that are dedicated to study social design education and the way it impacts communities and societies in general; in the area of design thinking some researchers have developed theoretical models (working closely with undergraduates in design) that support decision-making and the design process; and there are studies on teacher-student communication in the design studio.

### 2.2 *The research agenda of REDES*

The research agenda of REDes derives from its goals and presupposes several activities distributed along a timeframe. The reasoning behind the scheduling of research activities has to do with a set of priorities establish with the agreement of all group's researchers starting with:

1. Internal activities:
1.1 The diagnosis and assessment of the design education systems – including formation and investigation.
1.2 The connections between education and research, existing bridges and collaborations to be established;

2. Relationship between the education-research system and society:
2.1 Diagnosis of existing connections and gaps;
2.2 Bridging the two systems with a consistent and sustainable knowledge transfer strategy.

The two main points of the agenda address, on the one hand, internal issues concerning the effectiveness of design education, starting at the Lisbon School of Architecture but aiming to arrive at an understanding of the challenges facing Portuguese education in design in general; also, the group directs considerable efforts in establishing effective bridges between education in research with the conviction that improving one will necessarily have a positive impact on the other. On the other hand, the second point deals with the important issue of expanding beyond the confines of academia into society and question if the work conducted in the university is making the necessary links with society in general.

### 2.3 *Ongoing project: REDes diagnosis about design internal activities – the link between design research and design education at the 3rd cycle (the PhD course)*

Besides the individual research interests and projects of each member, REDes also has research projects that involve the whole group. The first one is an ongoing in-depth analysis of all the PhD and Master thesis in Design completed at the Lisbon School of Architecture (we describe the methodology supporting this study in another paper [Authors, 2019]).

In short, the research is a detailed analysis of PhD and Master thesis in Design detailed according to four levels of inquiry: research identification, territory, processes, and results. The analysis functions on two levels: a) the research reported in each thesis manuscript, and b) the aggregate of the analysis of each thesis results in meta-data about the research conducted in the PhD course as a whole.

We have completed phase 1 of the project which comprised the analysis of the PhD thesis completed between 2008 and 2019 (totaling 103 completed analysis out of 106).[3] Of course, the project is ongoing, meaning, having developed an analysis framework and a database structure the group will update the database every time a new thesis is defended and published, thus contributing to fulfil the aim of working as an observatory for education in Design at the university.

Phase 2 will comprise the analysis (using the same analytic framework) of master thesis in design; the goal is to look into the results in search of ways to connect these two cycles in terms of their research-education-practice interrelationship. At the moment, it is our assumption (from the experience with both courses) that the transition from Master to PhD is not a fluid move but rather a steep jump, that is, the experience in the Master course only marginally prepares the students for the challenges of conducting independent research in Design during their PhD.

This project produces a vast amount of valuable knowledge, both as raw data (straightforward metrics on author identification, method use, or research themes) as well as interpretative insights resulting from meta-analysis of all the thesis. The resulting database is thus

[3] Of the three theses missing from the analysis, two are being translated from Spanish to Portuguese and one is awaiting the results of a patent request and therefore is not available for analysis.

a valuable resource for academia which requires careful dissemination; possible uses of this pool of information are enumerated below:

2.3.1 *How to treat the knowledge created*

As content to enrich the teaching process and the design education system (across the three levels of higher-education);

As a research process/outcome to be critically analysed in the context of the research line consistency and development;

Internal improvement and transformation of the PhD and Master courses – specific studies; integration of empirical work in ongoing PhD research;

Critical assessment of the Centre for Architecture, Urbanism and Design of the Lisbon School of Architecture;

Assess the external impact of the project research.

2.3.2 *Knowledge transfer to education*

Engaging the PhD work with ongoing classes from design programs – undergraduation and masters

Translating the conclusions of the empirical studies into improvements in the curricula of the design courses (service design; ergonomics; sustainable design)

Creating pedagogical material to support the classes – exercise books; tools, models

2.3.3 *Expected outcomes*

The consolidation of a knowledge transfer system to the Design education field;

Operationalize ways to transfer that knowledge and to put it in practice

Publish an analysis of the research system in CIAUD namely in these 4 areas – themes/methodologies/tools/outcomes.

## 3 LIMITATIONS AND CONCLUSIONS

With this paper, we aimed to present the group to the design research community. Before we conclude, it is important to acknowledge the specific difficulties and limitations that derive from the group's mission and aims and how to operationalise these into concrete actions. Also, there are concrete challenges that any research group in design faces when trying to establish itself and develop meaningful work (such as funding and dissemination of results).

REDes is ambitious in its scope, that is, based on the vast territory covered by the studies that preceded it, the group hopes to improve educational practices (across the three levels of higher-education) in design and advance the design research conducted at the Lisbon School of Architecture. Such developments take time and require concerted efforts; the most important results of the first 24 months were the completion of phase 1 of the diagnostic of PhD thesis and the organisation of the first International Research and Education in Design Conference, as well as the advancement of several individual research projects of its members. Be that as it may, having established the group and stabilising its members into a promising mixture of experienced and novice researchers, the group expects to accelerate its output and productivity.

For the immediate future, the group aims to have at least two (self-funded) broad research projects that will involve the combined effort of several of its members. To achieve this, it is important to organise the members into smaller groups with effective coordination of a full professor that oversees the tasks to be completed, and plans and guides the research to be conducted. Only by implementing a systematic way of working can the group hope to produce timely results and achieve effective dissemination outputs.

Furthermore, henceforth the group will apply a dissemination strategy that focuses on less publications but with higher impact; to achieve this, we have created a list of high-impact conferences and journals for submission of the group's papers (partly based on the work on the impact factor of design journals developed by Gemser, De Bont, Hekkert, & Friedman (2012).

To conclude, the vision that inspired the creation of REDes is the ambition to have a research group that looks into the practice of research and education in design and uncovers ways to improve and advance the field. The materialisation of this vision works from the ground up, that is, by observing and analysing education and research in practice (meaning its actual results in students and in research outputs) question it, explore alternatives, and reach confident conclusions that reveal ways to move forward.

By sharing and publicly presenting the formation and functioning of the group with the research community we hope, on the one hand, to inspire the development of similar efforts elsewhere, and on the other hand to invite feedback, suggestions, and critiques that contribute to the group's development and consequently the advancement of design research.

## REFERENCES


Almendra, R. & Christiaans, H. (2009) Decision-making in the conceptual design: A comparative study. *Journal of Design Research*, *8*(1), 1–22.

Almendra, R. & Christiaans, H. (2010) Accessing decision-making in software design. *Design Studies*, *31*(6), 641–662.

Almendra, R. (2012) "Educating Critical Thinking in Design Research" Conference Proceedings 14th International Conference on Engineering and Product Design Education,2012, Antwerp.Belgium.

Almendra, R. & Veiga, I. (2014) Social design principles and practices. In Lim, Y.; Niedderer, K.; Redstrom, J.; Stolterman, E. & Valtoren, A. (Eds) Proceedings of DRS 2014: Design's Big Debates. Design Research Society Biennial International Conference, 6-19 June 2014, Umeå, Sweden.

Dias, A.; Almendra, R. & Moreira da Silva, F. (2016) The mismatch among design education-research-practice: How to strengthen the bridge? *The Design Principles and Practices Journal Collection (11)*4, 17–28.

Almendra, R. & Ferreira, J. (2019) A Framework to Analyse PhD Theses in Design In Almendra, R. & Ferreira, J. (Eds), *Proceedings of the International Conference on Research & Education in Design: People & Processes & Products & Philosophy, Lisbon, Portugal, November 14-15, 2019*. CRC Press.

Francis Almendra, R. (2013). Metodologias e métodos aplicados ao estudo dos processos de design. In T. Ramos (Ed.), *Arquitetura, Urbanismo, Design: Metodologias e métodos de investigação*. Lisbon: Caleidoscópio.

Best, M., & Neuhauser, D. (2004). Ignaz Semmelweis and the birth of infection control. *Quality and Safety in Health Care*, *13*(3), 233–234.

Buchanan, R. (1992). Wicked problems in design thinking. *Design Issues*, *8*(2), 5–21.

Cross, N. (2007). *Designerly Ways of Knowing*. Basel: Birkhäuser Verlag AG.

Gemser, G., De Bont, C., Hekkert, P., & Friedman, K. (2012). Quality perceptions of design journals: The design scholars' perspective. *Design Studies*, *33*(1), 4–23.

Rittel, H., & Webber, M. (1973). Dilemmas in a general theory of planning. *Policy Sciences*, *4*(2), 155–169.

*Design Thinking*

*Research & Education in Design: People & Processes & Products & Philosophy – Almendra & Ferreira (eds)*
 *ISBN 978-0-367-49296-0*

# Design at Stanford: The D.school's Daddy

S. McCarthy
*University of Minnesota, USA*

ABSTRACT: "Design thinking" has emerged as one the of the discipline's key movements, and is typically associated with having roots at the d.school at Stanford University. Its founder, and also the creative force behind global design company IDEO, is alumnus David Kelley, who has achieved legendary status. Prior to formation of the Hasso Plattner Institute of Design (the d.school's formal name), however, since the 1950s design education at Stanford consisted of many lesser known, but still influential, faculty members and an innovative curriculum combining engineering and art. This paper exposes the "founding fathers" of design education at Stanford (many of whom were David Kelley's professors), and reveals two curricular proposals that preceded the d.school.

"Design thinking" is often associated with the Hasso Plattner Institute of Design – the "d.school" – at Stanford University for its emphasis on teaching creative problem-solving methodologies and the iterative process to those from disciplines other than design. Design thinking has also been sold to the corporate world and other institutions through the consulting work of IDEO, a global design company. Both enterprises share a co-founder: David Kelley, himself a graduate of Stanford's Joint Program in Design. (McCarthy, 2017) But what preceded Kelley's global influence, and by association, that of Stanford in creating a fertile ground for creativity, expression, problem-solving, need finding and entrepreneurial ventures in design?

The Joint Program in Design at Stanford was founded in 1958 when faculty from mechanical engineering and art decided to collaborate on a joint graduate program in a "mercenary academic way." (Kahn, 2009) Each department facilitated a specific pathway through the program: mechanical engineering students earned a Master of Science degree and focused on product design; art students completed a Master of Fine Arts degree and created primarily visual designs. They shared coursework, faculty and a year-long thesis experience. Thesis projects were functioning designs that were experimental, human-centered and varied. Due to Stanford's integrated approach, it was not always clear who did their degree in which department.

Stanford design graduates went on to create Apple Computer's early products and graphics – besides Apple co-founder Steve Jobs, US Patent D268584 for the product design of a "personal computer" has three Stanford design alumni on it, including David Kelley. Other grads went on to design the water fountain at the base of the Burj Khalifa tower, design Google and Twitter's logos, create typefaces for Adobe, write a history of Buckminster Fuller, found and staff IDEO, patent and design the Koosh ball, serve as director of the Design Research Council at Northwestern University, and much else.

Besides the typical structures of academia – the campus resources, courses, faculty, facilities, studios, projects and so on – what was Stanford's "secret sauce"? There were two main ingredients: one, a philosophy of "creative disobedience" as embraced by key faculty, and two, tangible environments, programs and facilities that enabled creativity, collaboration and transformation. (McCarthy, 2019)

The notion of "creative disobedience" is attributed to the legendary professor Matt Kahn, a Cranbrook-trained designer who taught in Stanford's Department of Art from 1949 to 2009. Kahn maintained a wide creative practice, both professional and personal: interior design, graphics, photography, jewelry, fine art painting. His work was exhibited at the nation's top venues, like the Museum of Modern Art and at the Cooper Hewitt. As a teacher he was provocative, contrarian, inspirational and challenging, and through this transformed his students. In a hand-written note to Kahn from 1977, David Kelley wrote, "You have been such an inspiration to me. You see things the rest of us never will – and say things I only wish I could. You have changed me. Thank you." (Kelley, 1977)

Another unorthodox contributor to design education at Stanford was John Arnold, a mechanical engineering professor who was hired mid-career from the Massachusetts Institute of Technology (MIT) in 1957. He pioneered a concept called "creative engineering" – a blend of psychology, business, science fiction, invention and synthesis. Arnold used sci-fi prompts in his teaching, such as creating a fictitious planet and its inhabitants, to challenge his product design students to not take the "real world" for granted. He named Stanford's Department of Mechanical Engineering's "Design Division" by printing that term on letterhead (has was an amateur printer and had a press in his home basement). He was a maverick, he didn't wait for permission.

This philosophy of creative disobedience encouraged – even demanded – risk and experimentation, personal transformation, "bias through action" (physically creating designs in a studio or machine shop) and the consideration of the user at all scales, from individuals to humanity. Stanford's autonomous graduate student studio – "The Loft" – was the laboratory and playground for this approach. The culture at Stanford was permissive – *do anything*, just do it really well and cause no harm! Conversely, some university cultures value consensus, tradition and caution, with rules, policies and procedures that potentially circumscribe creativity.

Another stake holder to the notion of creative disobedience was mechanical engineering professor Bob McKim. His 1972 book *Experiences in Visual Thinking* was seminal in getting engineers to use sketching as an ideation tool. Thus, McKim asserted, analysis would combine with synthesis, objectivity with subjectivity, and rationality with expression in an integrated design education. (McKim, 1972) The book, which drew on perceptual psychology, neurology, semantics and art, became the bible of the program. McKim also exposed students to hallucinogens with the hope of increasing creativity, and he and other faculty became involved with the "human potential movement" as proffered by the Esalen Institute in Big Sur, California. He was also an active design consultant to technology companies in Silicon Valley.

Shockingly, neither Professor Kahn nor Professor McKim had a graduate degree (Kahn didn't even have a bachelor's!). Kahn finished three years at the Cranbrook Academy of Art before being hired by Stanford at age 21, while McKim earned a BS in industrial engineering from Stanford and a post-baccalaureate certificate in industrial design from the Pratt Institute in New York. Yet both men were professors, and key influencers, of David Kelley, who has gone on to be one of this generation's design giants.

Cranbrook faculty and students in the mid-twentieth century, however, included architects and designers Eliel Saarinen, his son Eero, Charles and Ray Eames, Harry Bertoia, Florence Knoll, Marianne Strengell, Jack Lenor Larsen and many others – these giants of design were Kahn's teachers, compatriots and the inspiration for his design practice. McKim worked for the famous industrial designer Henry Dreyfuss. Perhaps these exposures, to the practices of leading designers, were more important than graduate degrees then.

Of course, times have changed, and today, in most cases, graduate degrees represent increased knowledge and experience. A terminal degree is the typical minimum requirement for a faculty position at a research university. But graduate education is also a form of institutionalization and can limit one's creativity to the conventional expectations of the academy. For design faculty, the very process of going through tenure is anathema to creativity – judgment by staid senior colleagues, requirements to follow established procedures and policies, and pressure to conform to the institution's culture run counter to risky, innovative, truly creative behaviors. To this pressure, add today's campus milieu: an epidemic of mental health issues, student hyper-sensitivity, an unfortunate focus on identity politics, stringent financial limits, a bloated corporatist and bureaucratic approach to the "business" of education, and an uninspiring work environment. It is hardly surprising that some faculty and students come nowhere near their creative potential.

Another potential limitation is the traditional scholar's purely academic training – "book smarts" instead of "street smarts." Stanford professor David Kelley asks of this breed of faculty: "are they a designer or designologist?" (David Kelley conversation with author, June 9, 2016) Designers have applied creative experience making things for users; designologists merely study and write about design processes, artifacts and systems. While professional experiences are not the sole determinant in making designers superior faculty over designologists, the expanded experiential toolkit that comes from working in the field adds to the person's ability to solve problems creatively, to think laterally, and, it gives them professional credibility. Working materially in a studio (with a medium, tool, process, craft, etc.) gives the designer as professor additional haptic knowledge and enables them to better teach those methods to students. This is not to say that professional practice is necessarily intellectually enlightening; an engaged creative production of self-initiated, speculative works of critical design can also be a legitimate form of scholarly inquiry.

The faculty and interdisciplinary collaboration mentioned above helps provide context for Stanford's association with design thinking and doing today. Few are aware of Stanford's successful Joint Program in Design formula, that art + engineering = design. Many are aware of the popularity of Stanford's "d.school" but only in a superficial way (for example, it is not one of Stanford's seven official schools but a poke at the Graduate School of Business, the "b.school;" the d.school does not grant degrees, and did not even offer for-credit classes until 2017).

The d.school was founded by professors David Kelley and Bernie Roth in 2005 with a $35,000,000 gift from German industrialist Hasso Plattner. It is primarily a project-based connective tissue to students from all corners of campus. Students coalesce around topics, problems or opportunities in small, agile teams that are led by faculty facilitators. In this context, they are introduced to "design thinking": empathetic need-finding, brain-storming and conceptualizing, deferring judgment, sketching and visualizing, the iterative process, prototyping, user testing, refining and being entrepreneurial. Some teams' projects advance to attracting venture capital or to manufacturing and marketing.

After starting out in a trailer, the d.school is now housed in a new facility that is just off of Stanford's iconic quadrangle that was designed in the 1880s by the eminent landscape architect Frederick Law Olmstead. It is around the corner from the graduate studio, the Loft, and the Product Realization Lab. The exposed sandstone walls, black I-beams, glass ceiling, and red stairs and bannister envelope a modular and informal workspace. In one area, a 1950 Chevrolet panel van serves as d.school-branded mascot and mini-lounge, with its interior turned into a carpeted room with throw pillows. Post-It notes festoon its windows. A black and white banner with a John Cage quote hangs in the d.school atrium: "Nothing is a mistake. There's no win and no fail. There's only make." The places buzzes with creative energy.

Prior to the founding of the d.school in 2005, two competing curricular proposals were advanced. In 1999, mechanical engineering professor and director of the product design program Rolf Faste advocated for creating a college-level School of Design that would include engineering, art, architecture, human factors, cultural anthropology and psychology "as applied to design." (Faste, 1999) A devotee of Eastern philosophies and pioneer of a concept he called "Zenengineering," Faste sought to create a school that valued balance and "ambidextrous thinking": "The School of Design would value form as much as content. Thus concerns of aesthetics, symbiotics, [sic] etc. would be as important as utility." (Faste, 1999) It was an ambitious proposal that was not implemented, yet it opened the conversation about the future of design education at Stanford.

Professor Larry Leifer's research unit, the Center for Design Research's Design Theory and Methodology group, created a degree proposal titled "Comprehensive Design Engineering" (CDE) in 2003. It was admittedly influenced by Kelley's ideas related to design thinking; plus, the term "comprehensive design" harks back to John Arnold's concept of the creative engineer in the 1950s. The CDE curriculum proposal drew less from Stanford's longstanding engineering and art relationship, and more from the intersection of business, technology and human issues.

The Comprehensive Design Engineering plan borrowed from corporate hierarchy to equate bachelors students with being apprentices, masters students as mentors, and doctoral students as leaders. The apprentice would equate with being a "Designer, Engineer," the mentor with being a "Manager," and the leader with serving as "Educator, CTO [Chief Technical Officer], Entrepreneur." (Cockayne & Feland, 2003) Yet these types of pyramid-like organizational structures can be at odds with creative learning environments. Stanford mechanical engineering professor Jim Adams sums up this condition well in his book *Conceptual Blockbusting*: "The natural tendency of organizations to routinize, decrease uncertainty, increase predictability, and centralize functions and controls is certainly at odds with creativity, and conceptual blocks can abound." (Adams, 1974, p. 143)

For educational institutions anywhere to effectively use design thinking – both in teaching and in faculty research and creative production – they should encourage *creative disobedience*, evolve thinking into making through *bias towards action* (or "demo or die," the slogan attributed to the Media Lab at MIT), and balance its faculty *designologists* with *designers* – because hierarchy, inaction and theory alone will not do. Stanford's success in achieving renown through the d.school was predicated on decades of its faculty, curricula, facilities and students as engaged in world-influencing product and visual design.

Kelley's vision – as articulated in an email sent from his @ideo.com address (Kelley, 2002) – for establishing an institute that would emphasize design thinking across disciplines – was realized as the Hasso Plattner Institute of Design at Stanford University. Elements of Faste's and Liefer's curricular proposals influenced the d.school, directly and indirectly, through ideation and framing. The "founding fathers" (to borrow a cliché) of design education at Stanford from the 1950s forward – professors Matt Kahn, John Arnold and Bob McKim – must be acknowledged too.

## REFERENCES


Adams, J. L. (1974). *Conceptual Blockbusting: A Guide to Better Ideas.* San Francisco: W.H. Freeman.

Arnold, J. (1956/2016). *Creative Engineering: Promoting Innovation by Thinking Differently* edited by W. J. Clancey. Stanford Digital Repository.

Faste, R. (1999). School of Design document in Rolf Faste papers. Privately held. Unpublished.

Kahn, M. (2009). Stanford Historical Society Oral History Program interview. https://exhibits.stanford.edu/oral-history/catalog/rj210sm4070

Kelley, D. (1977). Hand-written note in Matt Kahn papers, Stanford University Libraries Special Collections and University Archives.

Kelley, D. (January 10, 2002). Kelley–Faste email in Faste papers. Privately held. Unpublished.

McCarthy, S. (2017). "Design Education at Stanford: The Formative Years" in *Creativity on the Line: Design and the Corporate World, 1950–1975* edited by Wim de Wit Cantor Arts Center, Stanford University. London: Lund Humphries. 85–93.

McCarthy, S. (2019). *Design at Stanford: A Visual History of Thinking and Doing* finished book manuscript.

McKim, R. (1972). *Experiences in Visual Thinking.* Pacific Grove, California: Brooks/Cole Publishing.

*Research & Education in Design: People & Processes & Products & Philosophy – Almendra & Ferreira (eds)*
*© 2020 Taylor & Francis Group, London, ISBN 978-0-367-49296-0*

# Industrial designers problem-solving and designing: An EEG study

S. Vieira
*Politecnico di Milano, Italy*

J.S. Gero
*University North Carolina at Charlotte, USA*

J. Delmoral, M. Parente & A.A. Fernandes
*Faculty of Engineering University of Porto, Portugal*

V. Gattol
*Austria Institute of Technology, Austria*

C. Fernandes
*Saint John Hospital, Porto, Portugal*
*Institute of Technology, Austria*

ABSTRACT: This paper presents results from an experiment to determine brain activation differences between problem-solving and designing of industrial designers. The study adopted and extended the tasks described in a fMRI study of design cognition and measured brain activation using electroencephalography (EEG). By taking advantage of EEG's high temporal resolution we focus on time-related neural responses during problem-solving compared to design tasks. The experiment consists of multiple tasks: problem-solving, basic design and open design using a tangible interface. The tasks are preceded by a familiarizing pre-task and then extended to a fourth open design task using free-hand sketching. The results indicate design cognition differences in the brain measurements of task-related power and temporal analysis of transformed power between the constrained problem-solving task and the open design tasks. Statistical analyses indicate increased brain activation when designing compared to problem-solving. Results of time-related neural responses connected to Brodmann' areas cognitive functions, contribute to a better understanding of industrial designers' cognition in open and constrained design spaces and how the problem statement can constrain or expand conceptual expansion.

## 1 INTRODUCTION

The study of the cognitive behavior of industrial designers while designing, based on methods such as protocol analysis (Ericsson and Simon, 1983, Kan and Gero, 2017), has produced important results covering foundational aspects of design cognition. The notions of problem space and solution space have been the ground of interpretations of the designing process (e.g., Kruger and Cross, 2006) in the last fifty years of design research (Jones, 1963). The problem-solving view of design claims that the designing process commences with an exploration within the problem space (Goel and Pirolli, 1992). Alternative perspectives assert that design thinking is primarily solution focused (Dorst, 2011; Darke, 1979). One of the initial and core research questions is whether designing as a cognitive process is distinct from problem-solving (Goel and Pirolli, 1992; Visser, 2009). Neurophysiological studies offer a new integrative perspective into how brain behavior progresses during the designing process, which makes them a robust tool for connecting to design cognition. Recent design studies based on functional magnetic resonance imaging (fMRI) (Alexiou, et al., 2009; Goucher-Lambert, et al., 2017), electroencephalography (EEG) (Liu et al., 2018; Liu et al., 2016; Liang, et al., 2017) and functional near-infrared spectroscopy (fNIRS) (Shealy, Hu and Gero, 2018) attempt to understand designing from a neurophysiological perspective. The present paper describes a study from a larger research project whose goal is to correlate design cognition with brain activation of designers across design domains. EEG's high temporal resolution makes it a more suitable tool than fMRI (Hinterberger, et al. 2014; Dickter and Kieffaber, 2014) to investigate designing as a temporal activity. The study reported in this paper is based on the analysis of industrial designers' brain activation using an EEG headset in the context of performing problem-solving and design tasks in a laboratory setting. The objective of the study is:

Investigate the use of the EEG technique to distinguish design from problem-solving in industrial designers.

We adopt and extend the tasks described in a controlled experiment of an fMRI-based design study (Alexiou, et al., 2009). That study suggested higher activation of the dorsolateral prefrontal cortex is consistent for design tasks and ill-structured problems and recruits a more extensive network of brain areas than problem-solving. We postulate the following hypotheses:

Hypothesis 1. Design neurocognition of industrial designers when problem-solving and designing are different.

Hypothesis 2. Neurocognitive temporal distributions of activations of industrial designers are significantly different across design tasks.

## 2 EXPERIMENT DESIGN

We have adopted and replicated two of the layout tasks described in the Alexiou, et al. (2009) fMRI-based study. We extended their experiment to a third open layout design task with the purpose of opening the solution space to produce a block experiment as depicted in Table 1 and Figure 1. The set of three tasks is preceded by a pre-task so that participants can become acquainted with the physical interface and headset. The three tasks are followed by a fourth open

Table 1. Description of the tasks.

| Task 1 Problem-solving | Task 2 Basic design | Task 3 Open design |
|---|---|---|
| In Task 1 the design of a set of furniture is available and three conditions are given as requirements. The task consists of placing the magnetic pieces inside a given area of a room with a door, a window and a balcony. | In Task 2 the same design set of furniture is available, and three requests are made. The basic design task consists of placing the furniture inside a given room area according to each participant' notions of functional and comfortable using at least three pieces. | In Task 3 the same design available is complemented with a second board of movable pieces that comprise all the fixed elements of the previous tasks, namely, the walls, the door, the window and the balcony. The participant is told to arrange a space. |

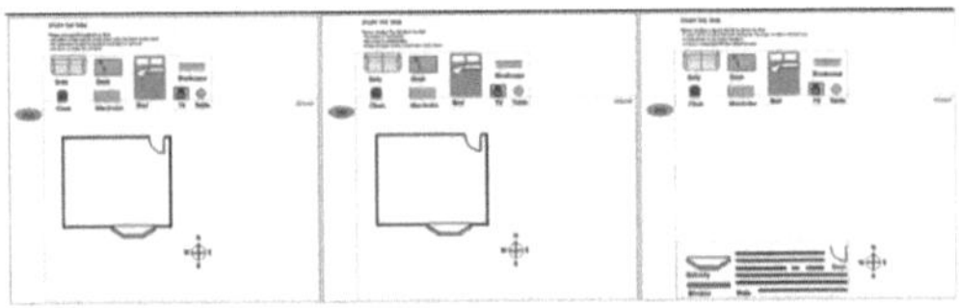

Figure 1. Problem-solving Task 1, basic design Task 2 and open layout design Task 3.

design free-hand sketching task. A tangible interface for individual task performance was built based on magnetic material for easy handling. The pre-task was designed so that participants can familiarize themselves with the use of the EEG headset, and necessary corrections can be made before advancing to the block experiment, manoeuvring the magnetic pieces that make up the physical interface and prevent participants from getting fixated in the problem-solving Task 1. The block experiment consists of a sequence of 3 tasks: problem-solving, basic design and open layout design, as illustrated in Figure 1. We have matched Tasks 1 and 2 with the problem-solving and design tasks from Alexiou, et al. (2009) in terms of requests, number of constraints, stimuli and number of instructions. The open layout design Task 3 provides an enlargement of the problem space and the solution space and the opportunity of evaluating and reformulating the previous design solutions. In Task 4, the participants are asked to propose and represent the outline design of a future personal entertainment system, which is an ill-defined and fully unconstrained task unrelated to formal problem-solving. The Mikado pick up sticks game was given to the participants to play in the breaks between tasks to break their focus on the tasks.

Differently from the original tasks (Alexiou, et al. 2009), the magnetic pieces were placed at the top of the vertical magnetic board to prevent signal noise due to eye and head horizontal movements. Two video cameras for capturing the participant's face and activity and the audio recorder were streamed in Panopto software (https://www.panopto.com/), Figure 2. One researcher was present in each individual experiment to instruct and record the participant performance. A period of 10 minutes for setting up and a few minutes for a short introduction were necessary for informing the participant, reading and signing of the consent agreement and discussing the experiment. The researcher sets the room temperature and draws each participant's attention to minimize the following actions as these affect the signal capture, namely: blinking, muscle contractions, rotating the head, horizontal eye movements, neck movements, pressing lips and teeth together in particular during the tasks. The researcher follows a script to conduct the experiment so that each participant is given the same information and stimuli. The researcher positioned the participants at the desk and checked for metallic accessories that could produce electromagnetic interference. Before each task, participants were asked to start by reading the text which took an average of 10s. Then the subjects performed the sequence of five tasks previously described. In the breaks between the tasks, participants played the Mikado game. The participants performed the tasks in a linear sequence as the objective of the study is the measurement of brain activation of designers through a sequence of tasks that gradually expand the design solution space from a problem-solving to basic and then open design tasks.

Electromagnetic interference of the room was checked for frequencies below 60Hz. The experiments took place between March and July of 2017 and June and September 2018 in a room with the necessary conditions for the experiment, such as natural lighting from above sufficient for performing experiments between 9:00 and 15:00 and no electromagnetic interference. The experiments took between 34 to 67 minutes. The EEG activity was recorded using a portable 14-channel system Emotiv Epoc+. Electrodes are arranged according to the 10-10 I.S, Figure 3.

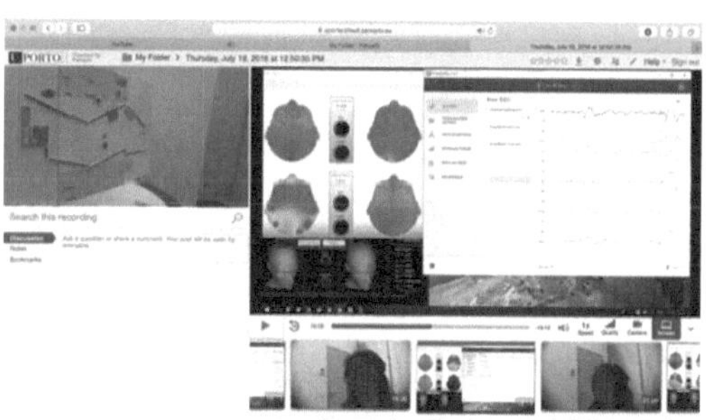

Figure 2. Audio, video and screen streaming in Panopto.

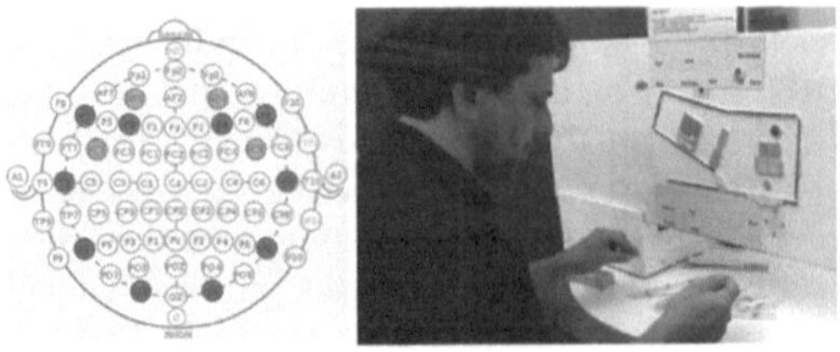

Figure 3. Emotiv Epoc+ Electrodes (10-10 I.S.) and experiment setup.

## 2.1 *Participants*

A total of 29 experiments were conducted with industrial designers. Due to EEG or video recording issues five experiments were excluded. The analysis then proceeded based on the EEG data recorded and processed for each of the 24 remaining experiments, and each of the 14 electrodes used for averaging, for each of the tasks. A z-transform was conducted to determine outliers. The criteria for excluding participants were based on the evidence of 6 or more threshold z-score values above 1.96 or below -1.96 and individual measurements above 2.81 or under -2.81. This resulted in a further two experiments being excluded leaving 22. After the division of the Pow into time deciles (which provides the basis for the temporal analysis) and based on the evidence of threshold values above two and a half average plus standard deviation per channel, a further 4 experiments had to be excluded leaving 18.

The analysis is based on the experimental data of 18 industrial designers, aged 25-43 (M = 31.7, SD = 7.3), 10 men (age M = 35.1, SD = 7.2) and 8 women (age M = 27.5, SD = 5.1), all right-handed. The study was approved by the local ethics committee of the University of University of Porto. Each participant was reminded to use the bathroom and spit out any gum before the start of the experiment. The researcher sat each participant at the desk, asking him/her to untie hair and remove earrings and other metallic accessories, check if they are using contact lenses as these may cause too much blinking and interfere with data collection. Time was given to the participants, in particular in Tasks 3 and 4 so they could find a satisfactory solution. Average time taken per task is as follows: Pretask, 101s, Task1, 90s, Task2, 97s, Task3, 373s and Task 4, 725s.

## 2.2 *Data processing*

For the present analysis, all the EEG segments of the recorded data were used for averaging throughout the entire tasks, from beginning to end. In order the remove spurious effects such those produced by eye blinks, jaw muscle contractions and speaking we adopt the blind source separation (BSS) technique based on canonical correlation analysis for the removal of muscle artifacts from EEG recordings (De Clercq, et al. 2006, Vergult, et al. 2007) adapted to remove the short EMG bursts due to articulation of spoken language, attenuating the muscle contamination on the EEG recordings (Vos, et al. 2010). The fourteen electrodes were disposed according to 10-10 I.S, with a 256 Hz sampling rate, a low cutoff 0.1 Hz, and a high cutoff 50 Hz. Data processing includes the removal of DC offset with the IIR procedure, and BSS.

## 2.3 *Data analysis*

We focus on the overall activation per channel, per task, per participant as the study aims to determine how the results for problem-solving and designing can be distinguished. We compare absolute values known as transformed power (Pow), and task-related power (TRP). The Pow is the transformed power, more specifically the mean of the squared values of microvolts per second (μV/s) for each electrode processed signal per task. This measure tells us about the amplitude of the signal per channel and per participant magnified to absolute values. We present Pow values on aggregates of participants' individual results, per total task and for each task deciles for the temporal analysis. The task-related power (TRP) is typically calculated taking the resting state as the reference period per individual (Rominger, et al. 2018, Schwab, et al. 2014). We analyzed the EEG recordings of the resting periods prior to the experiment of some of the participants and their results varied considerably, with some participants showing signals that can be associated with the state of being nervous and expectant and their cognitive effort and activity is unknown. As the focus of the study is to determine how well designing can be distinguished from problem-solving, we take the problem-solving Task 1 as the reference period for the TRP calculations. Thus, for each electrode, the following formula was applied taking the mean of the corresponding electrode i, in Task 1 as the reference period. By subtracting the log-transformed power of the reference period (Powi, reference) from the activation period (Powi, activation) for each trial j (each one of the five tasks per participant), according to the formula:

$$TRP_i = log(Pow_i,\ activation)_j - log(Pow_i,\ reference)_j \quad (1)$$

By doing this, negative values indicate a decrease of task-related power from the reference (problem-solving Task 1) for the activation period, while positive values express a power increase (Pfurtscheller, Lopes da Silva, 1999). TRP scores were quantified for total power and Pow temporal analysis was carried out by dividing each experiment session into deciles per task (power and activation refer to brain wave amplitude). Data analysis included Pow and TRP values on individual and aggregate levels using MatLab and open source software.

## 3 ANALYSIS AND RESULTS

Preliminary results of total task-related power (TRP) across the 18 participants indicate that the tasks can potentially be distinguished from each other using the TRP values. The open design Tasks 3 and 4 show higher TRP from the constrained Task 1. The transformed power (Pow), was calculated for each of the 5 tasks and electrodes. Results between the tasks for the industrial designers are depicted in Figure 4. Higher activation in the open design Tasks 3 and 4, particularly in the channels of the right occipitotemporal cortex (F8 to O1), translates the higher conceptual expansion in the problem and solution spaces.

To compare the TRP scores we performed an analysis by running a 4x2x7 repeated-measurement ANOVA, with the within-subject factors task, hemisphere and electrode. From the analysis of the 18 participants *we found a significant main effect of: task*, $p$=.02, and *hemisphere*, $p$=.02. There was no main effect for electrode, $p$=.60. A significant interaction effect between the factors *hemisphere and electrode* was found: $p$<.01. In addition, we conducted pairwise comparisons to check for differences among participants comparing electrodes, hemisphere and task. The pairwise comparisons revealed that Task 4 differs significantly from Pre-task ($p$=.02) and Task 2 ($p$<.01). The transformed power (Pow), was calculated for each of the 5 tasks, electrodes and deciles. To compare the Pow scores we performed an analysis by running a 5x2x7 repeated-measurement ANOVA, with the within-subject factors task, hemisphere and electrode. We found a significant main effect of: *task*, $p$<.001, *hemisphere*, $p$<.001,

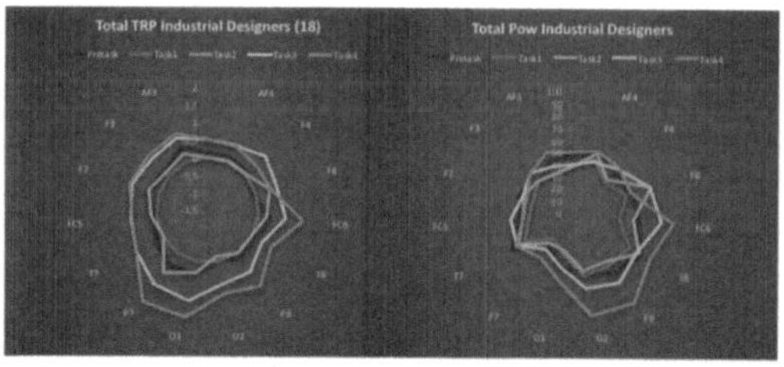


Figure 4. Task-related power (TRP) and transformed power (Pow).

and electrode, $p<.001$. The pairwise comparisons revealed that Task 4 differs significantly from Task 1($p<.001$) and Task 2 ($p<.01$), and Task 3 differs significantly from Task 1($p<.01$) and Task 2 ($p=.01$).

### 3.1 *Temporal analysis and Brodmann areas*

For a temporal analysis of the data, each experiment session is divided into ten equal segments called deciles. The transformed power (Pow) for the constrained Task 1, and the open design Tasks 3 and 4 across channels per decile is depicted in Figure 5. Problem-solving Task 1 has increased general activation in deciles one and seven. Task 3 shows increased general

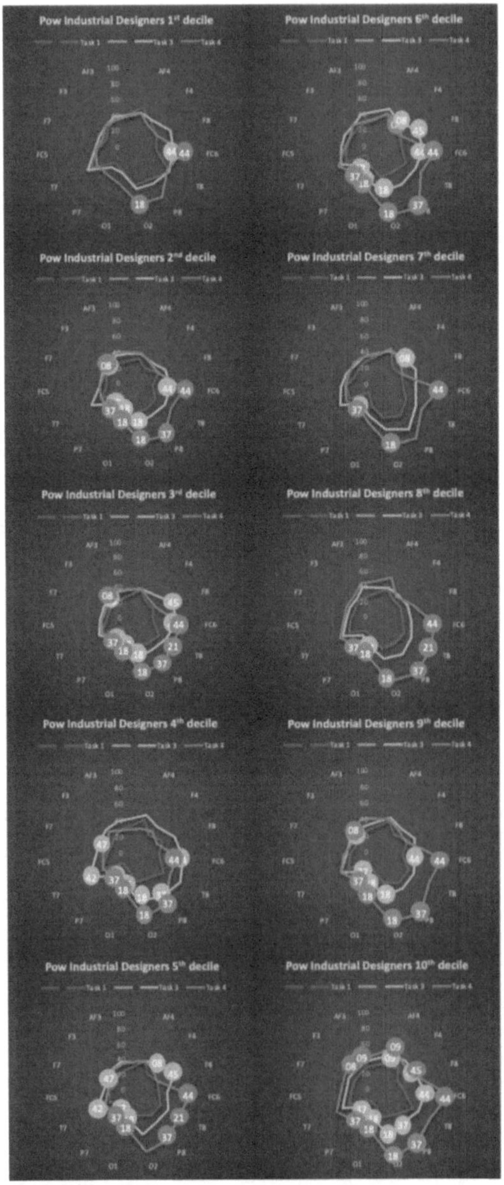


Figure 5. Circles indicate channels that differ from Task 1 to Task 3 and Task 4 by deciles correlated with their Brodmann areas (numerals inside circles).

activation in deciles one, four, six, seven and ten. Task 4 shows higher variation of temporal distributions of activations.

To compare the Pow scores for the deciles we performed an analysis by running a 5x2x7x10 repeated-measurement ANOVA, with the within-subject factors of task, hemisphere, electrode and decile. From the analysis of the 18 industrial designers we found a significant main effect of: *task*, $p<.001$, *hemisphere*, $p<.001$, *and electrode*, $p=.001$. A marginally significant main effect was found for decile, $p=.07$.

Significant interaction effects were found between the factors: *task and hemisphere*, $p=.01$, *task and electrode*, $p<.001$, *task and decile*, $p<.001$, *and hemisphere and electrode*, $p<.01$.

In addition, we conducted pairwise comparisons for hemisphere, electrode, decile and task. The pairwise comparisons revealed that Task 4 differs significantly from Task 1 ($p<.01$) and Task 2 ($p<.01$), Task 3 differs significantly from Task 1 ($p<.01$) and Task 2 ($p=.02$).

The pairwise comparisons also reveal significant differences between deciles: between the first two deciles from which it can be inferred that participants are sorting out how to tackle the tasks request; deciles four and six do not show differences with the others, from which it can be inferred that a more reflective and incubation stage while maturing thinking about the task request takes place; the third, fifth, seventh, eighth and ninth deciles differ from the last one as the refinement of the solutions may differ from searching how to tackle the request.

Statistical analysis indicates significant increased activation of channels placed on the left and right occipital and dorsolateral cortices in the open design tasks compared to the problem-solving task. These channels and their corresponding Brodmann areas (BA), are represented across the deciles in Figure 5. In Figure 5, the circles indicate significant differences and the numerals inside the circles are the Brodmann area number. Brodmann areas refer to unique regions of the cortex and are associated with particular cognitive activities. Brodmann's studies on brain cells' neuron structure and its cytoarchitectural organization in 52 areas (1909) have been refined and correlated to various cortical functions and cognitive activities by measuring blood flow in response to different mental tasks (Glasser, et al. 2016). Multiple magnetic resonance imaging (MRI) measurements have resulted in an extended map with 97 new areas, besides the 83 areas previously reported (Glasser, et al. 2016) with each discrete area containing cells with not only similar structure, but also function and connectivity. Various cognitive functions and connectivity have been identified in studies using fMRI and positron emission tomography (PET).

From the analysis of the open design Task 3, the time span for deciles is 36s. Single channel significant activation takes place in three deciles. In the first decile, channel FC6 shows increased activation of BA 44 whose cognitive functions are associated with inhibition actions, monitoring actions, goals, expressing emotions, working memory, episodic memory and object manipulation (Bernal and Altman, 2009). Such increased activation of FC6 takes place in seven deciles for the open layout design Task 3.

The left temporal cortex and secondary visual cortex have differences in the second, third, fifth, sixth and ninth deciles, with increased activation of BA37 associated with the functions of monitoring shape, intentions, drawing, episodes, familiarity judgments and visual fixation (Le, Pardo, Hu, 1998), and BA18, associated with the functions of spatial and emotional visual processing, on the right hemisphere and visual word form and mental imagery on the left hemisphere (Waberski, et al., 2008). Evidence for higher activation of the right dorsolateral prefrontal cortex happens in the fifth, sixth and tenth deciles in the open layout task. Single activation of channels in this region happen in the third and seventh deciles. No channel shows decreased activation compared to the constrained problem-solving Task 1.

From the analysis of the open design Task 4, the time span for deciles is 70s. For each decile of 70s, statistically significant differences between Task 4 and Task 1 take place in all the deciles. Channel FC6 increased activation of corresponding BA 44, whose cognitive functions are associated with inhibition actions, monitoring actions, goals, expressing emotions, working memory, episodic memory and object manipulation (Bernal and Altman, 2009) takes place in all the deciles as well. The right and left temporal and secondary visual cortices have differentiating contributions in the second to the sixth and in the eighth to the ninth deciles, with increased activation of BA37, associated with the functions of monitoring shape, intentions, drawing, episodes, familiarity judgments and visual fixation (Le, Pardo, Hu, 1998). As Task 4 is an open design free-hand sketching task, drawing activates BA37 (Le, Pardo, Hu, 1998), and other areas of the secondary visual cortex such as BA18 associated with the functions of spatial and emotional visual processing, on the right hemisphere and visual word form and mental imagery on the left hemisphere (Waberski, et al., 2008). Evidence for higher activation of the right dorsolateral prefrontal cortex just takes place in the tenth decile. Spatial memory, recall and planning among other functions attributed to BA09 (Slotnik, Moo, 2006) connected to channel AF4, just show increase in activation compared to Task 1 in the tenth decile. No channel shows decreased activation compared to Task 1. The co-activation of channels of significant differences have two moments of continuous and increasing engagement before and after the seventh decile.

## 4 DISCUSSION AND CONCLUSION

Results from this study demonstrate that EEG is both a practical and relevant technique to study differences in industrial designers while problem-solving and designing. The results of the analysis of the EEG data of the 18 participants show differences in the neurophysiological activations of these industrial designers across tasks and provide initial support for Hypothesis 1: the design neurocognition of industrial designers when problem-solving and designing is different, particularly in open design tasks, Task 3 and Task 4. Industrial designers show higher transformed power (Pow) and distinct task-related power (TRP) differences from the open design Task 3 and Task 4 to the constrained design Task 1. The neurocognitive temporal distributions of activations are non-uniform, providing initial support for Hypothesis 2: industrial designers show variation in the Pow between the problem-solving and design tasks, across the deciles. On a qualitative level the current study shows evidence of a distinct characteristic of increased Pow and TRP of Task 3 and Task 4. Increased activation is associated with conceptual expansion (Abrahams, 2019) from which we infer that the design space inherently expands as well in the designers' search for the problem and the solution.

Evidence for higher activation of the right dorsolateral prefrontal cortex across in design tasks (Alexiou, et al. 2009; Kounios and Beeman, 2009) is shown, particularly in the open layout design Task 3. Evidence from fMRI studies (Alexiou, et al. 2009) of a more extensive network of brain areas in designing than problem-solving can be inferred from these EEG results. Evidence for higher activation of the right occipitotemporal cortex is consistent for both open design tasks. We can propose that for open design tasks the co-activation of channels of significant differences, is consistent for the channels P7, O1, O2, P8 and FC6. In particular for the open layout design, F4 and F8 also integrate the co-activation of channels of significant differences, whose associated cognitive functions seem to be relevant for the design of spatial solutions. Results from the time-related neural responses connected to Brodmann areas' cognitive functions, contribute to a better understanding of industrial designers' cognition in open design tasks. These results can be correlated with previous cognitive studies that explore similar hypotheses (Jiang, Gero and Yen, 2014).

Further detailed analyses are being carried out to provide a more in-depth and comprehensive understanding of the neurophysiological differences between the tasks based on the temporal analysis of frequency bands and their relation to cognitive functions.

Neuroimaging studies (i.e. fMRI, EEG, fNIRS) are more advanced in creative cognition (Abrahams, 2019; Benedek, Jung, Vartanian, 2018; Gero, 2008; Gero, 2015; Kowatari et al., 2009; Martindale and Hines, 1975; Vartanian and Goel, 2005; Xue et al., 2018), and visual creativity, architecture and the arts (see review by Pidgeon, et al. 2016), than in design research. However, no consensus has been found as results do not converge among studies due to the different nature of the tasks and focus. Results from creative cognition studies with focus on insight and divergent thinking problems, may not be particularly central to understand creativity in the context of designing artifacts for the real world (Goel, 2014). Consequently, the design neurocognition field emerges as promising to further a better understanding of the acts of designing across domains and perhaps a more in-depth distinction of creativity in the mental processes associated with design.

Cognitive studies of designers commenced some 50 years ago (Eastman, 1968) with the bulk of the studies occurring in the last twenty years. Neuroimaging studies are a new approach to studying design cognition that have the potential to provide an objective measurement of brain behavior connected to cognition. The potential contributions of neuroimaging studies of design cover a large number of areas including studying the effects of: design domains, tasks, teams, tools and experience on design cognition. In particular, neuroimaging studies contribute to a better understanding of design cognition and have implications for design education, the development of design support and the management of design.

## REFERENCES


Abrahams, A. (2019). *The Neuroscience of Creativity*, Cambridge, UK: Cambridge University Press.

Alexiou, K., Zamenopoulos, T., Johnson, J. H., Gilbert S. J. (2009). Exploring the neurological basis of design cognition using brain imaging: some preliminary results. *Design Studies*, 30(6), 623–647.

Benedek, M., Jung, R., Vartanian, O. (2018). The neural bases of creativity and intelligence: common ground and differences. *Neuropsychologia*, 118, 1–3.

Bernal, B., Altman, N. (2009). Neural networks of motor and cognitive inhibition are dissociated between brain hemispheres: an fMRI study. Int J Neurosci. 119, 10, 1848–1880.

Darke, J. (1979). The primary generator and the design process. *Design Studies*, 1(1), 36–44.

De Clercq, W., Vergult, A., Vanrumste, B., Van Paesschen, W., Van Huffel, S. (2006). Canonical correlation analysis applied to remove muscle artifacts from the electroencephalogram. *IEEE Transactions on Biomedical Engineering*, *53*, 2583–2587.

Dickter, C., Kieffaber, P. (2014). *EEG Methods for the Psychological Sciences*. Sage.

Dorst, K., (2011). The core of 'design thinking' and its application. *Design Studies*, 32(6), 521–532.

Eastman, C. M., (1968). Exploration in the cognitive processes of design, *Defense Documentation Report* No. AD 671158, ARPA, Carnegie-Mellon University.

Ericsson, K. A., Simon, A. H. (1984). *Protocol Analysis: Verbal reports as data*. MIT Press.

Gero, J. S. (ed) (2008). *Studying Design Creativity*, Key Centre of Design Computing and Cognition, University of Sydney, Australia.

Gero, J. S. (ed) (2015). *Studying Visual and Spatial Reasoning for Design Creativity*, Dordrecht, Netherlands: Springer.

Glasser, M., Coalson, T., Robinson, E., Hacker, C., Harwell, J., Yacoub, E., Ugurbil, K., Andersson, J., Beckmann, C., Jenkinson, M., Smith, S., Van Essen, D. (2016). A multi-modal parcellation of human cerebral cortex. Nature. 536 (7615), 171–178.

Le, T., Pardo, P., Hu, X. "4 T-fMRI Study of Nonspatial Shifting of Selective Attention: Cerebellar and Parietal Contributions. The American Physiological Society. Vol.79 No.3. 1998: 1535–1548.

Goel, V. (2014). Creative brains: designing in the real world. *Frontiers in Psychology*, 8, 1–14.

Goel, V., Pirolli, P. (1992). The structure of design problem spaces. *Cognitive Science*, 16, 395–429.

Goucher-Lambert, K., Moss, J., Cagan, J. (2017). Inside the mind: Using neuroimaging to understand moral product preference judgments involving sustainability. *ASME Journal of Mechanical Design*, 139(4), 041103-041103-11.

Hinterberger, T., Zlabinger, M., Blaser, K. (2014). Neurophysiological correlates of various mental perspectives. *Frontiers in Human Neuroscience* (8): 1–16.

Jiang, H., Gero, J. S., Yen, C. C. (2014). Exploring designing styles using Problem-Solution indexes,. In J. S. Gero (ed), *Design Computing and Cognition' 12*, Springer, pp. 85 101.

Jones, J. C., Thornley, D. G. (eds) (1963). *Conference on Design Methods*. Oxford, UK: Pergamon Press.

Kan, J. W.T., Gero, J. S. (2017). *Quantitative Methods for Studying Design Protocols*. Springer.

Kounios, J., Beeman, M. (2009). The Aha! moment: The cognitive neuroscience of insight. *Current Directions in Psychological Science*, 18(4), 210–216.

Kowatari Y., Lee S.H., Yamamura H., Nagamori Y., Levy P., Yamane S., Yamamoto M. (2009). Neural Networks Involved in Artistic Creativity. *Human Brain Mapping*, 30, 1678–1690.

Kruger, C., Cross, N. (2006). Solution driven versus problem driven design: Strategies and outcomes. *Design Studies*, 27(5), 527–548.

Liang, C., Lin, C., Yao, C., Chang, W., Liu, Y., Chen, S. (2017). Visual attention and association: An electroencephalography study in expert designers. *Design Studies*, 48, 76–95.

Liu, L., Li, Y., Xiong, Y., Cao, J., Yuan, P. (2018). An EEG study of the relationship between design problem statements and cognitive behaviors during conceptual design. *Artificial Intelligence for Engineering Design, Analysis and Manufacturing*, 32, 351–362.

Liu, L., Nguyen, T., Zeng, Y., Ben Hamza, A. (2016). Identification of Relationships Between Electroencephalography (EEG) Bands and Design Activities. ASME 2016 International Design Engineering Technical Conferences and Computers and Information in Engineering Conference. Volume 7: 28th International Conference on Design Theory and Methodology. Charlotte, North Carolina, USA, August 21–24, 2016.

Martindale, C., Hines, D. (1975). Creativity and cortical activation during creative, intellectual and EEG feedback tasks. *Biological Psychology*, 3, 91–100.

Pfurtscheller, G., Lopes da Silva, F. (1999). Event-related EEG/MEG synchronization and desynchronization: basic principles. *Clin. Neurophysiol.* 110, 1842–1857.

Pidgeon, L., Grealy, M., Duffy, A., Hay, L., McTeague, C., Vuletic, T., Coyle, D., Gilbert, S. (2016). Functional neuroimaging of visual creativity: a systematic review and meta-analysis. *Brain and Behavior*, 6(10), 1–26.

Purcell, T., Williams, P. & Gero, J. S., Colbron, B. (1993). Fixation effects: Do they exist in design problem solving? *Environment and Planning B*, 20, 333–345.

Rominger, C., Papousek, I., Perchtold, C., Weber, B., Weiss, E., Fink, F. (2018). The creative brain in the figural domain: Distinct patterns of EEG alpha power during idea generation and idea elaboration. *Neuropsychologia*, 118, 13–19.

Shealy, T., Hu, M. and Gero, J. S. (2018) Neuro-cognitive differences between brainstorming, morphologivcal analysis and TRIZ, ASME IDETC.

Schwab, D., Benedek, M., Papousek, I., Weiss, E., Fink, A. (2014). The time-course of EEG alpha power changes in creative ideation. Frontiers in Human Neuroscience, 8, 1–8.

Slotnick, S., Moo, L. (2006). Prefrontal cortex hemispheric specialization for categorical and coordinate visual spatial memory. Neuropsychologia. 44, 9, 1560–1568.

Vartanian, O. and Goel, V. (2005). Neural correlates of creative cognition, in C. Martindale, P. Locher and V. Petrov (eds) *Evolutionary and Neurocognitive Approaches to the Arts*, Baywood Publishing Company, 195–206.

Vergult, A., De Clercq, W., Palmini, A., Vanrumste, B., Dupont, P., Van Huffel, S., et al. (2007). Improving the interpretation of ictal scalp eeg: BSS-cca algorithm for muscle artifact removal. *Epilepsia*, 48, 950–958.

Vos, D., Riès, S., Vanderperren, K., Vanrumste, B., Alario, F., Huffel, V., Burle, B. (2010). Removal of muscle artifacts from EEG recordings of spoken language production. *Neuroinform*, 8, 135–150.

Visser, W. (2009). Design: one, but in different forms. *Design Studies*, 30(3), 187–223.

Waberski, T., Gobbele, R., Lamberty, K., Buchner, H., Marshall, J., Fink, G. (2008). Timing of visuo-spatial information processing: Electrical source imaging related to line bisection judgements. Neuropsychologia. 46, 1201–1210.

*Beyond the Classroom*

*Research & Education in Design: People & Processes & Products & Philosophy – Almendra & Ferreira (eds)*
*© 2020 Taylor & Francis Group, London, ISBN 978-0-367-49296-0*

# Integration of novice designers into interdisciplinary teams

M. Zahedi
*University of Montreal*

ABSTRACT: It is accepted that "essential aspects of Design are better understood in its own terms". With this in mind, we attempt to understand the challenges of novice designers when they enter interdisciplinary collaborative design projects and try to tackle complex and ill-defined problems. We present a way of inquiry to collect data from novice designers themselves and their collaborators. By using the Designerly Activity Theory model, we examine how novices mobilize and apply their theoretical and practical design knowledge in the real practices of collaborative design. In this perspective, it becomes possible to assess the alignment of design education in relation to the aptitude of working with others and towards professional dynamics of collaboration. Identifying the gap will allow us to create training strategies and tools that would improve novice designers' integration into teams.

## 1 INTRODUCTION

Interdisciplinary collaborative design is widely practiced in industry, but design education –in most institutions–, only lightly prepares students to work as team members with non-designers. The increasingly complex society, the growing demand for more inclusive products and services, and the need to achieve results in short timeframes all point to the importance of effective interdisciplinary work. This, in turn, enjoins us to further reflect on how interdisciplinary collaboration affects design practice, and to find ways to bridge research on these matters with design education.

This exploratory project seeks to study the experiences of novice designers when integrating multidisciplinary teams during their early professional tasks. The project is guided by the following research questions:

In interdisciplinary situations, what challenges do novice designers typically encounter during the various phases of the project, and, in particular, during problem reasoning and problem framing?

What strategies do novice designers employ in order to partake in the process of framing problems (i.e. verbally, visually) during collaborative sessions?

How do novice designers enrich their design expertise through their collaborations with more experienced professionals, including ones from different disciplines than their own?

What training do novice designers require in order to effectively engage in problem framing, negotiating and working toward solutions with other experts in interdisciplinary settings?

## 2 THE CHALLENGE OF COMPLEX AND ILL-DEFINED DESIGN PROBLEMS

Design projects (and problems) are commonly characterized as wicked or ill-defined (Rittel and Webber 1973, Buchanan 1995, Cross 2006). Compared to tame problems which mostly have a single and well-defined goal, wicked problems are loosely formulated, they are subject to redefinition/reformulation, and "their formulation already depends on the viewpoints of those presenting them" (Coyne 2005, p 6).

Design projects are also characterized as uncertain, unstable, and unique (Schön 1983). Such problems tend to be difficult to define, and designers alone are seldom enough for successful innovative solutions (Valkenburg & Dorst 1998, Preece et al. 2002, Nelson & Stolterman 2003, Zahedi et al. 2011). As explained by Cross (2002), professional practice throughout the field of design must deal with "messy, problematic situations" where there are many conflicts of ideas and challenging power dynamics among team members. Efficiency of innovative human-centred design solutions requires analysis of problems and elaboration of solutions to progress in parallel involving the unique and diverging ideas of all stakeholders. The difficulties of generating a common understanding in these complex situations and working in collaborative manner are amplified when teams are composed of experts from different disciplines and backgrounds (Dorst & Cross 2001, Kleinsmann & Valkenburg 2008, Zahedi et al. 2011). Within collaborative design, although it might be clear for the team that they all work toward the same goal, they each enter the project with their own expert view. Each individual has a different language or ways of communicating, different ways of representing ideas, different type of reasoning, and different processes for decision-making, tackling projects, and responding to conflicting situations (Bucciarelli 1988, Kleinsmann et al. 2007). Moreover, design activities always aim at human needs and desires, and enriching user experience. These objectives are dynamic, multifaceted, always changing, and dependent on sociocultural and economic situations.

To deal with the complexity and wickedness of these projects, design is in need to use a systemic approach. Systemic approaches help generate a holistic view of the situation at hand in order to understand relations among the interlaced elements impacting the design project (Jones 2017). Design, as a coherent discipline of study, has its own ways of knowing, thinking and doing (Cross 2006, 2011, Dorst 2011) and often uses systemic approach for its reasoning and sense making (Dorst 2015a, 2015b).

During the past fifty years, designerly ways of thinking gained recognition as effective approaches to problem-framing and problem-solving. Also, there has been an ongoing quest for in-depth knowledge on co-creation and co-construction of multidisciplinary teams (Lloyd 2017). Research on collaborative design, particularly within multidisciplinary/interdisciplinary teams, has been a subject of study in itself (Bucciarelli 1988, Valkenburg & Dorst 1998, McDonnell 2015). As stated by McDonnell,

> *Efficiency and effectiveness in design collaboration rely on making it possible for collaborators both to exercise their skills and knowledge relevant to the design task and, at the same time, to be able to exercise their abilities to work collaboratively with others to achieve a shared goal (2012, p 44).*

Designers work in an increasingly varied field of activities and are involved in designing a wide range of products, services, environments, systems, policies and more (Sanders & Stappers 2008). Furthermore, the spread of "design thinking" – which insists on the aim of a systemic view of problems and iterative processes for designing – is changing the way other researchers and practitioners approach problems. In other words, design orients creativity towards innovative solutions and positive social and economic benefits (Nelson & Stolterman 2003, Valkenburg & Sluijs 2012). Based on this understanding, experts from other disciplines (i.e. management, health, economy) now often seek insights through designers' work, thinking and reasoning.

## 3 THEORETICAL FRAMEWORK

The underlying premise of this research is that design is a *social* activity (Bucciarelli 1988, 2002) which involves building agreement through negotiating uncertainty and ambiguity (McDonnell 2015). However, many researchers agree on the potential for functional problems at the outset of design projects due to the differences of disciplinary backgrounds and the context of some projects (for ex. Valkenburg & Dorst 1998, Preece et al. 2002). In other words, lack of understanding among team members and inadequate approaches at the conceptual level may occur, resulting in unsuccessful design (Bucciarelli 2002; Carlile 2004, Kleinsmann et al. 2007). In order to explore possibilities for a better preparation of novice designers in their integration into interdisciplinary design teams, the project draws its theoretical framework from two primary fields of research: (1) Collaborative design (for ex. Bucciarelli 1988; Zahedi et al. 2011, McDonnell 2015), and (2) Activity theory (Vygotsky 1978, Engeström 1987, Kuutti 1996, Engeström1999a).

### 3.1 *Collaborative design*

The guiding principles of innovative and successful design solutions are functionality, desirability, and feasibility. To achieve these characteristics in design projects, a human-centred approach to design –which, by definition, is interdisciplinary– is required to address problems from diverse knowledge perspectives (Kleinsmann et al. 2007, Carroll et al. 2014). Therefore, the engagement of diverse experts is needed from the outset (Chiu 2002, Sanders & Stappers 2008). Referring to Krippendorff, by working with stakeholders as collaborators, the design process changes from problem-solving to a "social process that relies on stakeholders with different and potentially conflicting interests" (Krippendorff 2008, p 65). Since design is purposeful and all activities are directed toward reaching particular goals (Weick 1995, Dorst & Cross 2001, Cross 2006), design teams need to develop consensus on objectives, courses of action and priorities, while taking into account both project context and constraints. In this respect, it is important to highlight that collaborative design is not the same as cooperative design: the latter address problems by drawing on disciplinary expertise as needed; experts of different disciplines engage in solving problems of their respective parts, and parts and solutions are later integrated. In collaborative design, team members work together to share perspectives on project issues and tasks, and take advantage of recurrent exchanges of knowledge. Team members are interdependent and accomplish together what could not be accomplished by each individual alone. Collaborators develop attitudes of openness, aptitudes in sharing information and skills for the co-construction of project-specific knowledge (Kvan 2000). According to Achten,

*Collaborative design is a process in which the participants work together in a meaningful way, not just working together efficiently, but stimulating each other to contribute to the design task. They act towards mutual understanding and maximizing outcomes that satisfy not only their respective goals, but also those of other participants. (2002, p 7)*

In collaborative projects, discipline-specific approaches, different mental models, varying individual aptitudes and abilities, project limitations, and lack of shared language and understanding are all potential barriers to efficient collaboration in design (Kvan 2000; Sanders & Stappers 2008). In summary, a collaborative design team need to cross traditional boundaries and engage in the processes, methods, and tools of other disciplines. The team also needs to develop critical vision, shared commitment, dialogue, and knowledge exchange on creative and innovative processes (Valkenburg & Sluijs 2012, Dorst 2015a). However, most novice designers enter the workforce without a deep understanding of these challenges, and their education often does not provide them with the adequate experiences, tools, and approaches to help them overcome these problems otherwise than by improvising strategies on the spot.

### 3.2 *Activity theory*

Activity theory is a framework that is based on the works of Vygotsky (1978), Leont'ev (1978) and their colleagues. The theory has been further developed until today by Yrjö Engeström (Engeström 1987, 1999a). Activity theory focuses on the development of ideas and the constructive role of humans. It is a framework that studies activity in the context in which it occurs (Engeström 2011, 2015; Igira & Gregory 2009). It describes work and human interactions as a system and examines the activities that people engage in by understanding who they are (subject), what their goals are (object) and what results from the activity (outcome). Other components of activity theory are instrumental and social mediators: tools, rules, community, and division of labor (Engeström 1987, 1999a). These seven interacting components are represented in a triangular model that aid in identifying the contradictions emerging over time in collaborative processes due to the diverging viewpoints, opinions, and needs of involved stakeholders. Activity theory looks at activities as collective dynamics in order to make the invisible visible (Engeström 1999b). Activity theory is widely used in the domains of psychology, ethnography, and education as "a philosophical and cross-disciplinary framework for studying different forms of human practices as development processes, both individual and social levels interlinked at the same time." (Kuutti 1996, p 23). In the context of design research, activity theory is less established (Kuutti 2009), but growing attention is noted from the community. However, activity theory provides a cohesive relationship between scales of activity and a holistic description of design work (Cash et al. 2015). It supports the integration of complex design work and promises a deeper understanding of design processes.

Previous research conducted by the authors confirmed the strong potential of the theory to gain an in-depth understanding of teams' progression in collaborative design projects (Zahedi, Tessier & Hawey 2017). Based on empirical observations and thorough coding of retrieved data, the authors developed an expanded model of activity theory (Zahedi & Tessier 2018, Zahedi, Tessier & Heaton 2018). This expanded model, based on real practice, lived experiences and design situations allows a more fine-grained interpretation of collaborative design activities by considering, observing and analyzing systemically all active components of design activity processes.

This Designerly Activity Theory model adds a second triangular layer of six components to the classic activity theory model, namely: *collective subject*, *object in context*, *signs*, *design criteria*, *imagined community* and *agreed process*. Each plays a role in a collaborative team's journey to framing, proposing ideas and decision-making in a design project. The central gray zone (Figure 1) depicts the original Activity Theory model, the whole figure represents the Designerly Activity Theory (d.AT).

Each component of d.AT is linked to its analog component from the original model, but is specifically connected to design activities. That is, "*collective subject*" translates an effort toward shared understanding in the team to work for the construction of a shared goal by a collective mental model. *Object in context* is associated with the collective understanding of anticipated ideas or solutions in the future context of use, socio-cultural and historical

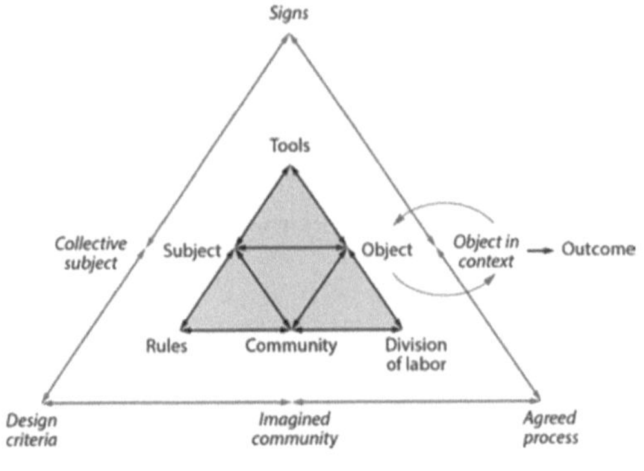


Figure 1. Designerly activity theory model.

settings. At the tip of the triangle, the new component *signs* exposes intangible and internal elements influencing team members' experiences and expertise supporting the team's reflection. Then, the last three designerly components are *design criteria*, *imagined community* and *agreed process*. The first identifies a team's collectively imposed constraints to frame the project. The second considers stakeholders through experiences, stories, scenarios, personas or journey maps to create sense of and interpret available information in a human-centered perspective. Finally, the last component identifies the adoption of agreed-upon ways of tackling a project within a team through the adoption of a design attitude (Boland et al., 2008, p 13). With the model, we seek to make visible taken for granted details of everyday design practice in order to improve it, adapt it and enrich it. We posit that this expanded model will even further enable us to disentangle the collaborative dynamics of interdisciplinary teams. During the present exploratory project, we apply the model to the analysis of real-world situations – the experiences of novice designers' integration into and contributions to– interdisciplinary teams addressing complex design problems. Working with the model to analyze and interpret collected data will also allow us to refine it and to render it more widely applicable to other collaborative activities.

## 4 RESEARCH STRATEGY AND METHODOLOGY

The research project involves the development of knowledge through practice-based inquiry related to real design activities: in other words, research-through-design (Frayling 1993; Findeli 2001, Jonas 2006, Durrant et al. 2015). Thus, the data collection occurs from within the active collaborative teams in which novice designers are engaged. While the research will be guided by the four questions listed in the introduction, it is expected that new questions emerge as the research processes in this type of qualitative inquiry (Findeli 2004; Zeisel 2006). This methodological approach thus allows for an iterative research process. The research collects data from multiple case studies each involving a novice and a design agency. The first case study that has been undertaken recently is considered a pilot. According to which the research team will be able to make any needed adjustments to the research strategies (in particular, in terms of reviewing and/or adding to the main research questions and adapting the data collection tools accordingly).

Data collection takes place during the collaborative design activities of six novice designers (0 to 2 years of experience) during the early stages of their professional tasks. Novice designers are selected amongst our graduating students (each year approximately 50 students graduate from our school of design). Some design agencies, on the other hand, have shown interest and agreed to participate in this endeavor. At the beginning of each case study, the participating novice designer receives a brief training on the principles of research-through-design in order to be prepared to maintain rigorous documentation of his/her design activities.

### 4.1 *Data collection*

In order to gain a strong grasp of the particular dynamics of collaboration, data collection is focused on verbal and visual communication (Lawson 2004). This is for the purpose of creating a profound understanding of design activities (i.e. approaches, explorations, methods), framing and generation of ideas (i.e. negotiations, tools, creativity, tensions, critics), and the roles of the team members who interact with the novice designer (i.e. questioning, reasoning, decision-making). Data collection is achieved through three strategies: (A) in-situ audiovisual recordings, (B) reflective journals, and (C) follow-up interviews. The expected results' outcome will bring clarity on the challenges of novice designers (first three questions of the introduction) and guide the research to propose required training (forth question). In the following paragraphs, we describe each data collection activity and its expected outcomes.

#### 4.1.1 *A. In-situ audiovisual recordings of collaborative design activity*

Periods of conversations during which designers interact with other team members (i.e., during activities of exploration, problems formulation, ideation, branding) will provide the research team with an opportunity to understand "what goes on" during early collaborative design sessions and to gain familiarity with the dynamics of practice among each design team. Close examination of verbal communications during the design discussion is used by many researchers as a fundamental way to build knowledge and understanding of design activity (Cross & Cross 1995; Cross 1999; Dong 2005; Oak 2010; Mohammed & Dumville 2011). As noted by Oak in her article *What can talk tell us about design?:* "through talk, the creativity and constraints of design are continually being managed and performed by participants in practice" (2010, p 214). In addition, attention to the visual and material dimensions of communication, including through the study of gestures, drawings, sketches, charts, prototypes, and other images or objects used in the collaborative process, will offer insight into how creative thinking does not only reside within mental images, but can also potentially be constructed in the material world (Murphy 2005). By focusing on novice designers in real professional situations, we access participants that are still grounded in their disciplinary formation (fresh in their memory), but that have integrated a new professional environment. It is likely that comparing both will let us identify gaps in the existing practices and methods of design education in order to propose paths to enhance it. Moreover, the analysis of interactions (with humans and objects) will contribute to the systemic analysis of these situations.

#### 4.1.2 *B. Detailed critical reflective journals*

We will ask novice designers to write journals (15 to 30 minutes daily) of their collaborative design sessions according to guidelines that focus on tasks, activities, discussions, and development of reflection-on-action (Schön, 1983, Valkenburg & Dorst 1998) and a detailed critical assessment of these experiences. They will share their journals with researchers through a secure online platform. They will also have the possibility to upload and share files and images of visual and material communication tools. As needed, researchers would seek answers to guiding questions that focus on significant events, discussions, actions, reactions, decision-making, etc. This method of data collection aims at recording developments in the design process as they occur, including problems encountered, breakthroughs, or shifts in the efficiency or effectiveness of the collaboration based on changes in the overall strategy, for instance. Reflections on recent lived experiences can raise insights on missing skills, aptitudes or knowledge.

#### 4.1.3 *C. Follow-up interviews*

Following data collection through strategies (A) and (B), interviews will be scheduled with each novice designer, their team leader, and one or two other team members, in order to discuss and reflect on their experiences of the collaborative process. Interviews will be semi-structured according to a set of themes and questions that will have surfaced in the data gathered through strategies (A) and (B) regarding their design activities. These interviews will provide added depth and precision to the previously collected data and its interpretation. By including other participants and practitioners, we may be able to generalize results for a wider range of disciplines.

#### 4.1.4 *Data analysis and interpretation*

All data will be coded using the components of the Designerly Activity Theory model described above. Cross-comparisons of randomly selected sections of the resulting coding will allow the verification of the analytical consistency (see Authors 2018). From this, researchers will proceed to a sequential analysis of the collaborative process in order to highlight the challenges faced by novice designers in interdisciplinary design activity, as well as the strategies they employ to overcome them. Finally, the in-depth analysis of key episodes in the design process will enable us to identify the skills, tools, approaches, and other training strategies that should be integrated into future design education curriculum.

## 5 CONCLUDING REMARKS

The *About* section of REDES2019 conference website states: "The skill-set required from a designer keeps expanding as the complexity of the contemporary world grows and with it the design challenges and situations become ever more ill-defined". Questions about challenges of novice designers, as we presented earlier, align with this statement. With the overarching goal of preparing students for successful design opportunities, we also question "what skills and knowledge future designers will need to cope with the hyper-complexity of the contemporary world" (REDES2019 website). And, how these designers will navigate the expectations that are placed on them within collaborative activities? Up-to-date knowledge is required in order to effectively reinforce educational design programs as well as to inform new models for the integration of novice designers into the workforce.

Indeed, many design programs include some form of service-learning or professional internships that often involve collaboration with stakeholders of varying disciplinary backgrounds and socio-demographic profiles. However, students are often left to learn how to go about such work on the ground. Likely, it is judged that they learned collaboration competencies from experience. This, in turn, can hinder optimal results in terms of learning and project outcomes for both the student and the organization or community they were working for (Bringle & Hatcher 1995, Angotti, Doble & Horrigan 2011, Himley 2004).

Having been celebrated for decades as adopting innovative practices, partly due to studio and project-based approaches, design education is now confronted with a need to adapt to emerging methods and practices of the work world. As mentioned by Jones (2017), we face new challenges related to our societies, migration, equitable economy, as well as climate change, housing and health care; systemic approaches are needed to tackle these complex problems. In this sense, the objective of systemic design is "to affirmatively integrate systems thinking and systems methods to guide human-centered design for complex, multi-system, and multi-stakeholder services and programs across society." (Jones 2017, p 157). The complexity of projects and problems which designers face collaboratively asks for extensive approaches and perspectives to guide their reflexions and progress. Activity Theory, because of its systemic and situated nature is one possible avenue to support the development of design practice accordingly. Moreover, by using the Designerly Activity Theory to analyse the collected data, and by introducing it in the practice of novice designers, the research proposed in this article seeks to identify new avenues in design education that are better aligned with the needs of students preparing to enter the workforce as design practitioners. Designerly Activity Theory will allow us to endorse a systemic point of view to scope relevant aspects of the situation, identify persistent contradictions, and evaluate the impact of active components. The expanded model will support the in-depth comprehension of the challenges encountered by novice designers, both visually and cognitively – allowing for thorough interpretation and comparison of the various case studies.

Taking a closer look at current professional contexts and practices will allow a deep dive and update of the tools that are used, the rules followed and introduced, the community involved, and the division of labour with different team members. Every day, novice designers are confronted with unique challenges, obstacles and opportunities, which they should be fully equipped to solve efficiently and confidently. The observation of diverse contexts and the input of many individuals should allow us to generalize research results to wider teamwork settings – on the basis that designers are involved. The project will also give the opportunity to refine the application of Designerly Activity Theory to interdisciplinary collaborative design. Of course, as the research is still in its preparatory stage, we can only hope for such results to materialize. Still, in coherence with the authors' past research, we know that the components of this project will bridge an existing gap between research, practice, and education in the context of collaborative design.

## ACKNOWLEDGEMNTS

Many thanks to Virginie Tessier for her active participation in this study, for her revision and invaluable comments. This work is supported by the Social Sciences and Humanities Research Council of Canada under Grant number 430-2019-01110.

## REFERENCES

Achten, H. H. (2002). *Requirements for Collaborative Design in Architecture*: Eindhoven University of Technology, Department of Architecture, Building and Planning, Eindhoven.

Angotti, T., C. Doble, & P. Horrigan, eds. (2011). *Service-Learning in Design and Planning: Educating at the Boundaries*. NYU Press, New York.

Boland, R. J., Collopy, F., Lyytinen, K., & Yoo, Y. (2008). Managing as designing: Lessons for organization leaders from the design practice of Frank O. Gehry. *Design Issues, 24*(1), 10–25.

Bringle, Robert G., and Julie A. Hatcher. (1995). "A Service-Learning Curriculum for Faculty." *Michigan Journal of Community Service Learning* 2 (1).

Bucciarelli, L. L. (1988). An ethnographic perspective on engineering design. *Design Studies, 9*(3), 159–168.

Bucciarelli, L. L. (2002). Between thought and object in engineering design. *Design Studies, 23*(3), 219–231.

Carlile, P. R. (2004). Transferring, Translating, and Transforming: An Integrative Framework for Managing Knowledge Across Boundaries. *Organization science, 15*(5), 555–568.

Carroll, J. M., Jiang, H., & Borge, M. (2014). Distributed collaborative homework activities in a problem-based usability engineering course. *Educ Inf Technol, 20*(3), 589–617.

Cash, P., Hicks, B., & Culley, S. (2015). Activity Theory as a means for multi-scale analysis of the engineering design process: A protocol study of design in practice. *Design Studies, 38*(C), 1–32.

Chiu, M.-L. (2002). An organizational view of design communication in design collaboration. *Design Studies, 23*(2), 187–210.

Coyne, R. (2005). Wicked problems revisited. *Design studies, 26*(2), 5–17.

Cross, N. (1999). Design Research: A Disciplined Conversation. *Design Studies, 15*(2), 5–10.

Cross, N. (2002). *Designing Design (Research) 3*. The Inter-disciplinary Design Quandary, De Montfort University.

Cross, N. (2004). Expertise in design: an overview. *Design Studies, 25*(5), 427–441.

Cross, N. (2006). *Designerly Ways of Knowing*. London: Springer.

Cross, N. (2011). *Design Thinking: Understanding How Designers Think and Work*: Berg Publishers.

Cross, N., & Cross, A. C. (1995). Observations of teamwork and social processes in design. *Design Studies, 16*(2), 143–170.

Dong, A. (2005). The latent semantic approach to studying design team communication. *Design Studies, 26*(5), 445–461.

Dorst, K. (2006). *Understanding Design*. Amsterdam: Bis Publishers.

Dorst, K. (2011). The core of 'design thinking' and its application. *Design studies, 32*(6), 521–532.

Dorst, K. (2015). *Frame Innovation: Create New Thinking by Design*. Massachusetts: The MIT Press.

Dorst, K. (2015). Frame creation and design in the expanded field. *She Ji The Journal of Design, Economics, and Innovation, 1*(1), 22–33.

Dorst, K., & Cross, N. (2001). Creativity in the design process: co-evolution of problem–solution. *Design Studies, 22*(5), 425–437.

Durrant, A., Vines, J., Wallace, J., & Yee, J. (2015). Developing a dialogical platform for disseminating research through design. *Constructivist Foundations, 11*(1), 8–21.

Engeström, Y. (1987). *Learning by expanding: An activity-theoretical approach to developmental research*. Helsinki: Orienta-Konsultit.

Engeström, Y. (1999a). Perspectives on Activity Theory (Learning in Doing: Social, Cognitive and Computational Perspectives). Cambridge: University Press.

Engeström, Y. (1999b). Expansive visibilization of work: An Activity-theoretical perspective. *Computer Supported Cooperative Work, 8*, 63–93.

Engeström, Y. (2001). Expansive Learning at Work: Toward an activity theoretical reconceptualization. *Journal of Education and Work, 14*(1), 133–156.

Engeström, Y. (2015). *Learning by Expanding: An Activity-Theoretical Approach to Developmental Research*. (Second Edition). New York: Cambridge University Press.

Findeli, A. (2001). Rethinking Design Education for the 21st Century: Theoretical, Methodological, and Ethical Discussion. *Design Issues*, *17*(1).

Findeli, A. (2004). *La recherche-projet: une méthode pour la recherche en design*. La conférence de recherche sur le design, Tenu à la HGK de Bâle sous les auspices du Swiss Design Network.

Frayling, C. (1993). *Research in Art and Design*. Research Papers, presented at the Royal College of Art (RCA).

Himley, M. (2004). "Facing (Up to) 'the Stranger' in Community Service Learning." *College Composition and Communication* 55 (3): 416–38.

Igira, F. T., & Gregory, J. (2009). Cultural Historical Activity Theory. In Y. K. Dwivedi, B. Lal, M. D. Williams, S. L. Schneberger, & M. R. Wade (Eds.), *Handbook of Research on Contemporary Theoretical Models in Information Systems* (pp. 434–454): IGI Global.

Jonas, W. (2006). *Research through DESIGN through research - a problem statement and a conceptual sketch*. Paper presented at the Design Research Society International Conference, Lisbon, Portugal.

Jones, P. (2017). The Systemic Turn: Leverage for world changing. *She Ji. The Journal of Design, Economics, and Innovation*, 3(3), 157–162.

Kleinsmann, M., & Valkenburg, R. (2008). Barriers and enablers for creating shared understanding in co-design projects. *Design studies*, *29*(4), 369–386.

Kleinsmann, M., Valkenburg, R., & Buijs, J. (2007). Why do(n't) actors in collaborative design understand each others? An empirical study towards a better understanding of collaborative design. *CoDesign*, *3*(59-73).

Krippendorff, K. (2008). *The Semantic turn*. Boca Raton, USA: CRC Press Book.

Kuutti, K. (1996). *Activity Theory as a potential framework for human-computer interaction research*. In B. Nardi (Ed.), Context and Consciousness: Activity Theory and Human Computer Interaction (pp. 17–44). Cambridge: MIT Press.

Kuutti, K. (2009). Artifacts, Activities, and Design Knowledge. In S. Poggenpohl & K. Sato (Eds.), *Design Integrations: Research and Collaboration* (pp. 67–85): Intellect.

Kvan, T. (2000). Collaborative design: what is it? *Automation in Construction*, *9*.

Lawson, B. (2004). Schemata, gambits and precedent: some factors in design expertise. *Design Studies*, *25* (5), 443–457.

Leont'ev, A. N. (1978). *Activity, consciousness and personality*. Englewood Cliffs, NJ: Prentice Hall.

Lloyd, P. (2017). From Design Methods to Future-Focused Thinking: 50 years of design research. *Design studies*, *48*(C), 8.

McDonnell, J. T. (2012). Accommodating disagreement: A study of effective design collaboration. *Design studies*, *33*(1), 44–63.

McDonnell, J. (2015). Gifts to the Future: Design Reasoning, Design Research, and Critical Design Practitioners. *she ji The Journal of Design, Economics, and Innovation*, *1*(2), 107–117.

Mohammed, S., & Dumville, B. C. (2001). Team mental models in a team knowledge framework: expanding theory and measurement across disciplinary boundaries. *Journal of Organizational Behavior*, *22*, 89–106.

Murphy, K. M. (2005). "Collaborative Imagining: The Interactive Use of Gestures, Talk, and Graphic Representation in Architectural Practice." *Semiotica* 2005 (156): 113–145.

Nelson, H. G., & Stolterman, E. (2003). *The design way*. Englewood Cliffs, New Jersey: Educational Technology Publications.

Oak, A. (2010). What can talk tell us about design?: Analyzing conversation to understand practice. *Design Studies*, *32*(3), 211–234.

Preece, J., Rogers, Y., & Sharp, H. (2002). *Interaction Design: beyond human-computer interaction*. New York: Wiley.

Rittel, H., & Webber, M. (1973). *Dilemmas in General Theory of Planning, PP 155-169*. Working Papers from the Urban & Regional Development, University of California, Berkeley.

Sanders, E. B.-N., & Stappers, P. J. (2008). Co-creation and the new landscapes of design. *CoDesign*, *4* (1), 5–18.

Schön, D. A. (1983). The Reflective Practitioner: How Professionals Think in Action. New York: Basic books.

Valkenburg, R., & Dorst, K. (1998). The reflective practice of design teams. *Design Studies*, *19*(3), 249–271.

Valkenburg, R., & Sluijs, J. (2012). *The world of the open innovator*. The Netherlands: The Hague University of Applied Sciences.

Vygotsky, L. S. (1978). *Mind in Society: the Development of Higher Psychological Processes*. Cambridge: Harvard University Press.

Weick, K. E. (1995). Sensemaking in organizations. Thousand Oaks, CA: Sage.
Zeisel, J. (2006). *Inquiry by design: environment, behavior, neuroscience in architecture, interiors, landscape, and planning*. New York: W.W. Norton & Company.
Zahedi, M., Guité, M., & De Paoli, G. (2011). Addressing user-centeredness: Communicating meaningfully through design. In P. Leclercq, A. Heylighen, & G. Martin (Eds.), *Designing Together, CAADFutures 2011* (pp. 513–524). Liège, Belgium: PUL.
Zahedi, M., Tessier, V., & Hawey, D. (2017). Understanding Collaborative Design Through Activity Theory. *The Design Journal*, *20*(1), S4611–S4620. doi:10.1080/14606925.2017.1352958
Zahedi, M., & Tessier, V. (2018). *Designerly Activity Theory: Toward a New Ontology for Design Research*. The Design Research Society (DRS) - Catalyst conference, University of Limerick, Ireland.
Zahedi, M., Tessier, V., & Heaton, L. (2018). *Designerly Activity Theory insights on the Design Processes of a Korean Company*: Design Thinking Research Symposium (DTRS12), UNIST, Korea.

*Research & Education in Design: People & Processes & Products & Philosophy – Almendra & Ferreira (eds)*


# Design Education for the 21st Century: The multiple faces of disciplinarity

S. Antunes & R. Almendra
*CIAUD, Lisbon School of Architecture, Universidade de Lisboa, Portugal*

ABSTRACT: Preparing designers for the 21st century working environments involves providing them with skills to quickly adapt to a world of continuous change, to face a time where flexibility is a must and where uncertainty is the most assured reality.

Societies and economies are now knowledge-based, due to globalization and technology; knowledge is then the key to succeed; in a knowledge-based word, professionals with different disciplinary backgrounds need to join efforts in teams where they interact and share their expertise to achieve common goals; collaboration, communication and critical thinking are some of the crucial skills they should possess and their education pathway should include opportunities to acquire those skills. The education process should then include experiences that mimic those environments, that some classify as multidisciplinary, while others prefer interdisciplinary, or even transdisciplinary. This paper contributes to clarify these terms and to assert which the correct level of disciplinarity might be.

## 1 INTRODUCTION

"Profession-specific world-views merely prepare individuals to work within their own profession, not to communicate with individuals from another profession" (Hall, 2005, p. 193).

Like in most other professions, the designers of today and tomorrow need to improve their skills (or even acquire new skills) to face the challenges of a fast-evolving world, in which knowledge-based economies predominate (MGI, 2017c; MGI, 2017a; MGI, 2017b).

Among the so-called skills for the 21st century, collaboration and communication stand out as essential tools to succeed in working environments including a diversity of professions (P21, 2016; WEF, 2016; Lai et al., 2017). Sometimes designated as multidisciplinary, those environments are often referred to as interdisciplinary, transdisciplinary or even cross-disciplinary. These terms are also used when describing teaching and learning situations, most of the times interchangeably, without a coherent distinction between them. In fact, in this almost promiscuous use of words, other terms such as pluridisciplinary or interprofessional are also used; in this research, we will approach the most common, and perhaps most discussed ones – multidisciplinary, interdisciplinary and transdisciplinary, in an effort to contribute to a clearer understanding of its meaning and applicability.

In fact, several authors (Devlin & Davies, 2007; Park & Son, 2010; Pirrie et al., 1999; Uwizeymana & Basheka, 2017) have tried to establish that distinction, and several approaches have been used during those trials; and although we could find a tendency or trend that points in a certain direction in doing that distinction, there is still a great lack of consensus, as their conclusions are based on diverse assumptions and using different approaches.

When discussing how to prepare future designers for the working environments of the next decades, a clarification of the terms used to characterize those environments may help understanding in which ways can we help those students in acquiring the essential 21st century skills (Komnenić et al., 2016).

## 2 TERMS AND DEFINITIONS

Before getting into the confusion and ambiguity created by the various views on what the terms multidisciplinary, interdisciplinary and transdisciplinary mean to different authors, it is important to analyze the basic words from which those terms originated, namely discipline and disciplinarity.

While a consensual meaning for the prefixes 'multi', 'inter' and 'trans' is easily found in dictionaries, as well as it seems easy to understand that the adjective 'disciplinary' relates to the basic word 'discipline', a less consensual opinion arises when several authors propose definitions for the composed prefix + adjective terminology.

Considerable efforts have been made to clear the confusion created by that lack of consensus, and one that's worth to mention is that of Bernard Choi & Anita Pak (2006), in their 1st article on *Multidisciplinarity, interdisciplinarity and transdisciplinarity in health research, services, education and policy*. Taking a methodic and deep approach to the terminology, they searched both hard-copy and online dictionaries, looking for a clearer understanding of wording and definitions.

From that profound research we acknowledge that ***'discipline'*** is defined in most dictionaries as 'a branch or field of knowledge', as well as 'learning', 'teaching', 'instruction', 'education' and 'activity' while the meaning of ***'disciplinary'*** is anything that relates to disciplines; '***multi***' is defined as 'many' or 'more than one', '***inter***' means 'among', 'between', 'mutual' and '***trans***' is explained as 'across', 'over', 'beyond', 'through' or 'on the far side of'. While these basic words are then easier to define and understand, there seems to exist less consensus when the composed terms are referred to, and as a matter of fact they are even missing in older dictionaries, indicating that its origin and use is somewhat recent: the term *multidisciplinary* is defined as 'involving several disciplines', *interdisciplinary* as 'involving two or more disciplines' and *transdisciplinary* as interdisciplinary, showing almost no difference between the understanding of the three words (Choi & Pak, 2006).

It should be mentioned that the fact of an existing common root – the word 'discipline' – this does not actually help much in clearing the confusion of the disciplinary variations being discussed. As Olga Pombo (2013) points out in *Epistemología de la interdisciplinaridad. La construcción de un nuevo modelo de comprensión*, even the root word 'discipline' may have at least three different meanings: one is the aforementioned field of knowledge, such as Mathematics, Chemistry or Biology; a related second one would describe a discipline as a specific part of a curricula, as Applied Mathematics or Biochemistry; and a third and well distinct meaning is when using the word discipline to define a set of rules applied to a certain activity or group – as in military discipline or in school discipline.

Several other authors have also struggled to find a clear distinction between the terms used to describe the various forms of disciplinarity, in contexts mostly related to education, as well as in working environments. When investigating this terminology within the context of higher education, Drs. Marcia Devlin & Martin Davies (2007), researchers at the Centre for the Study of Higher Education of the University of Melbourne, found interesting considerations made by other authors about what is traditionally known as an 'academic discipline', such as those of Beyer & Lodahl (1976)[1] that describe it as providing a structure of knowledge, of Becher (1981)[2], who classify it as discrete and autonomous, but not homogeneous, and of Squires (1992)[3], who defines it as a distinct area of study within an institution, including theories, methods and content. Devlin & Davies (2007) consider that the features usually used to define the nature of an 'academic discipline' include a community of scholars, a traditional mode of inquiry to treat data, the requirements to define new knowledge and

[1] Beyer, J. & Lodahl, T. (1976). A Comparative Study of Patterns of Influence in United States and English Universities. *Administrative Science Quarterly, 21*, 104 - 129.

[2] Becher, T. (1981). Towards a Definition of Disciplinary Cultures. *Studies in Higher Education, 6*(2), 109-122.

[3] Squires, G. (1992). Interdisciplinarity in Higher Education in the United Kingdom. *European Journal of Education, 27*(3), 201-210.

a communications network; and as such, the term 'disciplinarity' is related to that traditional academic interpretation of 'discipline', and is used to describe academic disciplines as "autonomous and discrete areas of study within which academic communities rarely cooperate or coordinate their academic efforts" (Devlin & Davies, 2007, p. 2).

However, academic disciplines evolve over time and they also change with cultural differences, meaning that are not historically or culturally fixed. Devlin & Davies (2007) also mention the problem posed by that traditional view on what is an academic discipline, and describe some attempts to redefine that notion, such that of Squires (1992)[4], who stated that all disciplines are "multidimensional spaces in which define, protect and enlarge themselves along those dimensions, and in so doing, come into conflict or cooperation with other disciplines" (as cited in Devlin & Davies, 2007).

From the above we conclude that the basic term 'discipline', both in the traditional view and in the more recent notion, is strongly related to the academic terrain, although it can, and is also used, in the professional environment, due to the intrinsic connection between education and work. In both environments the term usually relates to a field of knowledge, although it may also be used to describe a set of rules.

Due to the strong 'academic nature' associated with the term 'discipline', is not surprising that considerably more education-related disciplinary approaches have been made than those linked to the professional environment; but as just mentioned, education is usually acquired as a preparation to the future working arena and this link between both worlds suggests that a similarity between approaches would exist, when analyzing discipline-related terminology. Let's then have a brief look on some of those approaches.

## 3 DISCIPLINARY APPROACHES TO EDUCATION

Education: Should learning be multidisciplinary, interdisciplinary or transdisciplinary? Is there any significant difference in using one of the terms and not another, when exploring different approaches to education? If yes, which term and approach should we use in design education?

At a practical level, as there is a lack of consensus about possible distinct meanings, these terms are used interchangeably, most of the times with the same intention. What most authors mean when using either of them is essentially that more than one discipline is involved. But when discussing which are the relevant issues to design a course of any kind, all possible theories must be explored, even if only at the theoretical level. Not doing so could lead to a less solid foundation on the knowledge base that must support any well-designed course.

As such, it not only interesting, but in fact important, that we explore the different views on how a distinction can be made between multidisciplinary, interdisciplinary and transdisciplinary, in order to establish a coherent position about our own interpretation of each of them.

One simple approach is that referred by Wilson & Pirrie (2000) in *Multididisciplinary Teamworking Indicators of Good Practice*. They based their interpretation of the terms in the assumption of the existence of three dimensions, numerical, territorial and epistemological. They argue that '*inter*' relates to the involvement and reciprocal interaction of only two disciplines or only two professions; still having in mind the numerical dimension approach, when more than two disciplines are involved, *'multi'* would then be the more appropriate term. When applying the territorial dimension, discipline boundaries are the important matter; in this view, those boundaries become blurred in an interdisciplinary environment, while in a multidisciplinary one, there is little interaction between disciplines. The epistemological dimension relates to the creation of new knowledge that results from the interaction of different disciplines, well beyond the blurring of their boundaries.

[4] Squires, G. (1992). Interdisciplinarity in Higher Education in the United Kingdom. *European Journal of Education, 27*(3), 201-210.

It is interesting to notice that other authors also tend to agree on a certain *crescendo* of this level of discipline interaction, although sometimes approaching it from different perspectives, like when reflecting about the cognitive integration associated with that interaction:

During the plenary talk from the First Global Conference on Research Integration and Implementation: *Interdisciplinarity Then, Now, and Into Networked Futures* (2014), and in what concerns discipline interaction, Julie Thompson Klein, Faculty Fellow and Professor at the Wayne State University, explains her view on the degree of cognitive integration associated to each of the commonly used terms, which she synthetizes in the diagram bellow:

| UNIDISCIPLINARY | MULTIDISCIPLINARY | INTERDISCIPLINARY | TRANSDISCIPLINARY |
|---|---|---|---|
| Focusing | Juxtaposing | Integrating | Transcending |
| Concentrating | Sequencing | Interacting | Transgressing |
| Analyzing | Coordinating | Linking | Transforming |
| Segmenting | | Focusing | Overarching |
| | | Blending | |
| | | Hybridizing | |
| | | Synthesizing | |

Figure 1. Degree of cognitive integration (ANU, 2014).

Klein (ANU, 2014) considers that in the 'multidisciplinary' case, only a juxtaposition of disciplines is provided in an additive, rather than integrative process, thus where disciplinary perspectives are retained. For example, a painting may be analyzed under different and not interacting disciplinary views or perspectives, such as geometry or art history.

However, in an 'interdisciplinary' process, Klein (ANU, 2014) says that two or more disciplines may be involved in a deeper interaction, and from this interaction integrated knowledge may appear, such as when nuclear physics and medicine combined may lead to a new treatment for a certain disease; or the appearance of a new discipline as a result from combining, such as the combination of math and physics that originated mathematical physics. A lot of other examples exist, such as astrophysics or biochemistry.

And in what concerns 'transdisciplinarity', in Klein's opinion (ANU, 2014), what matters are the dynamics of whole systems and as such, holistic schemes are created having in mind those dynamics.

According to the holistic view, "transdisciplinarity defines a resarch focused on problems that cross the boundaries of two or more disciplines" (Zaman & Goschin, 2010, p. 7). Zaman & Goschin (2010) refer that a holistic or transdisciplinary approach to a problem such as pollution, may use concepts or methods that are not exclusive of one discipline, although might have initially been developed by that discipline.

Despite the confusion still surrounding the three terms (some authors even use others, such as cross-disciplinary) we tend to agree with Choi & Pak (2006), who concluded that there is a trend to converge towards certain meanings; most authors use these terms to describe "the involvement of multiple disciplines to varying degrees on the same continuum" (Choi & Pak, 2006, p. 351) and those degrees of involvement can be summarized as follows:

Multidisciplinary – several disciplines, working side by side with little or no interaction (basic involvement);

Interdisciplinary – two (sometimes more) disciplines interacting to a level where disciplinary borders are crossed and new knowledge is generated, in the form of new perspectives, methods or disciplines;

Transdisciplinary – involves a deeper involvement level, where members from the different disciplines sometimes expand or exchange roles, looking for the dynamics of a whole system (holistic approach).

Choi & Pak (2006) further propose that those terms can be translated into everyday words, and present a table where a salad bowl represents multidisciplinary, interdisciplinary is represented by a melting pot, and the transdisciplinary example is a cake; the respective ingredients either remain totally distinguishable, partially distinguishable or completely undistinguishable:

Table 1. Views on multidisciplinary, interdisciplinary and transdisciplinary (Choi & Pak, 2006, p. 359).

| | Multidisciplinary | Interdisciplinary | Transdisciplinary |
|---|---|---|---|
| Keyword | Additive | Interative | Holistic |
| Mathematical | 2 + 2 = 4 | 2 + 2 = 5 | 2 + 2 = yellow |
| Food example | A salad bowl | A melting pot | A cake |

Park & Son (2010) also present an interesting comparison of terms, focusing however in the learning modes associated with each of them, and in the type of interactivity involved in each mode, as well as in the roles of student and teacher:

Table 2. Learning modes and interactivity (Park & Son, 2010, pp. 82-93).

| Learning mode | Interactivity | Student identity | Teacher identity |
|---|---|---|---|
| Subject<br>Disciplinary | Topic driven<br>Subjects driven | Knowledge receiver | Knowledge deliverer |
| Multidisciplinary | Discipline to discipline driven | Knowledge consumer | Knowledge facilitator |
| Interdisciplinary | Learner collaboration driven | Knowledge collaborator | Learning designer |
| Transdisciplinary | Learner participation and new knowledge creation driven | Knowledge producer | Interactive learning designer |

Approaches that are mostly education-focused tend, as seen, to agree on an increasing level of knowledge integration of discipline interaction, related to the three most common terms used to describe disciplinarity; and as we will show, the same tendency exists on professional-related approaches.

## 4 DISCIPLINARY APPROACHES TO WORKING ENVIRONMENTS

"The importance of interprofessional teamwork is becoming increasingly recognized" (Hall, 2005, p. 194).

Current societies and economies are based on a vast and deep knowledge foundation, mainly because of the unlimited access to information brought by constant and unstoppable advances in technology.

The vast amount and diversity of information available makes the task of individual professionals or of smaller teams extremely difficult execute, as it would require each of those individuals to acquire a considerable amount of specialized knowledge and training in a broad variety of disciplines; not an easy task at the least, but not using all available information would pose the risk of losing competitivity or of creating a less appealing product or service. This dilemma is probably the reason for the urgent need for an increasing collaboration in today's working environments, allowing professionals from different fields to contribute with their specific knowledge area to the common goal of the team.

This diverse composition of most working groups is important, when trying to achieve a goal that requires the use of a broad field of knowledges; and, such as when discussing approaches to education, it brings about the question of which is the appropriate approach (and term) to use, when analyzing how teams ought to be formed and how teamwork should be accomplished: should it be done using a multidisciplinary, interdisciplinary or transdisciplinary perspective?

If the simple concept of 'team' may vary from any group of persons brought together to pursue a certain goal to a number of specialists sharing their knowledge to accomplish a complex task, we see how challenging it might be to define the degree of disciplinarity that should characterize the interaction within those groups. More elaborated definitions such that of Lorimer & Manion (1996)[5] describes 'team' as "a small number of consistent people committed to a relevant shared purpose, with common performance goals, complementary and overlapping skills, and a common approach to their work" (as cited in Choi & Pak, 2006). But how small should that group be, and to which degree should their skills complement or overlap? Should group members work in parallel but separately in their specific field, then putting individual results together or should they all mix their ideas and roles from the very start of task execution? In fact, by posing these questions we are trying to assess which the adequate level of interaction is, or, in other words, the optimum degree of disciplinarity that should be involved; however, just like in learning & education, this is not an easy task.

The same difficulty was felt by Uwizeymana & Basheka (2017) when trying to assess the multidisciplinary, interdisciplinary or transdisciplinary nature of public administration, both as a discipline (PA[6]) and as a professional career (managing public affairs); as per the discipline, they opted by characterizing it as multidisciplinary, as PA curricula is constituted by a mixed combination of subjects, recognizing, nonetheless, the influence of disciplines among and between themselves, as no discipline is an island. But in relation to the professional activity, they preferred not to agree on using only one of the three terms, but mostly, all at the same time, or even a more general one, 'multiple disciplinary'. They argued that due to its political character and to the complex and sometimes contradictory nature of public demands, public administration needs to continuously evolve and adapt to the changing environment where it is practiced; as such, it often requires varying degrees of the same continuum (Uwizeymana & Basheka, 2017).

In their research on collaborative practice in health professions *Multidisciplinary education: some issues and concerns*, also troubled with the striking terminological lack of consensus, although recognizing a general trend to accept that there is a varying degree of disciplinary interaction associated with the three most commonly used words (multidisciplinary, interdisciplinary and transdisciplinary), Pirrie, Hamilton & Wilson (1999) decided on a more conservative approach: using multidisciplinary as the most ambiguous one, and as such allowing for a broader interpretation. This is to prevent an eventual misunderstanding on which degree of multidisciplinarity should exist in health education to prepare their professionals to deliver their service, as they think that individual practitioners should be able to work collaboratively, while retaining unique areas of skill and knowledge. Increased degrees of discipline interaction, such as those usually associated with interdisciplinarity or transdisciplinarity, would pose the risk of not being able to retain those unique areas of skills and knowledge, as opposed to the vaguer or less strict concept of multidisciplinarity (Pirrie et al., 1999).

Also analyzing the composition and work of health care teams, Hall & Weaver (2001) refer the dilemma clinicians face as a result of the complexities of patient care: an constant increase in specialization is required by the increase in complexity of knowledge and skills necessary to provide comprehensive care to patients; this decreases interdisciplinary exchange, as communication is eased by specialized vocabulary, similar understanding and common methods, thus challenging collaboration between specialists. The approach taken in many instances is grouping different professionals in health care teams, that will "function somewhere along a continuum of degrees of interaction among team members and their degrees of responsibility for patient care" (Hall & Weaver, 2001, pp. 867-868).

In our view, the conservative approach that prefers the less strict term 'multidisciplinary' is adequate, especially in the initial phases where different professionals collaborate to a common goal, and the desired level of interaction is not yet well defined; in a more advanced stage of collaboration, they should, and naturally will, evolve from the multidisciplinary to the interdisciplinary approach, as the integration level deepens; however, they should

[5] Lorimer W. & Manion J. (1996). Team-based organizations: leading the essential transformation. *PFCA Rev.*, 15-9. Spring.

[6] Uwizeymana & Basheka refer that the capital letters 'PA' are used to refer public administration as a discipline, to diferentiate from the public administration activity.

probably avoid a further evolution to the transdisciplinary approach, as that implies a blurring of knowledge and roles and this might pose a serious risk to the necessary deeper field-specific knowledge and experience that each one shall retain.

## 5 DESIGN EDUCATION

"Education on how to function within a team is essential if the endeavour is to succeed" (Hall & Weaver, 2001, p. 868).

Just as in other fields of activity, the design profession has its own specific language, its understanding on how to approach and solve a problem, often being more based on a design-related view (although including perspectives from several design areas) than on an integration of disparate opinions of people diverse professions. But the working arena where designers of today and tomorrow have to contribute is characterized by an increasing need of skills such as collaboration and communication within a multidisciplinary environment, here understood as including other professions; when preparing future designers, special attention must then be taken in providing them with those skills, so they are able to easily integrate multidisciplinary teams, minimizing the adaptation process needed to start contributing to common goals and projects.

The beneficial effects of multidisciplinary exposure are show through research, such as that of Tang & Hsiao (2013), who observed teams that included students with different backgrounds and origins working on three design projects over a period of one year; the projects originated from a design studio and students were required to continue the design process until achieving fully functional prototypes. The researchers closely followed the work of the different teams and their final conclusions indicate that multidisciplinary collaboration had improved the student's communication skills, their understanding of the collaborative process, their professional abilities and brought a sense of collective achievement (Tang & Hsiao, 2013).

Since our preferred approach to the way a team of diverse professionals should interact is to start as a multidisciplinary team that may later evolve to an interdisciplinary group, where the level of interaction should not jeopardize the need to retain certain field-specific characteristics, design education should also, in our opinion, follow the same rule: design students must be exposed to a certain level of multidisciplinary experiences, where they will learn how to work with colleagues, and if possible also teachers and industry professionals from knowledge fields and activities other than design-related, to facilitate their entry in the future working arena; the number and characteristics of these academic multidisciplinary experiences will determine the easiness and speed they will need later, when being part of multidisciplinary or even interdisciplinary working teams.

## 6 CONCLUSIONS

There is an increasing need to empower design students with crucial skills that will enable them to easily integrate the multi-professional working teams that characterize today's and tomorrow's businesses and societies; an effort must be made to provide them with opportunities to acquire those skills, during their educational process.

Having in mind the terminology aspects discussed earlier in this article, from the several possible approaches on how that effort should be made during education, our preference is to use the multidisciplinary path, as it is a more cautious practice that may prevent an excessive blurring of knowledge and roles, that is, in our view, undesirable, even later, when in working environments; that is the risk posed by the transdisciplinary approach. By starting with multidisciplinary experiences during education students will learn the fundamentals of collaboration and communication, while retaining the field-specific knowledge and language; later on, during their working career, an increasing level of discipline interaction may occur, such as that described as interdisciplinarity.

The specific characteristics and implementation methods of such experiences or activities require further investigation and research.

## REFERENCES


Australian National University – ANU (2014, October 8). *Interdisciplinarity: an overview*. Retrieved from: https://i2s.anu.edu.au/resources/interdisciplinarity-overview.

Choi, B. & Pak, A. (2006). Multidisciplinarity, interdisciplinarity and transdisciplinarity in health research, services, education and policy: 1. Definitions, objectives, and evidence of effectiveness. *Clinical Investigative Medicine*, *29*(6), 351–364.

Devlin, M. & Davies, M. (2007). Interdisciplinary higher education: Implications for teaching and learning. Melbourne: CSHE.

Hall, P. (2005). Interprofessional teamwork: Professional cultures as barriers. *Journal of Interprofessional Care, Supplement 1*, 188–196.

Hall, P. & Weaver, L. (2001). Interdisciplinary education and teamwork: A long and winding road. *Medical Education, 35*, 867–875.

Komnenić, B., Borovnjak, I., Filek, S. & Velinovski, A. (2016). *Practicing Design: Rethinking Design Education*. Zagreb: Designaustria - Croatian Designers Association & Public Room Skopje.

Lai E., DiCerbo, K. & Foltz, P. (2017). *Skills for Today: What We Know about Teaching and Assessing Collaboration*. London: Pearson.

McKinsey Global Institute - MGI (2017a). *A Future that Works: Automation, Employment and Productivity* (Executive Summary). McKinsey & Company.

McKinsey Global Institute - MGI (2017b). *Jobs Lost, Jobs Gained: Workplace Transitions in a time of Automation* (Executive Summary). McKinsey & Company.

McKinsey Global Institute - MGI (2017c). *Technology, Jobs, and the Future of Work* (Briefing Note). McKinsey & Company.

Park, J. & Son, J. (2010). Transitioning toward transdisciplinary learning in a Multidisciplinary environment. *International Journal of Pedagogies and Learning*, *6*(1), 82–93.

Partnership for 21st Century Learning - P21 (2016). *Framework for 21st Century Learning*. Washington, DC: Partnership for 21st Century Learning.

Pirrie, A., Hamilton, S. & Wilson, V. (1999). Multidisciplinary education: some issues and concerns. *Educational Research*, *41*(3), 301–314.

Pombo, O. (2013). Epistemología de la interdisciplinariedad. La construcción de un nuevo modelo de comprensión. *Interdisciplina I*, *1*(1), 21–50.

Tang, H. & Hsiao, E. (2013). The advantages and disadvantages of multidisciplinary collaboration in design education. *5th International Congress of International Association of Societies of Design Research*. Tokyo, Japan.

Uwizeymana, D. & Basheka, B. (2017). The Multidisciplinary, Interdisciplinary and Transdisciplinary Nature of Public Administration A Methodological Challenge?. *The African Journal of Public Affairs*, *9*(9), 1–28.

Wilson, V. & Pirrie, A. (2000). *Multidisciplinarity teamworking Indicators of good practice*. Edinburgh: SCRE.

World Economic Forum - WEF (2016). *New Vision for Education: Fostering Social and Emotional Learning* through Technology. Retrieved from: http://www3.weforum.org/docs/WEF_New_Vision_for_Education.pdf.

Zaman, G. & Goschin, Z. (2010). Multidisciplinarity, Interdisciplinarity and Transdisciplinarity: Theoretical Approaches and Implications for the Strategy of Post-Crisis Sustainable Development. *Theoretical and Applied Economics, Volume XVII*, *12*(553), 5–20.

*Research & Education in Design: People & Processes & Products & Philosophy – Almendra & Ferreira (eds)*
*© 2020 Taylor & Francis Group, London, ISBN 978-0-367-49296-0*

# *Parergon* in *K* Magazine

P.C. Viegas
*CIAUD, Lisbon School of Architecture, Universidade de Lisboa, Portugal*

ABSTRACT: Under the fields of design theory and practice, we propose a critical comment on *K* magazine (1990-1993), under Derrida definition of *parergon* to understand the publication design solutions. The survey is the result of two crossing methodologies: the empirical analysis of the magazine issues and the critical review of its theoretical frame, where the device *parergon* is rescued to grasp the magazine's graphic design strategies, providing us a close reading of *K* visual layout through postmodern graphic design scope. The essay will identify and analyse 4 case studies of *parergon* in the magazine layout to demonstrate the significant role of graphic design in the text interpretation, as a postmodern graphic design feature. The pertinence of the study concerns the ability of design to become a subjectivation device in communication process.

## 1 INTRODUCTION

This a critical comment on *K* magazine (1990-1993) – study object of a design doctoral research – about the unfolding of *parergon* in its own design process to conceive the graphic layout. We borrowed the term from Jacques Derrida deconstruction theory to help us observe, explain and understand graphic design operations before digital advent: its ways of thinking, modes of work, uses of visual communication elements and rules.

The research main purpose was to grasp the *K* magazine as an expression of post-modernism in visual culture, during the early 90's in Portuguese editorial panorama. The survey was drawn upon an empirical analysis of all magazine editions, its historical context and a close reading of critical review about post-modernism literature on graphic design. Among others references, Ellen Lupton and J. Abbott Miller text "Deconstruction and Graphic Design" (1996) provided a clear reading of deconstruction theory upon the graphic design and typography domain, introducing the key concept of *parergon* as the frame of design, a way to solve the problem: a determined number of operations in graphic design to tell a story, to reproduce a solution, to become a mean of communication. This essay acknowledges how *parergon* is revealed and used to solve the magazine visual layout, confronting the magazine graphic features with design theory, allowing us to comprehend it under a wider perspective than the practice.

More than a set of tools, *parergon* is a way of thinking in graphic terms, utterly linked to design process and methodology. Despite the general use of digital software to drawn graphic objects nowadays, this alliance between processes and philosophy is important to introduce students into the design processes, skills and tactics, enabling them to think, to make questions, to work with intellectual autonomy towards digital platforms. The essay aims to demonstrate how the graphic layout of selected articles altered, changed and interacted with editorial contents in order to enrich the message, providing a more complex object of communication.

### 1.1 *K magazine, 1990-1993*

The magazine was published monthly between 1990 and 1993, under the direction of a young dauntless group of journalists, photographers, writers and artists. Held by private investors, its editorial leading was generalist-targeted, with articles mostly about culture, politics, fashion, mundane life. The only editorial rule was freedom, assigning a variety of themes and visual references into the magazine pages, from *high* to *low* culture. The magazine was an elegant publication: it was well-printed in glossy paper, it looked sophisticated, similar to a fashion magazine, with good photography, elegant typography and attractive spreads.

The magazine graphic layout was planned and composed always by hand by Luis Miguel Castro. Each number was drafted in paper models, composed with text blocks, images, photos and illustrations, photocopies, acetates, etc... and then sent to offset print. Castro imported, from Elle internship, the method *chemin de fere*, which allowed the representation of all pages on the same plan. The magazine layouts were designed under the logic of *the flow*, managing the visual continuity throughout all pages as one single object, like cinema editing. He also asked the collaboration of several and acknowledged artists, photographers, illustrators, which imprinted great quality into the magazine art direction. The graphic editor also used to copy other magazine layouts, adapting previous these visual schemes to new contents.

## 1.2 *Parergon*

After a complete review of all 32 editions, the survey proceeded to inquiry the graphic filiation of *K* design under a theoretical frame close to post-modernism. Post-structuralism theory and post-modernism were not coincident, but ideas and objects shared a common concern: to question the modes of representation and the way technologies reproduced social and institutional relations.

Rick Poynor set the guidelines to post-modern graphic objects in his book "No more rules: graphic design and post-modernism" (2003), recognizing the deconstruction theory influence in graphic design objects in the 90's, as part of the process of visual communication. Lupton and Miller text "Deconstruction and Graphic Design" (1996) reinforce the same argument where deconstruction theory is as part of the continuing development of design and typography. The authors claim Derrida main argument of *On Gramatology* (1967) as the basilar fundament of graphic design substance, considering writing as distinctive form of representation, independent from speech and able to transform and change the message. Derrida promotes deconstruction as mode of research, of questioning the binary relation between representation and reality. It concerns modes of form-making: spacing, framing, punctuation, type style, layout and non-phonetic elements in the communication process. "Design and typography work at the edges of writing, determining the shape and style of letters, the spaces between them, and their placement on the page" (Lupton and Miller, 1996. p. 14).

The term *parega* (plural of *parergon*) is rescued in Lupton and Miller essay as a reinforcement of Derrida thesis against the idea of writing as secondary copy of speech. In his text *The truth in Painting* (1978) the author reclaims the function of *parerga* in opposition of Kant's definition: *parerga* as the frame of a picture, an element outside or around the work, an ornamental appendix that touches the work of art but remains outside of it. For Derrida, *parergon* is about framing, but even as an ornament or accessory device, it is fundamental to the constitution of the work of art, is communicates with its essence, allowing the work to be materialized and admired. *Parergon* operates in the work margins: "it put pressure on it, press against it, seek contact, exert pressure on the boundary" (Derrida, p. 20), as a secondary work, it facilitates the appearance of work. Derrida notion of *parergon* included all non-phonetical forms present in writing. Later, in their essay, Lupton and Miller reclaim:

*Spacing and punctuation, borders and frames: these are the territory of graphic design and typography, those marginal arts which articulate the conditions that make texts and images readable. (...) Design and typography work at the edges of writing, determining the shape and style of letters, the spaces between them, and their positions on the page. Typography, from its position in the margins of communication, has moved writing away from speech. (Lupton and Miller, 1996, p. 14)*

In short, Lupton and Miller assign *parergon* as subjectivation device in design, where the graphic editor becomes an active agent in communication process, in line with postmodern graphic design definition provided by Poynor. Subjective in terms of narrative itself, where the multiple combinations of text and image and options as type selection, spacing and orientation allow unlimited ways to tell a story.

The authors propose a study of graphic design and typography history informed by deconstruction focused on "a range of structures that dramatize the intrusion of visual form into verbal content, the invasion of "ideas" by graphic marks, gaps, and differences." (Lupton and Miller, 1996. p. 17). Modes of deconstruction in graphic design are expressed in interventions that reveal, challenge or transform the rules of communication, but also social, technological, and aesthetics conventions. Again, the subjective dimension of graphic design is asked to materialize the message, adding more layers of signification into the final object. In conclusion, deconstruction theory and *parergon*, according to Lupton and Miller, reclaim functions of repetition, quotation and fragmentation into writing conditions as permanent into human communication. These are the work operations which enable graphic design to intervene and transform material media of visual and verbal writing. These are also the graphic features *K* design appealed to materialize visual contents and we shall inquire next.

### 1.3 *Frame*

The magazine has several examples of frames, composed by different means and visual elements. In Figure 1 is presented one of its most obvious forms: the spread has one block of text place in a white rectangle centered at each page, the background is full black with red figures displayed as a pattern. The composition is quite balanced and recalls a conventional book layout, oriented in a symmetrical axiality with generous margins. The background pattern reminds the marbled paper in book guards, assigning the article a credible context. However, the text is about series of jokes and silly stories. This case shows how graphic design can transform the eminent humorous and light tone of the text, into a formal appearance and more serious and worthy context.

Other case of framing is patent in Figure 2, where all page elements, body text, titles and imagens seem to struggle in constrained and irregular frames. Contents are under visual pressure, margins almost lack between forms, titles type are rotated, leading is irregular and colorful frames compete with text. The asymmetric composition provokes some vibration to contents related to politics and classic literature, making this spread an opposite attitude towards the previous example.

### 1.4 *Quotation*

*K* graphic was always aware other publications and visual references. In a pre-digital era, copying and appropriation were common operations to solve the work. Quotation can be understood as a direct reference to something and in *K* design was used to both pay homage or playfully mock

Figure 1. *K* n. 12 (September 1991).

Figure 2. *K* n. 30 (February 1993).

other publications. One of many cases of quotation is the cover of *K* n.4 (Figure 3): an instant mention to a Portuguese vanguard arts magazine, "K4 O quadrado azul" (Figure 3) conceived by plastic artist José de Almada Negreiros in 1917, fully immersed in modernism esthetics and ideas. This appropriation of cover layout expresses the *K* editorial and cultural influences but also settles the magazine editorial ambition to become a landmark in Portuguese panorama.

On the opposite side, the next case (Figure 5) is a parody of feminine fashion magazine *Marie Claire*. An exercise of *pastiche* is used to compose a humorous sample of articles dedicated to male audience into a feminine targeted magazine, called *Mário Cláudio*: its style sheets and layouts were made a *simulacrum* to confuse the reader and play with publishing conventions.

### 1.5 *Repetition*

In *K* n.7 (Figure 5) there is a double rubric called "Parabéns" (congratulations) and "Pragas" (plagues) where each subject is both praised and criticized. The graphic layout is exactly the same for both sections, only with different body texts. This the most obvious and deceptive example of repetition in the magazine: challenging the reader between good and bad reviews under the same plan of composition. Repetition of elements was also use do compose patterns and backgrounds, like the next example (Figure 7) which is also a frame.

### 1.6 *Fragmentation*

By fragmentation, we assume all kind of manipulation of text and images. Still in an analogic way, Castro improved this technique through photocopy to "destroy", alter and decompose

Figure 3. *K* n. 4 (January 1991). Cover (left) and *K4. O quadrado azul* (1917) (right).

Figure 4. *K* n. 29 (January 1993).

Figure 5. K n.7 (April 1991).

graphics elements. On images, fragmentation changes the original meaning and assigns alternative atmospheres to the page layout (Figure 7).

Other form of fragmentation is present in typography, where the leading would disappear and letter titles would lose some legibility but they would become strong graphic elements in the page. The excessive space between letters would also be fragmentation, compromising the legibility as well. (Figure 8)

Figure 6. K n.5 (February 1991).

Figure 7. K n.15 (December 1991).

Figure 8. K n.31 (March 1993).

## 2 DISCUSSION

We can assume that the result of these operations as the magazine visual grammar, developed under analogical processes, informing us what technologies were available at that time to produce such graphic features. But they also disclose the graphic editor interpretation of the texts: he is not just a maker, he is also a thinker, a visual thinker: he introduces his comments and reactions in the magazine graphic layout, as another layer of communication, the graphic design one.

Derrida considered *parergon* not just framing, it communicates with the work (text), it can change, manipulate and transform work itself. Probably Castro was not aware of this "mechanism", but as designer he simply applied the discipline operations. Lupton and Miller argued:

*Design can critically engage the mechanics of representation, exposing and revising its ideological biases; design also can remake the grammar of communication by discovering structures and patterns within the material media of visual and verbal writing. (Lupton and Miller, 1996, p. 23)*

When we review *K* magazine and identify all these visual strategies, we may conclude that its design had this ability to modify contents, to add meaning to them. Operations like framing, quotation, repetition and fragmentation were used to access another communication plan, they did not just materialize text and contents, they constitute a wider and richer message. Assuming this post-modern feature in the magazine, as a place of *parergon*, a place of tension and confrontation between text and graphics. The importance of such device as *parergon* is related with the emerging authorship of graphic design in media and editorial context. As a subjectivation device, *parergon* displays the way designers think and react to contents, allowing them to actively participate in communication process.

## REFERENCES

Lupton, E. and Miler, J. (1996) *Design Writing Research*. New York: Phaidon Press.

Poynor, R. (2003) *No more Rules: Graphic Design and Post modernism*. New Haven: Yale University Press.

Derrida, J. (1978) *The Parergon*. In October. Vol. 9 (Summer 1979). MIT Press.

*Design & Society*

*Research & Education in Design: People & Processes & Products & Philosophy – Almendra & Ferreira (eds)*
*© 2020 Taylor & Francis Group, London, ISBN 978-0-367-49296-0*

# Design & entrepreneurship – a reflection on the approximation of areas

B.R. Moreira, A.C. Dias, N. Plentz & R. Almendra
*CIAUD, Lisbon School of Architecture, Universidade de Lisboa, Portugal*

ABSTRACT: This paper discusses the relationship between design and entrepreneurship in the theses of the PhD Program in Design of the Lisbon School of Architecture, Universidade de Lisboa. We analyzed 72 theses defended between 2010 and 2017 and the results indicate that there is a little relation between the areas. So, we discuss the negative effects that this context can produce in the design competences and Portuguese economy. Moreover, we explore how design can benefit from entrepreneurial skills and present some scenarios that propose ways to approach each other in design education.

## 1 INTRODUCTION

Design is being practiced at more complex levels, therefore, its activities have gone beyond the development of artefacts and started to include the development of services, processes and systems as well. The results of design practice are diverse and, recently, market movements indicate the coming together of design and entrepreneurship. This phenomenon may be a reflection of different factors, such as: the recognition of economic potential the creative economy has, which, according to different researches, grows day by day (British Council, 2010; Governo de Portugal, 2013); and the increase of Startups[1] that use creativity as a primary input. Whether motivated by a number of successful cases or even by the difficulty of getting a job in the area, designers are approaching entrepreneurship in order to make their ideas come true. This relation uses creativity and intellectual capital as primary inputs to create, to produce and/or to distribute consumer goods that are, above all, inserted in the creative economy market.

However, to merge two areas which historically have such distinct natures often becomes a challenge for designers and for design education itself. For entrepreneurial designers, the challenge is to understand the dynamics of a "hard area" (Administration) and how to launch a business in the market. For design education, on the other hand, the difficulties can be related to different factors, such as: curricular structures that do not offer the space to discuss design through other perspectives; teachers focused more on technique than strategy; project disciplines that do not go beyond the prototype phase (such as insertion of the product or service in the market and it is application idea in a profitable business), etc.

Thus, this study aims to analyse how the design-entrepreneurship relationship is manifesting itself on the theses of the design PhD Program from the Lisbon School of Architecture (ULisboa). To turn it possible, we used the database of the Program and the University that contemplates 72 Theses, presented between 2010 and 2017.

The results indicate that the design-entrepreneurship approach is not yet evident in the PhD in design, which corroborates with the idea that, in general, market movements tend to be

[1] According to data from Startup Portugal, 584 new startups were created between October 2016 and October 2017, and each one creates on average 2.2 jobs in during their first year (Startup Portugal, 2017).

more agile than academia. Due to the results we had, we broadened the discussion of the harmful effects that this context can produce and also present some scenarios that reflect on ways to strengthen the relationship between design and creative entrepreneurship in design education.

## 2 THE ORIGINS OF THE ENTREPRENEUR DESIGNER

In the design postindustrial phase, the designer became a professional capable of also making feasible the management and strategic planning in competitive markets. With this, designers began to pay more attention to the consumer goods environment and not only the product itself. This responsibility gave a new function for design, demanding new skills and abilities from designers. Among them, are the competencies related to entrepreneurship that emerge from the possibility of designers transforming ideas into creative economy businesses. Thus, we suppose the design-entrepreneurship binomial can be a result of several factors, such as: (a) few formal job opportunities have encouraged designers to create new businesses; (b) the growth of faster business models, such as Startups, may be encouraging them to build their own paths; (c) the generations that are seeking jobs with a purpose and not just a job; and, above all, (d) the significant growth of the creative economy in Portugal and in the world.

According to the "Culture and creativity in the Portuguese economy," launched by the Government of Portugal (2013, p. 42), "Design products are one of the most popular exports in the country, encompassing the protagonists of the internationalization of the Portuguese economy, from furniture to textiles". Regardless of the reason that stimulates the relation between design and entrepreneurship, we need to recognize that there is a movement of convergence between the two areas. For this reason, the next item presents some data regarding creative entrepreneurship in Portugal.

## 3 CREATIVE ENTREPRENEURSHIP AND THE PORTUGUESE CONTEXT

We cannot talk about creative entrepreneurship without first of all defining the environment that entrepreneurship in design is inserted: the creative economy. The report made by UNCTAD (United Nations Conference on Trade and Development, 2012) refers to Creative Economies as a cycle of creation, production and distribution of products and services using creativity and intellectual capital as primary inputs. In this sense, it encompasses: (a) the set of knowledge based in activities that generate sales revenues and intellectual property rights; (b) those activities that constitute tangible products and intellectual or artistic services with creative content, economic value and market goals; (c) processes that are between the artistic, service and industrial sectors; (d) the results that constitute a new dynamic sector in world trade (UNCTAD, 2012).

One of the important characteristics of this type of economy is that it is often linked to a specific geography. That is, it exploits local characteristics of certain cities or regions in order to generate value for the companies and, consequently, to the places where they are inserted (Howkins, 2013). In this way, unlike the one intended by an automobile factory, for example, which aims to operate in countries where labor costs are lower and tax regimes are favorable, the Creative Industries cannot be moved from one place to another due to the correlation with specific localities.

In order to map Lisbon's potentialities and provide ideas for the expansion of the city's competitiveness, the Cross-Innovation Project, made in partnership with the Lisbon City Council, presented a "Blue Print" of the local creative economy (Lisbon Municipal Chamber, 2013). The document points out that 30% of the country's creative resources are concentrated in Lisbon and there are almost 22,000 companies of the sector in the Portuguese capital. In addition, the research also shows that Lisbon has more than 100 establishments of higher education and many of them are related to artistic education. This information becomes relevant because, according to the British Council research (2010), the relationship between university and creative economies must be symbiotic. That is, educational institutions play an important role in this type of economy as they are able to nurture the skills for the development and maintenance of local creative businesses and local job markets.

At the end of the presentation of the research developed by the Lisbon City Council (2013), the report cites the axes considered strategic and possible to be developed in Lisbon: (a) internationalization; (b) events: competitiveness anchors; (c) creative territories and neighborhoods; (d) creative entrepreneurship; (e) talents; (f) spaces and equipment: new and functions; (g) workshops and residencies for artists. However, even if the numbers indicate the insertion of the Arts area in Portuguese higher education establishments, it is necessary to understand how this relation is actually manifesting in the Universities. Thus, the following section presents the results obtained with the focus object of this study.

## 4 CREATIVE ENTREPRENEURSHIP IN DESIGN: ANALYSIS METHOD

### 4.1 *Theses overview*

With regard to the discipline that the theses are related, we found that the most common are: Communication design (26), followed by product (20), design (11), interior (8) and fashion (7). For this reason, the first keywords most cited in the theses are, respectively: design (8), visual communication (4), communication design (4), product design (3), graphic design (3) and furniture (2).

Open your old file and the new file. Switch between these two with the Window menu. Select all text of the old file (excluding title, authors, affiliations and abstract) and paste onto bottom of new file, after having deleted the word INTRODUCTION (see also section 2.5). Check the margin setting (Page Setup dialog box in File menu) and column settings (see Table 1 for correct settings). After this copy the texts which have to be placed in the frames (see sections 2.3 and 2.4). In order to avoid disruption of the text and frames, copy these texts paragraph by paragraph without including the first word (which includes the

Table 1. Overview of the PhD theses of ULisboa.

| Criteria | Answers | Count | % |
|---|---|---|---|
| Type of thesis | Theoretical | 48 | 66% |
| | Practical | 24 | 33% |
| Discipline | Communication Design | 26 | 36,11% |
| | Product Design | 20 | 27,77% |
| | Design | 11 | 15,27% |
| | Ambient Design | 8 | 11,11% |
| | Fashion Design | 7 | 9,72% |
| First keyword | Design | 8 | 11,11% |
| | Visual communication | 4 | 5,55% |
| | Communication design | 4 | 5,55% |
| | Product design | 3 | 4,16% |
| | Graphic design | 3 | 4,16% |
| | Furniture | 2 | 2,77% |
| Methods used | Literature review | 70 | 97,22% |
| | Questionnaire | 69 | 95,83% |
| | Interview | 45 | 62,5% |
| | Project | 37 | 51,38% |
| | Artefact analysis | 37 | 51,38% |
| | Case study | 25 | 34,72% |

*(Continued)*

Table 1. (*Continued*)

| Criteria | Answers | Count | % |
|---|---|---|---|
| | Usability testing | 20 | 27,77% |
| | Focus group | 13 | 18,05% |
| | Tools tests | 11 | 15,27% |
| | Field observation | 11 | 15,27% |
| Type of research | Praxeology | 22 | 30,55% |
| | Phenomenology | 18 | 25% |
| | Praxeology/Phenomenology | 12 | 16,66% |
| | Others | 20 | 27,77% |

In the Methods Used category the percentage refers to the number of theses that used a specific method; for instance, 97,22% of theses used the literature review method.

Regarding the type of thesis, we observed that 48 of them are theoretical and only 24 are practical. This may explain why the least used methods for analyzing the theses were: field analyzes (11), tool tests (11), and focus groups (13). Different from the bibliographic review (70), questionnaire (69) and interview (45) that were among the most usual methods. Regarding the type of research, the theses were more classified as: praxeology (22), phenomenology (18) and praxeology/phenomenology (12). The others, however, are situated in epistemology. Finally, the scope of research relates to: history of design (10), inclusive design (5), sustainability (4), museology (4), typography (4), lighting design (2) codesign (2), and design teaching (2).

### 4.2 *Title, Comparison between theses completed between 2010 and 2017*

The metrics presented differ when we analyze the publications of 2010 and 2017 separately (types of theses and disciplines), as illustrated on Table 2.

Regarding the discipline, Communication design remained as the main subject in the general average in both years, however, the diversity of theses increased in 2017 (communication, product, design and fashion) compared to 2010 (communication and product). Such facts may demonstrate the evolution of the lines of research of the Doctoral program, the expansion of the research interests of the students themselves and/or the extension of the understanding of the design areas. When looking at the comparative data, we noticed that in 2010 only theoretical theses were defended. Unlike in 2017, which had a 50% theoretical and 50% practical balance, a fact that shows a significant increase in applied research. However, this fact is not related to the applicability of research related to entrepreneurship, as it is presented in the following section.

Table 2. Overview of the 2010 and 2017 PhD theses of ULisboa.

| | Total | | 2010 | | 2017 | |
|---|---|---|---|---|---|---|
| Theses | Sum | % | Sum | % | Sum | % |
| Theoretical | 48 | 66 | 3 | 100 | 3 | 50 |
| Practical | 24 | 33 | – | – | 3 | 50 |
| Communication | 26 | 36,1 | 2 | 66,6 | 3 | 50 |
| Product | 20 | 27,7 | – | – | 1 | 16,6 |
| Design | 11 | 15,2 | 1 | 33,3 | 1 | 16,6 |
| Ambient | 8 | 11,1 | – | – | – | – |
| Fashion | 7 | 9,7 | – | – | 1 | 16,6 |

### 4.3 *Creative design and entrepreneurship*

In order to analyze the theses related to design and creative entrepreneurship, we filtered them first by the following keywords: 'entrepreneurship', 'undertake' and 'entrepreneur' because they are directly related to the purpose of this paper. However, we did not find any theses that had such links. Therefore, we made a second filter with the words 'business', 'startup' and 'business model', since entrepreneurship means the attitude of those who assume financial risks inherent to the creation of a business with a view to profit. However, again we did not get returns with such words.

Finally, we chose to look for theses that presented 'Innovation' as a keyword, as it relates to one of the elements of entrepreneurship. This time, we achieved more success and we found three theses with the following characteristics: (a) two theoretical/practical and one theoretical; (b) the theses were defended in the years 2012, 2014 and 2016; (c) the theses are focused on the areas of web design, sustainability and innovation; (d) two of them have theoretical and conceptual results and another one has practical results. The three theses are characterized below:

1. "Innovation, design et cetera"; 2012; Theoretical/practical research; Webdesign; keywords (innovation, design, collaborative network intelligence, inspiration, inspédia); research question: "Is it possible to develop a conceptual model of access to knowledge platform in innovation and design that allows acceleration of non-linear thinking and inspires collaborative network intelligence, exploring the potential of web 2.0?"; main result: A conceptual model and prototype Inspédia: a collaborative network platform to inspire innovation and design processes, which can be used in design.
2. "Design contributions to an innovation strategy: Technological centres in Portugal"; 2014; Theoretical; Innovation; keywords (Technology transfer, radical innovation, strategic design, design thinking, culture of innovation); research question: "Can strategic design meet the innovation needs of the national transforming industry?"; main result: Theoretical discussion about the potential for expansion that puts Design as a strategy to generate innovation.
3. "The importance of biotechnology in the textile industry"; 2016; theoretical/practical; sustainability; keywords (Natural dyes, sustainability, biotechnology, bio design, innovation, dying process); research question: "What are the possibilities of new manufacturing methods and innovative materials?"; main result: New industrial dying techniques using lycopene.

As illustrated in the points above, even though some Theses have the term Innovation as a keyword, we note that they are not directly related to entrepreneurship. Thinking indirectly, we can assume that theses 1 and 3 have the potential to become businesses in a future, but it was not mentioned in the studies. Because we did not find direct connections between design and entrepreneurship, we decided to discuss about the possible reflections of the academic distance between the areas, as we show in the next section.

### 4.4 *Reflections of the academic gap between design and creative entrepreneurship*

Keeping design education away from (or not connected to) entrepreneurship can generate impacts in the medium and long term which were not measured in the study, since the focus are the outputs of the ULisboa. However, in order to elucidate the discussion, we bring data obtained in a recent research (Dal Moro, 2017), carried out with graduates in a design Course located in the South of Brazil. Inspired by the high number of ideas generated in disciplines of design bachelor's degrees versus the low number of business born of such ideas, researcher Dal Moro (2017) conducted a survey to understand: "Do designers turn their project into business ideas?" And," Why do designers not convert good ideas into businesses?"

In seeking such answers, he found that young designers often have creative and potential ideas, but do not launch them into practice. Moreover, the survey showed that less than 10%

have launched ideas into the market. According to Dal Moro (2017), the inertia between the "knowing" and "doing" has multiple reasons, but one of the main problems is the discouragement induced by the lack of knowledge about entrepreneurship. Even though collected in a country which has big social and economic differences when compared to Portugal, the data is worrying and brings about a discussion on the competences that seem to be lacking in designers to undertake their ideas. In addition, we can think about the negative impacts that this scenario can bring to the market due to the many potential ideas created in Bachelors of Portugal that do not leave the 'drawers'. In order to illustrate this discussion, we looked for data and statistics of the Higher Education in Portugal (República Portuguesa, 2019). We found that, in April 2019, the survey indicated that there existed: 61 Degrees in design in the Portuguese territory; the courses have, in average, 3 years of duration; and, the courses incorporate, on average, 30 new students per year.

Based on these data, we estimated - just for illustration purposes - that there are approximately 90 students per degree in design (30 new students per year x 3 years) and, therefore, Portugal has around 5,490 students in design courses (90 students per course x 61 Courses in the whole country). If we consider that such students perform 1 project per year, whether in elementary, intermediate or advanced level disciplines, we could conclude that only the degree in design generates more than 5,000 ideas per year. In addition, we looked at the same body of research to find out the number of graduates per year. We found that this data varies greatly from course to course, but considering that, on average, only 2/3 of the original class graduates (20 students), we find that, on average, Portugal has 1,220 new designers per year (20 students x 61 degrees). In other words, there are more than a thousand people whose skills enable them to create products or services with creative content. However, we ask ourselves how many of these professionals are able to undertake their ideas in the market?

With the data gathered so far, we believe that it is necessary to think about viable paths that seek an approximation between design and entrepreneurship with the aim of Portugal benefiting even more from the creative potential of design professionals. Here is a reflection on this.

### 4.5 *Possible alternatives*

There are different ways to start a more effective approach between entrepreneurship and design in the academic universe and they are all legitimate in some way. Therefore, we emphasize that we do not intend to bring ready recipes, but rather, we use this space to reflect on some ways to increase the relations between the areas in question. Assuming that the meeting of competences between design and entrepreneurship can occur in a more exogenous way (when the incentive comes from outside the University, by means of edicts of development agencies, for example) or endogenous (incentive from within the University), we seek to focus on the internal actions of the university, since they are more manageable and do not require curricular reforms to be put into practice. These are: (a) partnerships with companies from the creative economy; (b) partnerships with Technology Parks; (c) partnership between administration and design disciplines, as shown below.

Partnerships with companies of the creative economy: We believe that establishing partnerships between the design education network (whether with the 1st, 2nd or 3rd cycle) with market companies can be a healthy alternative. Such partnerships can occur through different connections. One of them is through disciplines focused on design Projects that could develop, prototype, and launch business ideas for partner companies. This initiative has already been developed satisfactorily in an institution in the south of Brazil (Moreira et al., 2014). According to the study, there are partnerships between Project disciplines with real companies with the goal of bringing students closer to the market, allowing them to apply theory in practice and also have room for experimentation. Moreover, Moreira et al. (2014) emphasize that this practice is carried out along the program and that the complexity of the partner companies increases as the disciplines progress along the course curriculum. In order for such a partnership to be made feasible, each student starts to immerse themselves in the partner companies in order to broaden their knowledge regarding their market positions. At the end of the semester, the academics present their ideas to the managers and/or representatives of the companies themselves, a fact that, in our opinion, presents itself as an opportunity for expansion of student networking with a view to future career opportunities.

Partnerships with technology parks: We think that another way to foster entrepreneurship in design education is through the partnerships established between disciplines and the technological parks of universities. To support this idea, we use the teaching experience of one of the authors of this article who, for two years, taught a Project discipline in partnership with TechPark (Feevale University Technology Park). Such discipline aimed to bring students to the Park to encourage them to incubate their academic projects in order to turn them into business. This integration took place through the accompaniment of a TechPark teacher who acted as a provocateur of ideas and feasible and profitable projects in the market. Thus, in addition to the academic view of the teacher of the discipline, the students were also guided by a market look that led ideas for a business status applicable in the market. At the end of the semester, the project proposals were evaluated to indicate if they had the potential to be incubated by the technology park or not.

Partnership between administration and design disciplines: Because entrepreneurship is a branch of the administration course, we believe that fostering partnerships between disciplines from both areas (design and administration) is salutary, since both can learn from each other. In this perspective, students are encouraged to exchange experiences among themselves, indicating points of view through design's optics and, also, the Administration. In addition, the contacts between such students can promote networks of collaborations between them once they add knowledge. Also, we believe that such action collaborates with the construction of the visualization of complexity in both courses, especially design. This becomes particularly relevant when we realize that, as complexity changes society, the role of the designer also changes. Complexity, which is inherent in the current context, entails a series of developments for the area of design whose competences are also connected with several areas, becoming an articulator of complexity (Moraes, 2010).

#### 4.5.1 *Basic pre-requisites for the approximation between design and entrepreneurship*

We know that making academic changes in the day-to-day disciplines, schedules of teachers and courses is not an easy task because universities are embedded in processes, bureaucracies and people with different levels of knowledge and commitment. Within this universe, we consider that investing in design courses teachers' knowledge is a prerequisite of the first order. Therefore, regardless of the action taken in the university, we believe that the faculty of the University design courses should be, firstly, sensitized to the importance of the agenda and, also, trained in the theme of entrepreneurship. We understand that without this action, efforts may be in vain or drastically minimized since the teachers themselves are oblivious to the importance of the design-entrepreneurship relationship. However, in order to make this activity feasible, it is necessary that high academic management is sensitive to the importance of approaching both areas. Without this support, all the paths previously discussed may not be feasible or their political strength weakened within educational institutions.

## 5 FINAL CONSIDERATIONS

We cannot generalize the results obtained in this study because of the low number of data analyzed but we believe that they give rise to a very important indication: the need to approximate the competencies of entrepreneurship to design, whether at the Bachelor, Master or PhD level. We believe that, in mediating this approach, the creative economy of Portugal can benefit immensely from the large number of creative professionals that graduate annually in the country. We understand that this approximation must occur with parsimony and respecting the essence of both areas in order to avoid confusion or the exchange of roles between them. Therefore, the proposals presented in this study are not intended to transform designers into Entrepreneurs or vice versa, but facilitate a more profitable exchange between both areas for the sake of the country's economic growth. Finally, we believe it is essential that universities provide opportunities to converge knowledge between design and entrepreneurship, since, according to the Portuguese Government (2013, p.10), "the future of European economies depends decisively of their ability to put culture, creativity and knowledge at the center of economic activities".

## REFERENCES


British Council. (2010). Creative Economy: an introductory guide. Available from: <http://creativeconomy.britishcouncil.org/media/uploads/files/Intro_guide_-_Portuguese.pdf>. Retrieved on 14.08.2017.

Câmara Municipal de Lisboa. (2017). Lisboa: Economia Criativa, 2013. Available from: <https://issuu.com/camara_municipal_lisboa/docs/lisboa_economia_criativa>. Retrieved on 14. 08.2017.

Dal Moro, R. (2017). *CADI: rede de compartilhamento de recursos para inovação em projetos de design.* Monografia. (Graduação em Design) - Universidade Feevale, Novo HamburgoR.

Governo de Portugal. (2013). Secretaria de Estado da Cultura, Gabinete de Estratégia, Planejamento e Avaliação Culturais. *A cultura e a criatividade na internacionalização da economia portuguesa.* Relatório final.

Howkins, John (2013). The Creative Economy: How People Make Money from Ideas. London: Penguim Books Ltd,.

Moraes, D. de. (2010). *Metaprojeto*: o Design do Design. Blücher: São Paulo.

Moreira, B.; Camargo, C.; Soares, L.; Giongo, M. (2014). A relevância da multdimensionalidade das disciplinas de projeto na graduação em Moda. In: Colóquio de Moda. Caxias do Sul, 10, 2014, Rio Grande do Sul. *Anais*.... Caxias do Sul: Universidade de Caxias do Sul, 2014, Available at: <http://www.coloquiomoda.com.br/anais/Coloquio%20de%20Moda%20-%202014/COMUNICACAO-ORAL/CO-EIXO2-ENSINO-E-EDUCACAO/CO-EIXO-2-A-relevancia-da-multidimensionalidade-das-disciplinas-de-projeto-na-graduacao-em-moda.pdf>

República Portuguesa (2019). *Dados e estatísticas de cursos superiores.* Available from: <http://infocursos.mec.pt/>, Retrieved on: 24. 04.2019.

UNCTAD (2012). Conferência das Nações Unidas sobre Comércio e Desenvolvimento. *Relatório de economia criativa 2010: economia criativa uma, opção de desenvolvimento.* Brasília: Secretaria da Economia Criativa/Minc; São Paulo: Itaú Cultural.

*Research & Education in Design: People & Processes & Products & Philosophy – Almendra & Ferreira (eds)*


# Teaching of design for social innovation in Portugal: Perspectives for its improvement

N. Plentz, C. Miolo & R. Almendra
*CIAUD, Lisbon School of Architecture, Universidade de Lisboa, Portugal*

ABSTRACT: The paper discusses the teaching of design geared towards social innovation processes, identifying competencies required to work with this type of project. To arrive in these competencies, specialists were interviewed, crossing their answers with literature. To evaluate whether these competencies are being adequately taught, students from a social innovation-based course in a Portuguese graphic and product design undergraduate program were interviewed. The paper concludes the students find difficulties in working with projects of this nature, mainly for due to the fact that they spend most of the program working with a design process focused on results. We propose the competencies identified here should be worked on from the beginning of the program, stimulating earlier on practices like the interaction with communities and the problem identification process.

## 1 INTRODUCTION

In order to adapt to the current context, design has been shifting the way it acts in the last years (Manzini, 2009; Findeli, 2001). Two of the main characteristics of this change are moving from product projects to product-service systems projects and the transformation of linear design processes into networked processes (Manzini, 2011). These networked processes can be tied to the search for solutions of complex problems, like social problems, or the generation of new opportunities, which encompass social innovation processes. In the realm of design education, teachers are required to educate students in positioning themselves when faced with social and environmental problems we deal with today, in a way that allows these young people to participate in the change in society and design solutions that contribute to a more sustainable and socially just society (Manzini, 2009; Papanek, 1995).

However, what we see most of the time is design education focused only on the market, not giving enough attention to current problems and populations in need (Margolin & Margolin, 2002). In this context, a change in design education is urgent, with a larger focus on social innovation. Social innovation processes tend to be more complex than commercially based design processes, requiring from designers specific competencies, like the ability to work in a group or even empathy. Therefore, we started to question whether students are being prepared to work with social innovation. Moreover, which competencies need to be acquired in this context? To answer these questions, we used the case of a course in a Portuguese university's undergraduate design program.

## 2 DESIGN FOR SOCIAL INNOVATION

The social innovation thematic has been described as an approach to processes that aim to meet needs of vulnerable parts of society or to provide for needs which are not being met by the responsible public sector body (Murray et al., 2010). However, this view is being reformulated with the analysis of practical projects, which verify social innovation problems may include even State actors. Regardless of the actors involved, social innovations can be seen as discontinuities in the way society acts and lives, searching for more sustainable options (Manzini, 2008). As well as gains resulting from problem solutions or generation of new opportunities for common good, social innovations end up developing the social capital of people involved (Chick, 2012).

With the referenced authors in mind, we understand social innovation as a process that aims to meet social needs and generate new opportunities for common good. For this, efforts are made to change community behaviour, providing tools to enable members to work collaboratively in the search for solutions. As a result of the process, improvements must be obtained aiming at the common good and developing the social capital of those involved. Many different disciplines can act in the process of developing social innovation. Amongst them is design, which for a long time and still today is very attached to the development of products which meet market needs, decrease production costs and make products more attractive (Walker & Dorsa, 2001), as well as the visual communication of such feats.

This restricted view of design is not adequate to current society anymore, with its serious environmental and social problems. Besides searching for production processes which do not affect the environment, designers can be important creative agents, guiding social innovation processes and helping to solve complex social problems. In this context, designers act to generate innovative solutions, not only centered in a product but an integrated product-service strategy (Zurlo, 2010). Such solutions must provide improvements that consider the interests of the different actors and their interaction with the environment (Zurlo, 2010; Meroni, 2008). Therefore, the design process requires the systematic involvement of the interested parts, in such a way that an amalgam of disciplines, techniques, technologies and knowledge can interfere and collaborate; making the process intensely participative (Meroni, 2008).

Due to design's capacity to develop innovation processes in a participative way, which considers the interests, collective values, resulting socio cultural impacts, it is linked to social innovation (Meroni, 2008). In this context, designers, who have always created bridges between society and technology, must now act in another direction, which is social innovation (Manzini, 2008). More than facilitating the social innovation process, through work on multiple project levels, the designer also has a catalyst role in this journey (Chick, 2012). The sense of catalyst mentioned by the authors means the act of accelerating actions and enabling relationships, acquiring a fundamental role in stimulating the process to happen. To catalyse the process, designers must connect different disciplines, techniques and methods in order to facilitate the conception of products and services necessary to the social movement being developed by the community (Chick, 2012).

An important design competence to work in social innovation processes is working with empathy with the community members (Cantú, 2012). This involves the not always easy job of placing oneself in their place, in order to perceive what motivates them, the context in which they live, their habits and day-to-day experiences (Ouden, 2012). With the development of empathy, it is possible to learn from observation and reflection and reach hidden knowledge about their needs and desires, which will enable project insights (Cantú, 2012). A way of establishing empathic relations is through dialogue (Cantú, 2012), or empathic conversations, (Friedman, 2002).

From the literature review, it is clear that the role of design for social innovation is not a simple process, as it demands as a prerogative the ability to treat problems systemically, considering the consequences to the community as well as environmental, economic and political impacts. Its process demands the interpretation of community needs in an empathic manner, habilitating relationships between its members, so they can work collaboratively, using different techniques, technologies e methods to build innovative scenarios, prototypes and finally implement the co-created ideas. Therefore, considering the different types of actions e capacities demanded from a designer to stimulate social innovation, we return to the questions brought at the beginning of the paper, questioning how to teach designers to work in social innovation processes.

## 3 DESIGN EDUCATION

Since the middle of the twentieth century, the academic structure of traditional teaching in design, as in other fields, has been basically the same: it initially presents basic science, followed by applied science and finally practical education, requiring students to learn to apply the knowledge assimilated in the first two to solve everyday problems (Schön, 2000). However, a possible gap in this curricular structure is that this late implementation of practical assignments makes it difficult for students to learn since they are confronted with real design problems only at the end of the program. These problems, in practice, do not present themselves as well-delineated structures, or even as problems per se, but as chaotic yet indeterminate structures that need to be initially understood and defined (Schön, 2000).

The design process is traditionally seen as the bridge between identifying a problem and proposing a possible solution. But in doing so, it is necessary to take into account the environment where it is placed, not only its biophysical but also social, material and symbolic aspects. Thus, the purpose of design, when taking all these aspects into account, must consist of a horizon or even a set of values and not a single final goal, where there is a concern for the natural environment and also a search for a fairer society that coexists in harmony with the environment (Findeli, 2001). This goes through the new ideas of wellbeing as well as production systems which facilitate better living without as much compromise to the environment and social and cultural quality (Manzini, 2009). In order to guarantee the survival of design today, changes such as the systematic questioning of the design problem and the context in which it is inserted must occur (Findeli, 2001).

In this context, it can be affirmed that sustainability, including its social aspect, should become the goal of any research or practice in design, not being a specialty as it has often been in recent years (Manzini, 2009). Even because design for the market and design for social innovation are not necessarily opposites and their differences lie in priorities, not in the methods used. However, there are some specific skills that are necessary for designers to work with social innovation, such as those that facilitate the relationship with vulnerable populations and the work with people from other areas, since social projects often involve social assistants or health professional (Margolin & Margolin, 2002).

The article analyses the teaching of design specifically in Portugal. Currently there are a considerable number of design programs in Portugal (Guerreiro, 2012), only in Product Design and Industrial Design there are 29 graduations and 22 masters (Dias et al., 2013), not counting other areas such as fashion and communication. But the insertion of design education in the country was quite late, having only begun in the late 1960s (Guerreiro, 2012). Thus, the methodology of the work consisted of a review of the literature, where we raised competencies that different authors point out as necessary to work with social innovation. Specific skills needed to work with social innovation were also raised interviews with specialists. These interviews were done with designers who have already worked on social innovation projects and who are researchers in the field. Subsequently, students from a course focused on social innovation of the third year of the design degree in a Portuguese university were interviewed. This discipline seeks to create a social service in communities, identifying possible problems and opportunities that can be developed. Seven students were interviewed, trying to understand the main difficulties they face when participating for the first time in a process of social innovation.

## 4 COMPETENCIES OF DESIGN FOR SOCIAL INNOVATION

In order to arrive at the competencies, the answers of two specialists were compared with current literature. The first specialist lives in Brazil, graduated in design, with a specialization and master's degree in the area, she is doing her Ph.D. at the moment. The interviewee investigates collaborative processes and possibilities of building collective imagination from the perspective of design anthropology. The second specialist lives in Portugal, has a master in design, also currently doing his Ph.D., he works for a strategic design company. His research focuses on design for social innovation, with the study object being processes for the development of creative communities. Both doctoral and master's thesis projects were carried out within communities in situations of social vulnerability.

The following questions were asked of the experts: What competencies were learned at the undergraduate level that were useful in your work with social design? And what skills did you miss learning when starting your work with social design? With these questions we obtained specific competencies of social innovation projects, which were compared with the literature, arriving at three macro-categories of competencies: collaboration capacity, empathy and systemic vision. These categories and their specificities will be described below.

### 4.1 *Collaboration capacity*

This competency was reached from the intersection of the literature with the answers of the specialists. The capacity for collaboration is the ability to work with people from different cultures and different disciplines (Papanek, 1995). In social innovation processes, this ability also goes through the designer's ability to enable those involved to work collaboratively. As pointed out by Specialist 1, social innovation must "involve people in the project and empower them for various things, we shouldn't see this consumer and this user as passive agents in the process." Corroborating with this, Specialist 2 states feeling "a lack of understanding of how to involve these people, what tools I should use to empower them in a natural way, without it being imposed".

It is considered that this ability of collaboration passes through the ability to empower actors, and the designer should have a role of catalyst (Chick, 2012). The designer should also have the ability to communicate with non-designers, contributing to the dissemination of knowledge among those involved (Manzini, 2008). This capacity can improve the process of social innovation, since it contributes to the empowerment of the communities, an exchange of ideas between the actors, allowing the emergence of innovations that bring benefits to the common good.

### 4.2 *Empathy*

Empathy is the second most important competency to act in the realm of social innovation, and there is a need to learn from observation of communities and reflection, to achieve hidden knowledge about their needs and desires (Cantú, 2012). This involves a job of perceiving their motivations, their context, habits and experiences (Ouden, 2012). Corroborating with the authors, Specialist 1 affirms that the issue of empathy is important since designers do not have the habit of being close to who will benefit from a project. Specialist 2 also stated that he had not learned methods of how to develop empathic relationships, pointing out that he understands empathy "in the sense of trying to understand the real needs of this community and to understand the context of their lives, quite different from my own".

Thus, it is believed that empathy also involves a capacity to interpret the needs of communities, with the development of tools to interpret those needs.

### 4.3 *Systemic vision*

The third competency identified as fundamental is the systemic vision, which helps when constructing scenarios in codesign with the communities, more than arriving at solutions, the designer should seek being able to visualize and understand the interaction (Cantú, 2012). The process should not be linear, starting from a problem to a solution, one must constantly question the problem, thinking about the possible negative impacts (Meroni, 2008). That is, one must work on a broad view of the system while thinking about its parts. Experiments, prototypes, and tests can help create scenarios, anticipating the environmental, ecological, economic, and political consequences of design interventions (Papanek, 1995).

## 5 DATA COLLECTION AND ANALYSIS

With the competencies raised, we started to collect data from the study of a course focused on social innovation of the degree in design of a Portuguese university. This course is given in the last year of the degree (3rd year) and aims to develop a social service, through work in groups, consisting of approximately six students each. Interviews were conducted with one member of each of these groups midway through the semester. The purpose of the interview was to discover the main difficulties they were facing when working with social innovation and whether they had previously had contact with similar topics.

The interviews were open, with different questions being made, but the main issues present in all of them were: Is this discipline different from other project disciplines? How? Have you ever worked with sustainability or social design in other disciplines or is it the first time you are having this contact? Is this project more challenging than the ones done until now? What are the main difficulties? With the analysis of the interviews, it was possible to perceive a series of questions and difficulties that students have in working with the process of social innovation, such as difficulty in formulating problems, working in groups and having contact with people outside the university environment. Two questions emerged: difficulties encountered in the process of social innovation itself and problems identified in the curricular structure of the university.

### 5.1 *Difficulties in the social innovation process*

The first insight was that working in groups and in a collaborative way is not customary for students, they mentioned difficulties in this process, since most projects were carried out individually. Also, interacting with a community is very different to the university context and poses a challenge, they mentioned going out into the real world as a new and challenging experience. Another point that was raised is that these projects are more complex and time-consuming because they involve students and the community.

Analyzing the difficulties faced by the students, two main difficulties of working with social innovation were revealed: the ability to work collaboratively, identified in this article as one of the competencies of design in this field; and the difficulty in working with more complex, long-standing problems that require the third identified competency, which is to develop a systemic view.

### 5.2 *Problems in the curricular structure*

The first problem identified in the curricular structure starts from the difficulty of disciplines that contribute to the processes of social innovation, such as sociology, to establish a relationship with design, one of the students mentioned that "disciplines such as sociology and anthropology turn out to be a general introduction to these subjects, without talking about design. It is up to each one to make these relationships, but not everyone does".

The second problem identified was the late proposition of a course which requires students to identify the problem to be solved, it was pointed that "In the first two years of the course, the subjects are very focused on creating, for example, making a chair. One does not think of the people who will use the product, we create from what we think. Now (in the final year of the course) all disciplines are being focused on projects where problem identification is the first step".

The student further states that "(...) there is no direction from the beginning, we spend some time doing the projects in a certain way and suddenly the form of work changes, becoming much more difficult. There should be a concern with identifying problems from the start. "Corroborating this view, another student stated that "they usually give us a problem and we have to solve it, here it was the opposite, right, we had to look for a problem, to observe, to notice where there was something wrong."

Still in regards to the problematization required in this course, it was revealed that "we had never worked like this, it's a discipline that makes us think about things in a more adult way, it's not just that project-making for the sake of it, and in the workshops. And we get out of that sphere of the commercial, of making to sell, and I think that's great, it opens our eyes".

Therefore, two structural problems are observed in the disciplines: to be able to establish relations between design and other disciplines necessary to work with social innovation, such as sociology; and the late insertion of projects that require a problematization, to identify the problems to be solved or opportunities for improvement.

## 6 CONCLUSIONS

The purpose of this article was to investigate design education focused on processes of social innovation, identifying the main skills needed to work with this approach. The three design competencies identified as key to working on social innovation processes were, firstly, the ability to collaborate, work with people from different cultures and different disciplines, encouraging them to use the skills and wisdom they possess to work collaboratively and actively. The second competency the study revealed was empathy, which is often not practiced in regular design work and was considered of great importance, especially by the experts. It refers to the need to learn from observation and to put oneself in the place of the people of the community, in order to be able to understand their difficulties, their needs, their context and their perspectives on life. The third competency, which can be identified as necessary for any design project, is a systemic vision. This should be present throughout the process of social innovation, as it aims to predict all possible social, environmental, political impacts of the carried-out scenarios, behaviours and actions.

From the triangulation of bibliographic review data and interviews with the experts with the interviews with the students, it was possible to perceive that they had more difficulty with two of the competencies identified in the present article. The first is to work in a collaborative way, pointed out by them as a competency not explored enough in curricular activities. The second is the systemic vision, not said in these words, but revealed in the discourse, when they explained that they had difficulties in identifying problems, being able to work with diverse contexts and more complex problems. The third competency identified in this article, which was to work with empathy, was not mentioned by the students, but by the experts, who as students did not acquire this skill, which proved very important in the practical work of design for social innovation.

Thus, it is believed that the main gain of this study was to reach these three competencies, which can be better discussed, understood and formulated within the academic curriculum of universities. These skills can be the starting point for organizing the content brought in the classroom, as well as for the search for tools, methods and methodologies that can be understood and taught to design students.

Regarding design education, it was revealed, when interviewing students and design specialists for social innovation, that there are difficulties in working academically with projects of this character mainly for three reasons. The first one, because students spend most of their academic life working on object-oriented design processes, that is, more related to the creation of services or graphic materials. These projects only consider the participation of users remotely and start with an already pre-defined problem, which is completely different from what happens in practice, where problems are diffuse and need to be understood and revealed (Schön, 2000). The second reason is that the disciplines involving practical work focused on social innovation are brought near the conclusion of the program, or, as mentioned by one of the experts, not even clearly addressed. This form of work is outdated and maintains what was proposed in the middle of the $20^{th}$ century. It is clear that our life context has evolved a lot in the last 70 years and that teaching needs to be reviewed and improved.

The third reason was related to the curricular structure because there is a gap in the relationship between disciplines important to social innovation, such as sociology, to design. That is, the disciplines to assist design for social innovation exist, but are worked in a poorly integrated way. Therefore, it is proposed that the three competencies identified in this article be discussed and worked from the beginning of the course, stimulating earlier work in contact with communities. As well as processes aimed at identifying problems and opportunities, without there being only a rigid problem proposition, defined by the teacher himself.

As suggestions for future studies, we point out, firstly, to test how one can work on the design competencies identified in didactic practice. It is also relevant to analyse the curricula of different universities to find out if they bring social innovation into their curricular structure and how it works in practice.

## REFERENCES


Cantú, D. (2012). Ideas Sharing Lab. Community Centred Design for Multifunctional and Collaborative Food Services. Doctoral Thesis, Politécnico Di Milano, FEB.

Chick, A. (2012). Design for social innovation: Emerging principles and approaches. Available at: <http://iridescent.icograda.org>. Acesso em: 20 jan. 2015.

Dias, A.; Almendra, R.; Moreira da Silva, F. (2013).Teaching of sustainability in industrial/product design courses in Portugal. In A Insustentável Leveza do ter. Porto Alegre, Brasil: SBD + ISSD.

Findeli, A. (2001). Rethinking design education for the 21st century: Theoretical, methodological, and ethical discussion. Design issues, 17(1), 5–17.

Friedman, K. (2002). Design curriculum challenges for today's university. In International Conference from CLTAD (Vol. 1, pp. 27–63).

Guerreiro, M. (2012). A relação pedagógica entre docente e aluno no ensino do Design. Em Ensino Superior: Inovação e Qualidade na Docência. Porto: CIIE- Centro de Investigação e Intervenção Educativas. FPCE, Universidade do Porto.

Manzini, E. (2008). Design para inovação social e sustentabilidade: comunidades criativas, organizações colaborativas e novas redes projetuais. Rio de Janeiro: E-Papers.

Manzini, E. (2009). New design knowledge. Design studies, 30(1): 4–12.

Manzini, E. (2011, May). Design Schools as Agents of (sustainable) Change. In 1st International Symposium CUMULUS//DRS for Design Education Researchers. http://www. designresearchsociety. org/docs-procs/paris11.

Margolin, V., & Margolin, S. (2002). A "social model" of design: issues of practice and research. Design Issues, 18(4): 24–30.

Meroni, A. (2008). Strategic design: where are we now? Reflection around the foundations of a recent discipline. Strategic Design Research Journal, v.1, n.1, Dec 1, p.31–38.

Murray, R; Caulier-Grice, J; Mulgan, G. (2010). The open book of social innovation. London: The Young Foundation & NESTA.

Ouden, E. (2012). Innovation Design: Creating value for people, organizations and society. London: Springer-Verlag.

Papanek, V. (1995). The green imperative: Natural design for the real world (p. 234). New York: Thames and Hudson.

Schön, D. (2000). Educando o Profissional Reflexivo. Porto Alegre: Artmed.

Walker, S.; Dorsa, Ed. (2001). Making design work–Sustainability, Product Design and Social Equity. The Journal of Sustainable Product Design. 1(1) 41–48.

Zurlo, F. (2010). Design strategico. In: AA. VV. Gli spazi e le arti. Roma: Enciclopedia Treccani. v. 4: Opera XXI Secolo. van Wezel W. and Jorna W.J. Paradoxes in planning. Engineering.

*Research & Education in Design: People & Processes & Products & Philosophy – Almendra & Ferreira (eds)*
*© 2020 Taylor & Francis Group, London, ISBN 978-0-367-49296-0*

# Design education for theatre regarding craft-design alliance

L. Soares & E. Aparo
*Polytechnic Institute of Viana do Castelo (IPVC) & CIAUD, Lisbon School of Architecture, Universidade de Lisboa, Portugal*

R.A. Almendra & F.M. da Silva
*CIAUD, Lisbon School of Architecture; Universidade de Lisboa, Portugal*

ABSTRACT: This paper discusses the collaboration between design education and theatre. Currently, due to financial and social issues, some theatre companies are forced to eliminate different topics from the creative process, such as the culture of drawing, sketching and craft-work. On one side, this situation contributes to the stagnation of answers for theatre shows, devoid of a process, unconnected with the culture of project. On the other hand, manufacturing world loses an occasion to reinvent itself, building new business areas for a sector, also, with problems. Methodologically, the authors present three case studies that crosses those fields. With this paper the authors demonstrate that a craft-design collaboration contributes to the improvement of theatre audiences, design education and local entrepreneurs. The paper highlights the importance of connecting design and enterprise as a common learning system. Thus, design-business networks can reinvent local contexts and create opportunities for entities, design education and human beings.

## 1 INTRODUCTION

Today it is reasonable to translate the idea of modern as an allegory for the accelerated mode and changeableness of our time. An era that for being so fast appears to uncover a human being who is not set in space or time; reducing all pretensions in instant hedonism; remaining for ephemeral answers simply outdated by different ones. The concept of 'liquid' (Bauman, 2005) is interpreted as the substance without a shape as it used to be. Never remaining the same, the state of substance it is adaptable to any container. This singularity in design field is reflected in changes and in performing to transform society. When it is manifested, design redefines itself and enlarges its territories in rapport with reality, including persons, space, time, causalities and circumstances. It is possible to understand the current life cycle as a cross-venture process. It means, a rational outcome of philosophical thinking (Latour, 2008) which includes the identification and the comprehension of the *genius loci* as a sign of a particular place. As Ampelio Bucci states (2003) this means everything that is related to the only thing that cannot be moved: the place, its history, its physical and cultural landscape. This methodology emphasises how design education may act, interpreting the spirit of today's reality. In addition, a place strongly connected with handicraft means taking advantage of craft-design alliance. This phenomenon demonstrates that design stimulates connections between all contributors. Craft-design alliance focus on today's reality can reach a new life creating links with the Academy and the entrepreneurial reality. Therefore, design for theatre regarding craft-design alliance highlights the importance of creating links between design and enterprise as a mutual learning system, in particular interdisciplinary working in changing contexts, creative and cultural economic development and the role of design and craft in theatre.

## 2 BACKGROUND

### 2.1 *Call for interdisciplinarity*

Collaborations between fields have been a recurring theme since the second half of the twentieth century, intuiting a world of new connections, of several coincidences, as medium like process and not of origin or end. This phenomenon manifests itself in a significant way as thinking on these changes means acting to transform reality and not to be a product of it. Design is an agent that produces contemporaneity, it redefines itself and expands its territories. This action is related to people, space, time, causalities and environments. If this is an occasion to describe design as science and art, as thinking and inventing (Boradkar, 2010), for master design education it is an occasion to found a permanent debate, highlighting design students' interests. This means that design methods is understood as a principle of a new foundation, supporting the singularity of "designerly ways of knowing" (Cross, 2006). Regarding networks between design and theatre, not all contexts are prepared for liquid modernity (Bauman, 2005) that focuses the importance of creating connexions between design and enterprise as a common learning system. On the other hand, considering the present economic situation of a place, business connections may stimulate master design students to perform as entrepreneurs of their own enterprises. That is, for design education the theatrical practise becomes a pioneer experience that can be applied in the business context. Master in Design studies must expand the students' skills in a holistic process, intersecting political, economic, cultural, technological and social possibilities. The proposition of connecting design with theatre reveals a system of patterns that joins products to a larger context, which goes from the community to the territory. Therefore, design field accepts and proposes multidisciplinary, transdisciplinary and interdisciplinary interactions (De Moraes, 2006).

### 2.2 *Craft-design alliance creating links with design education*

In Portugal, the relationship between design and craftwork gave rise to solutions characterized by stylized ethnographic patterns and popular shapes of daily life. In the first half of the 20th century this craft singularity characterized Portuguese authors such as Fred Kradolfer (1903-1968) or Raul Lino (1879-1974). During that historical moment the goal was to reach an iconic image of the country, but this process revealed a country able to be characterized with different images focused on old processes and ancestral techniques. Today, as other countries, Portuguese reality has changed to become complex with the challenges of introducing new technological capabilities and with contradictory persons asking for multifaceted products and services. Although, this globalised reality reveals some places in the country that are characterized with tradition, handicraft and ancient processes, which guarantees the identity of the nation. Therefore, craft-design alliance may be an answer to add value to craftwork. Thus, creating links with design master students may be an answer to develop sustainable project propositions (Cross, 2006). Design education should create the basis of a creative process between all participants, introducing experimental projects that include materials, methodologies, process, and business. Therefore, this paper supports the idea developing entrepreneurial skills in design education in order to stand up for handicraft sector, micro-enterprises and cultural entities, such as, the theatre companies of a place. In the north of the country this methodology makes sense as the numbers of artisans is relevant (see Table 1).

Table 1. Number of artisans by region and sex, 2018. Source[1].

| Region | Sex | | Total |
|---|---|---|---|
| North | M (365) | F(405) | 770 |
| Centre | M (206) | F(318) | 524 |
| Lisbon | M (222) | F(394) | 616 |
| Alentejo | M (100) | F(138) | 238 |
| Algarve | M (57) | F(89) | 146 |
| Madeira | M (12) | F(25) | 37 |
| Azores | M (167) | F(390) | 557 |
| | M (1129) | F(1759) | 2888 |

## 3 DISCUSSION

### 3.1 *Design for theatre and the local context*

In 1986, Portugal entered in European Economic Community and 40,0% per 1000 residents used to go to theatre, but in 1993 the values decreased to 19,3%. Only in 2002 the values would be higher than 120,0%. Since 2016 values increased to 240,0%. This trend has been growing and in 2017 reached 244,0%. For instance, in 2015 some theatre companies captured an average of about 1,336,920[2] spectators. In order to expand the number of spectators and, above all, to develop the number of theatre activities in the country, it is necessary to undertake research initiatives, highlighting the singularity of the place. This means to create a network with all actors. This quote also means that it is urgent to educated a new generation of designers, preparing then to work with their own skills in different scenarios like theatre. The connection between theatre and the small businesses of a territory can be a way to build experiences of knowledge and sustainability. Nevertheless, due to financial and social issues, there are some Theatre Companies that remove different fields from the creative process. Among them, drawing, sketching or crafts. This contributes to the stagnation of solutions for theatre shows. Handicraft loses an occasion to reinvent itself, building new business areas for a sector also with economic problems. With this philosophy in mind, the authors intend to answer to the following research questions: To what extent can design - which combines drawing, sketching with craftsmanship and which is always an act of redesign - contribute to the effective improvement of the performances produced by Theatre Companies? What is the impact of this action on the transmission of knowledge to society?

### 3.2 *The theatre company as an active agent of a place*

The Teatro do Noroeste – CDV[3] is the professional theatre company based at the Sá de Miranda Municipal Theatre of Viana do Castelo (in the North of Portugal), since 1991 and with more than a hundred productions. This group is made up of different performers, including the director, ten actors, an artisan dealing with the construction of scenarios, a light and multimedia expert, a sound and assembly technician, a graphic designer, a video specialist, a locksmith, two art consultants and an accountant. Concerning physical ant technical terms, it is an Italian theatre from the end of the 19th century with a horseshoe-shaped audience and three orders of cabins, with a capacity of 400 seats. In addition to the showroom, there is a rehearsal room and a garage (Figure 1) located in the limit zone of the city, where costumes, scene props and scenarios are tested and produced and later applied in the theatre or in a specific place, according performing characteristics.

[1] https://www.dgadr.gov.pt/images/docs/div_meiorural/INFORMACAO_ESTATISTICA.pdf
[2] https://www.pordata.pt.
[3] http://centrodramaticodeviana.com

Figure 1. From left to right: Company's office space. Company's warehouse.
Source: Serena Russo.

Regarding the process, there is an open and empirical practise, anticipating advances and retreats throughout the research. All contributors take part in every performance. The actors read the text to the other participants - such as the director, the designers, the actors and the artisans - which analyse the data's collection. In addition, the company supports connections with national and international professionals, for instance, with actors, directors, musicians, set designers or researchers. This company also develops community partnerships in order to promote social integration supported by artistic practices. The Theatre Company also provides educational activities for the local community, creating links between groups that usually do not interconnect with each other. One of the activities develops skills and knowledge about the culture of theatre focus on young people. The other activity is concentrated on pensioners. On the other hand, the company also works with other professionals through partnerships or joint-ventures in order to create innovation. This action includes the development of projects with national and international theatres, generating theatre 's festivals or research projects in partnership with institutions of higher education, such as, the Master in Design of the place.

### 3.3 *The competence of drawing and sketching in theatre*

During the process of creation and learning at the theatre, several drawings and sketches are made to freeze and store ideas that provide a reasonable result. In this process, all participants - actors, director, technicians, artisans, designers - of the company contribute with their knowledge and practice. Scenario drawings should bring together the creative component and the rational component in order to respond to practical and concrete problems. Therefore, draws and sketches foundations are based on personal experience and must be versatile and adaptable to the needs of the theatre team. As John C. Jones states, "the traditional design method is to draw, and to draw, successive alterations, either on different parts of a large piece of paper or else on a series of tracing from the original sketch or layout." (Jones, 1991, p.22). This means that the design hypotheses are increasingly being developed as the play is advanced. Today, it is necessary that the process of insertion of values is "projectable" (Mari, 2002) in a way that allows the increase of the meaning of the product (the concept) and its significance (the value). In addition, scenography, staging or the art of representing, it is possible to create links with areas such as carpentry or the textile sector. It is possible to states that there is a cross-fertilization between areas that expands to other activities. In particular, the design of costumes and set design can be an occasion to create connections with professions that are hibernated to need of innovation, such as, sewing or carpentry. The link between entities and the Academia of the same region may be the key to sustaining all stakeholders.

## 4 A PROJECT BETWEEN DRAWING, MANUFACTURING AND PERFORMANCE

In this research project, the authors present three case studies, describing the experience of a class of Master students and the dissertation of 2 Master design students, who developed their experiences using different methodologies, connecting craft sector with the culture of theatre. The first case was developed in 2016 was performed in an original way in different spaces of the city, bringing together academic, business and cultural partners. The second case was developed between 2016 and 2017 and was positioned to create costumes and accessories for a play to be performed outside the theatre, in a specific space of the city. The third case was developed between 2017 and 2018 and was focus oriented to create scenarios and accessories for a historical piece to be taken in scene on stage, inside the theatre.

### 4.1 *Case study 1*

This project emphasized the creation of an event in the city of Viana do Castelo (in the north of Portugal) as a consequence of the synergy between entities, companies and institutions of the same place, providing a cultural journey of a day with the added value brought by the partners, transferring them in a cultural event on an urban scale. With this project it was intended to create a synergistic action at different levels such as: design, music, creative spaces, representation and ceramic production.

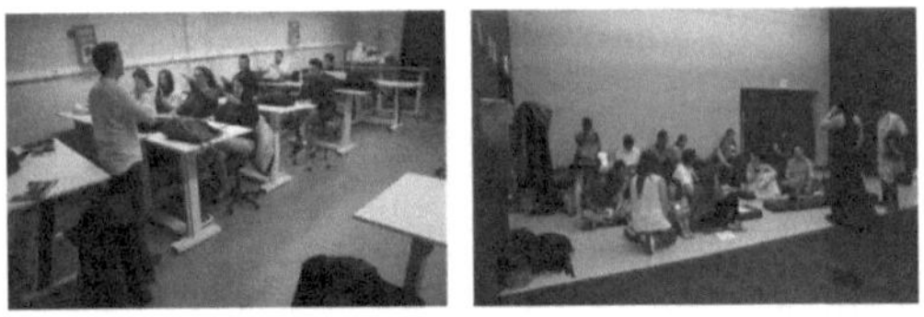

Figure 2. From left to right: Workshop with the master students and the Director of the local Theatre in the School. Workshop with the students, the actors and the Director of the local Theatre in the Theatre (eliminated for blind review).
Source: João Pontes.

The project focused on different actions of the story "Alice in Wonderland" and in partnership with different actors of the territory, namely, the Master Degree in Integrated Design of the Polytechnic Institute of Viana do Castelo, the productive company Fábrica de Cerâmica Artística da Vale do Neiva, the cultural agents of the city, namely, Inauguro[4], the Teatro do Noroeste – CDV and the Viana do Castelo Academy of Music[5]. In methodological terms, the creation of the event was a consequence of the interaction between the diverse players, taking the idea of a creative platform. In this sense, Master's students were organised in groups. Each unit created a performance that was physically associated with one of the cultural spaces of the project. As storytellers, each group of students developed ceramic objects carriers of "Alice in Wonderland". Throughout this process the ceramics factory and the theatre company provided specific workshops to the master students. Finally, the whole creative process was published on social networks to publicize and communicate the action.

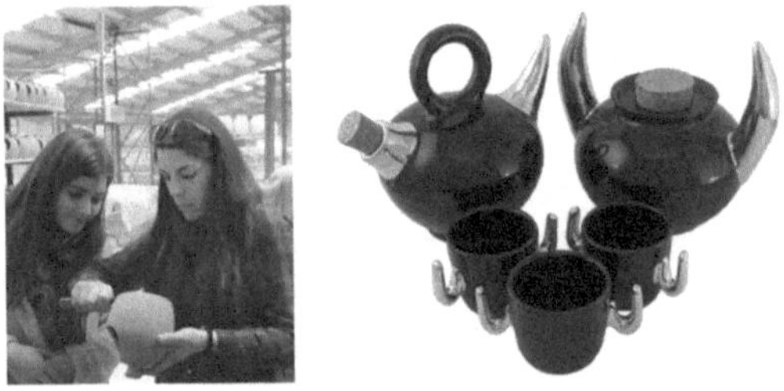

Figure 3. From left to right: The students' experiences in the ceramic factory. Set of tea designed by students.
Source: João Pontes.

[4] http://inauguro.pt
[5] http://www.amv.pt

## 4.2 *Case study 2*

In this project the stage was the ship "Gil Eannes", also named "White Angel", that is anchored in the river of the city. As a limit, the director proposed a costume to be used by an Angel, representing the women of the place, expression of their distinct realities - love, mourning, hatred, engagement, mothers, wives, revolt (Lopes, 2018). During design process it was created a palette of colours and materials of the culture of the place. Then, it was generated a palette of colours related with the colours of the river and the sea, using different textiles such as silk, polyester, cotton and fabric on canvas. In practice, student progressed directly into a three-dimensional scale phase, testing materials, mechanisms, and proportions.

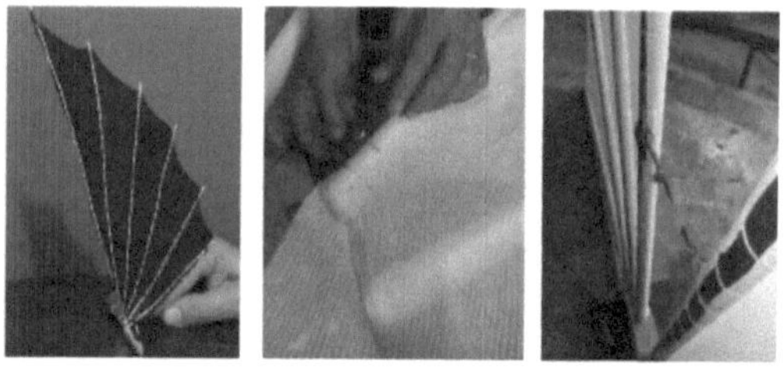

Figure 4. From left to right: Model. Testing components. Application of a steel cable to calculate the weight of the pipes.
Source: Andreia Lopes.

Then, photographic record and drawing were proceeded. This process was presented to the director and the supervisors. The next phase consisted on the materialization of the real-scale project. This phase would be subject to a series of limits like the existing machineries and the technical know-how of the theatre company's artisans. This stage was very important because it was the first contact between the student and artisans. In some cases, project required other materials, other techniques and other artisans, requiring the connection with other partners. This dynamic frame would create a system of companies, inciting innovation for other sectors that had never worked in theatre. "Connection between cultures of making and design through research and academic ventures, may be the key to sustainability and thus to the territory's effectiveness." (Aparo, Soares, Moreira da Silva, 2017).

Figure 5. The theatre support incorporated into the scenario and developed by the student Andreia Lopes.
Source: Andreia Lopes.

## 4.3 *Case study 3*

In this proposal student design the scenarios and supports for a play of Modernist period. There was no direct support of any professional set designer in the theatre (Russo, 2018), therefore the student worked with the director, the light technician and the artisan. The director combined the initial ideas with the proposals that came after the meeting with the designer and the technicians.

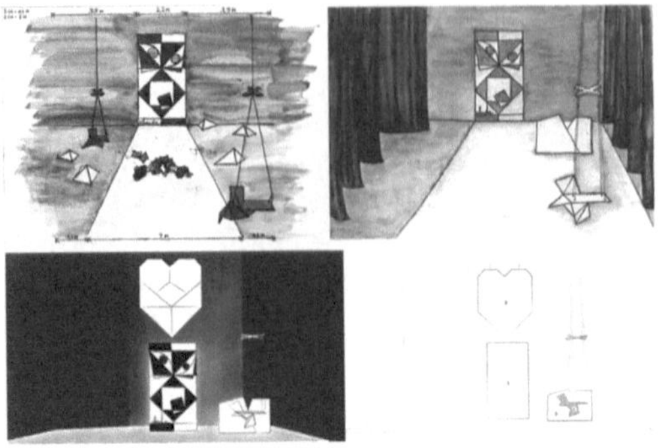

Figure 6. From left to right and from top to bottom: Sketches for the scenario (pen and watercolor on paper) and numbered scheme of the process of construction of set design highlighting part 4, developed by student Serena Russo.
Source: Serena Russo.

At this stage, the role of the designer was essential, regarding the creation of sketches and drawings of the stage, which show the ideas taken into account by the director and the totality of the supports, as well as their details of construction. During creation and learning process with the creative team, the student had made several sketches to solidify and store ideas that provided a satisfactory result, going through several stages of creation. Drawings demand for versatile in order to be readapt on personal experience and according to the needs of the theatre's team, responding to practical and concrete problems. Methodologically, it was created a scheme that indicates the operations that were required (Munari, 1983). This scheme was organized in a logical and chronological order, guiding the craftsman into the constructive process of each part that constituted the scenography as a whole.

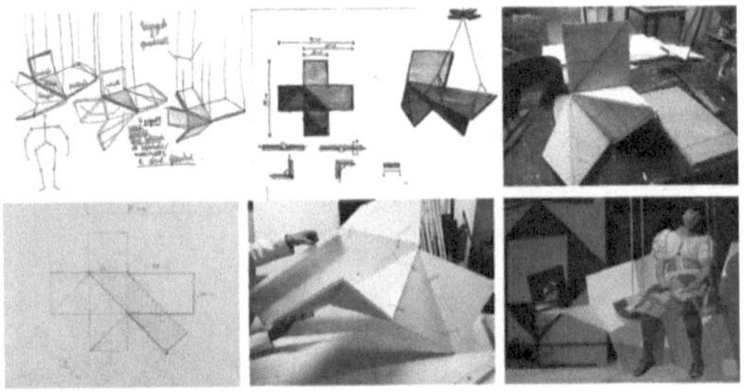

Figure 7. From left to right and from top to bottom: Sketches for the scenario (pen and watercolor on paper) and numbered scheme of the process of construction of set design highlighting part 4, developed by student Serena Russo.
Source: Serena Russo.

## 5 CONCLUSIONS

Today, the complementary attitude of designers and researchers in design process applied to theatre contributed towards the permanent debate and development of scenography as a discipline. The scenic project is not bound by a single method or style. It involves different directors and projects, letting external factors such as culture, communication, politics, people and time to define philosophies, products and processes, developing regenerative projects. This also means that the three case studies' presented in this paper had a common philosophical basis, which may contribute to the creation of a master design education, which strength is the system established among parts. It means the notion of a phenomenological school related to its own environment, always changeable. In the scope of Design, this phenomenon is significantly reflected in metamorphoses; in acting to transform society instead of being a by-product of society. Considering

Design as an agent to produce modernity, when it is manifested, it redefines itself and expands its territories in rapport with reality, people and other disciplines. Design profits from momentary environments. Subjective individualism substantiates this thesis. It characterized scenic arts teaching in the North of Portugal – exactly at the Polytechnic Institute of Viana do Castelo - as analysed through different cases in the same school and with the same mentors. The scenic project is not bound by a single method or style. It involves different directors and projects, letting external factors such as culture, communication, politics, people and time help define the process, developing regenerative projects. For design research, this article defends the open and empirical process, stimulating new links between the local productive culture and the territory and addressing new disciplines that are usually not associated with design. On the one hand, by stimulating links between the local productive culture and the territory, it is possible to restructure not only the offices but also a process that brings social and economic added value, capable of modernizing, promoting and sustaining a place or region. This article aims to contribute towards the study of set and stage design for theatre in our time, challenging designers, scenographers, handicrafts, architects and theatre directors to understand the concept of change as synonymous of the current life cycle. Liquid reality is a reference for this research since it allows building a self-critical course, assuring dialectics and subsequently self-knowledge and innovation. From this perspective, design education requires the practice of prudence (Howard, 2009) as a virtue, entailed by the use of reason, managing and motivating groups of people of different professional backgrounds - scenic designers, lighting designers, actors, carpenters, graphic designers - towards a design process that is always an act of redesign (Latour, 2009) old things to become modern.

## ACKNOWLEDGEMENTS

This work was carried out with the support of CIAUD - The Research Centre for Architecture, Urbanism and Design of Faculty of Architecture, University of Lisbon (FA/ULisboa).

## REFERENCES

Aparo, E., Soares, L., Moreira da Silva, F. (2017). "Either/or: reflecting design thesis orientation" In. Berg, Arild; Bohemia, Erik; Buck, Lyndon; Gulden, Tore; Kovacevic, Ahmed; Pavel, Nenad (Editor). DS 88: Proceedings for the 19th International Conference on Engineering & Product Design Education (E&PDE17), Building Community: Design Education for a Sustainable Future, Oslo, Norway, 7 & 8 September 2017. Section: Design Education for the General Public. United Kingdom. Pages: 674-679. ISBN: 978-1-904670-84-1.

Bauman, Z. (2005). *Modernità liquida*. Bari: Laterza.

Boradkar, P. (2010). Design as problem solving. *The Oxford handbook of interdisciplinarity*, R. Frodeman, Oxford University Press. 273–287.

Bucci, A. (2003) L'impresa guidata dalle Idee - Lezioni di management creativo dalla moda edal design. Milão: Ed. Arcipelago.

Cross, N. (2006). *Designerly ways of knowing*. London: Springer-Verlag AG.

De Moraes, D. (2006). *Metaprojeto: o Design do Design*. São Paulo: Blucher.

Howard, P. (2009). *What is Scenography?* London: Routledge.

Jones, J. C. (1991). *Design Methods*. New York: McGraw-Hill.

Latour, B. (2009). A Cautious Prometheus? A Few Steps Toward a Philosophy of Design (With Special Attention to Peter Sloterdijk). *Proceedings of the 2008 Annual International Conference of the Design History Society, Falmouth*, 3-6 September 2009, *e-book, Universal Publishers*, 2–10.

Lopes, A. F. O. (2018) O design na construção de figurinos para uma peça de teatro: o caso do projeto comunitário, "Anjo Branco - Gil Eannes", no âmbito do Noroeste Comunitário - Projeto Comunitário do Teatro do Noroeste – CDV. DES - Dissertação de Mestrado em Design Integrado. Instituto Politécnico de Viana do Castelo. http://hdl.handle.net/20.500.11960/1983

Mari, E. (2002). *Autoprogettazione?* Milano: Edizioni Corraini.

Munari B. (1983). *Como nacen los objectos: apuntes para una metodologia proyectual*. Barcelona: Gustavo Gili.

Russo, S. (2018) Relações metafóricas entre o design e o teatro: um caso de estudo de design de cenários no âmbito do Teatro do Noroeste – CDV. DES - Dissertação de Mestrado em Design Integrado. Instituto Politécnico de Viana do Castelo. http://hdl.handle.net/20.500.11960/2026

*Research & Education in Design: People & Processes & Products & Philosophy – Almendra & Ferreira (eds)*
 *ISBN 978-0-367-49296-0*

# Gender and territories of design research

A.C. Dias, B.R. Moreira & N. Plentz
*CIAUD, Lisbon School of Architecture, Universidade de Lisboa, Portugal*

ABSTRACT: This paper aims to map and discuss the relationship between researcher genders and their chosen research territories, using the PhD program in Design at the Lisbon School of Architecture as a case study. Based on the PhD thesis database and using the REDES' methodology for analysis, we identified some search tendencies on specific areas and some preferences for specific research processes, respectively by women or men. Thus, we reflected about how gender can influence the choice of research territories of a PhD student in Design.

## 1 INTRODUCTION

Observing the numbers of the resident population in Portugal in 2017, we can verify a slight preponderance of the female gender (52.7%) compared to the male gender (47.3%) (Pordata, 2018). On the other hand, there is a very large gap between the number of women with higher education (61.8%) and men (38.2%) (Pordata, 2019), being the percentage of the active population (employed and unemployed) with higher education 26.3% (Pordata, 2019).

These data lead us to reflect on the increasing access of women to education, namely to higher education, which was once elitist (Rodrigues, 2015) and mostly accessible to men. Today we see a massification of women in the labour market as a result of the right to a more complete education, with the direct consequence of recognizing the (active) role of women in society over the last decades, even though the positions of leadership are still reserved for men (APGICO, 2013).

In fact, the number of women on corporate boards or management positions has increased during the last years, but Portugal is still far from the 40% expected quota of each gender by the European Union (European Union, 2019) with just 14,6% of women as board members of the largest publicly listed companies in the EU, below EU average of 23.3% (European Union, 2016). The study *Report on equality between women and men in the EU* (European Union, 2019, p. 20) shows those numbers and says that "women tend to get stuck at middle or senior management level" and "...are less likely to be hired into manager-level jobs and they are far less likely to be promoted into them". Also, "women tend to earn less per hour than men do for the same profession, regardless of whether it is a highly skilled white-collar profession or a lower-skilled job" (European Union, 2019, p. 19).

We should note that the number of women with PhD level in Portugal grew significantly during the last decades. According to Pordata (2017), in 1970, 22 men and 1 woman completed their PhD studies in Portugal, comparing with the 1092 men and 1259 women in 2015. Since the year 2006 the number of female PhDs has exceeded the number of male PhDs every year, with 2007 being the only exception.

However, in a century that should be governed by the equality of rights and duties between women and men, we are interested in realizing if the social values that instilled in us and that we continue to instil, even unconsciously, in our children about gender differences have a repercussion on the thematic choices we make. Thus, we observed the Doctoral Program in Design of Lisbon School of Architecture, which encompasses four disciplinary areas (product, communication, fashion and spatial) branched out into several lines of research of a transdisciplinary nature and defined as the main objective to map and discuss the relationship between gender and the chosen research territory. For this study we used the database developed by the research group REDES (Research & Education in Design) that includes the analysis of the 72 PhD theses in Design defended between the years 2010 and 2017 as the object of study.

This doctoral program was sought by both men (34) and women (38), but it is important to understand the relationship between the gender of the researcher and the choice of study territory. This raises the main question we want to reflect on: Can gender influence the choice of research territories of a PhD student? leading us to other questions such as: What are the topics most explored by women or by men? Are men more involved with finding new ways to manage a project? Who is more awakened to societal problems?

The results obtained from this study are discussed in order to map the current scenario of thematic choices made by men or women in Lisbon School of Architecture's Design Doctoral Program and to awaken new researchers to a certain selective tendency, provoking them, on the one hand, to leave their comfort zones and, on the other hand, to discover new possible approaches throughout the doctoral program.

## 2 METHODOLOGY

This study used the database developed by the research group REDES (Research & Education in Design) and its methodology for analysis that includes criteria such as identification of the type of the thesis (theoretical or theoretical-practical), the taxonomy of the work, the scope and discipline explored, and the research methods used. Those data were organized and also other information about the researcher were included like the gender, the name, the name of the supervisors and also the year when the work was concluded.

Thus, we decided to structure our study in two moments: first we mapped the total information in tables, adding three different columns - one for women, one for men and another for the total information. Secondly, we identified some tendencies about each gender and discrepancies between choices of both to analyse in detail.

## 3 RESULTS

In a general overview, in terms of type of thesis and taxonomy approach, there are not many discrepancies between women or men choices.

A first observation of the sample reveals that 2/3 of the 72 PhD thesis defended between 2010 and 2017 are theoretical and just 1/3 theoretical-practical (Table 1) and, regarding the research taxonomy perspective, we can verify that 57 thesis (79,2%) are related with design phenomenology and/or praxeology - being phenomenology slightly more explored by women and praxeology by men - and less with epistemology or ontology (Table 2).

However, when it comes to the chosen research disciplines, scopes or methods the setting is somehow different and we can identify some (small) gender tendencies.

As we can see on Table 3, the major chosen disciplines are communication design (26) and product design (20). In terms of gender choices, on one hand we can perceive that women tend to prefer fashion design or spatial design when compared than men, and on other hand,

Table 1. Type of thesis.

| Type | Female | Male | Total |
|---|---|---|---|
| Theoretical | 25 | 23 | 48 |
| T/Practical | 13 | 11 | 24 |

Table 2. Taxonomy.

| Areas | Female | Male | Total |
|---|---|---|---|
| Epistemology | 0 | 2 | 2 |
| E/Phenomenology | 1 | 1 | 2 |
| E/Praxeology | 6 | 0 | 6 |
| Ontology | 1 | 4 | 5 |
| O/Epistemology | 0 | 0 | 0 |
| O/Phenomenology | 0 | 0 | 0 |
| O/Praxeology | 0 | 0 | 0 |
| Phenomenology | 11 | 7 | 18 |
| Praxeology | 12 | 10 | 22 |
| Pr/Phenomenology | 7 | 10 | 17 |

Table 3. Discipline.

| Discipline | Female | Male | Total |
|---|---|---|---|
| Communication | 14 | 12 | 26 |
| Design | 5 | 6 | 11 |
| Fashion | 6 | 1 | 7 |
| Product | 7 | 13 | 20 |
| Ambient | 6 | 2 | 8 |

compared to women, twice as many men choose product design - which is probably related to the researcher background.

Regarding the chosen research scopes (Table 4), and linked to the research taxonomy previously narrated, design history is the scope more sought (10 in 72 thesis), mostly by women (7). We also can verify that there are 47 different scopes, being 39 of them chosen just by one person – which means that 54,2% of the researchers choose a unique topic. This reveals that the other 8 scopes are chosen by more than one researcher.

So, we decided to look at that in a more detailed way. Those 8 scopes are co-design, lighting design, inclusive design, design education, design history, museology, sustainability and typography, there being a preponderance of the female gender in some areas such as history, collaboration, inclusion, museology or sustainability, compared to the more fragmented men's choices. When looking at the isolated topics, we can perceive that men prefer topics more related to materials, methods, perception or interaction, and women go for topics more related to affection, emotions or interiors.

Regarding the ten most used research methods, we can verify that literature review is used by almost all of the researchers, followed by the questionnaires (69) and interviews (45). With this, we can understand the need and importance to contextualize and justify the research problem and scope, but also to fundament with opinions of the stakeholders (users, clients or other people involved with in). The empirical research is crucial to the PhD work, keeping the impartiality of the researcher with the investigation. Other data we should consider are that exist a ratio of 2 women per 1 man choosing interviews as a process; but also, the artefact analysis, usability testing or observation are mostly chosen by women instead of men; and men prefer project, case study or tool testing as process possibilities.

## 4 DISCUSSION

This paper presents a holistic overview about the relationship between research genders and their chosen research territories, studying the PhD thesis database from the program in

Table 4. Research scope.

| Scope | Female | Male | Total |
|---|---|---|---|
| Affection | 1 | 0 | 1 |
| Agro-food | 1 | 0 | 1 |
| Authorship | 0 | 1 | 1 |
| Branding | 0 | 1 | 1 |
| Scenography | 0 | 1 | 1 |
| Co-Design | 2 | 0 | 2 |
| Japanese graphic communication | 0 | 1 | 1 |
| Colour | 0 | 1 | 1 |
| Human body | 0 | 1 | 1 |
| Observation drawing | 1 | 0 | 1 |
| Automotive design | 0 | 1 | 1 |
| Experience design | 0 | 1 | 1 |
| Identity design | 1 | 0 | 1 |
| Lighting design | 1 | 1 | 2 |
| Pattern design | 1 | 0 | 1 |
| Ceramics | 1 | 0 | 1 |
| Clothing | 0 | 1 | 1 |
| Emotional design | 1 | 0 | 1 |
| Inclusive design | 4 | 1 | 5 |
| Parametric design | 0 | 1 | 1 |
| Medical devices | 0 | 1 | 1 |
| Ecodesign | 0 | 1 | 1 |
| Education | 1 | 0 | 1 |
| Packaging | 1 | 0 | 1 |
| Drawing | 1 | 0 | 1 |
| Design Education | 0 | 2 | 2 |
| Geometry | 0 | 1 | 1 |
| Design history | 7 | 3 | 10 |
| Innovation | 1 | 0 | 1 |
| Tech innovation | 1 | 0 | 1 |
| HCI | 1 | 0 | 1 |
| Interaction | 0 | 1 | 1 |
| Design research | 1 | 0 | 1 |
| Materials | 0 | 1 | 1 |
| Design methods | 0 | 1 | 1 |
| Furniture | 1 | 0 | 1 |
| Urban furniture | 1 | 0 | 1 |
| Museography | 0 | 1 | 1 |
| Museology | 3 | 1 | 4 |
| Design perception | 0 | 1 | 1 |
| Pictograms | 0 | 1 | 1 |
| Design process | 1 | 0 | 1 |
| Signage | 0 | 1 | 1 |
| Sustainability | 3 | 1 | 4 |
| Typography | 1 | 3 | 4 |
| Urbanism | 0 | 1 | 1 |
| Webdesign | 0 | 1 | 1 |

Design at the Lisbon School of Architecture. The sample of 72 PhD thesis reveals that design research is focused on products (phenomenology) and processes (praxeology), respectively chosen by women or men, which leads us to think there is a tendency to choose topics directly related to the understanding of phenomena and human behaviour (with/facing to them) than to philosophic or ontological issues.

Table 5. Research methods.

| Scope | Female | Male | Total |
|---|---|---|---|
| Literature Review | 38 | 32 | 70 |
| Questionnaire | 35 | 34 | 69 |
| Interview | 31 | 14 | 45 |
| Project | 15 | 22 | 37 |
| Artifact Analysis | 33 | 4 | 37 |
| Case study | 11 | 14 | 25 |
| Usability Testing | 16 | 4 | 20 |
| Focus group | 7 | 6 | 13 |
| Tool Testing | 0 | 11 | 11 |
| Observation | 11 | 0 | 11 |

When considering the chosen research scopes, we can perceive that women choose topics more related to affection, emotions or interiors, and men prefer topics like materials, methods, perception or interaction. So, can we understand this as a reflex of the female essence (take care of the house and protect their children) or the origin of women's entrance on labour market, such as teaching or working as secretaries?

When it comes to the preferred research processes, we consider women are more interested in working with data collection tools, so, more focused on the development of empathy with the studied universe – preferring to conduct an interview, usability test, or observation. Perhaps, women designers are focusing more on the human being, that is, on results even more qualitative than the men, who have had more quantitative results (from project, tool test, case studies, or prototyping).

That lead us to reflect about the gender differences in temperament that Else-Quest, Hyde, Goldsmith, & Van Hulle (2006) elucidate in a meta-analysis. The authors – academics related to the area of psychology and understanding of human behaviour - studied the magnitude of gender differences in temperament in children. In general, they focused on childhood explaining that "temperament reflects biologically based emotional and behavioral consistencies that appear early in life and predict" (Else-Quest, Hyde, Goldsmith, & Van Hulle, 2006, 33), assuring the idea that "gender differences in temperament cannot be discussed without considering the evidence on gender differences in personality in adulthood" (idem, 2006, 62), because some dimensions like susceptibility to emotional stimulation, strength and speed of response and peculiarities of fluctuation and intensity of mood, shaped by biological and environmental factors, contribute to form the basis for adult personality. In an overall perception, authors found small gender differences in temperament between girls and boys. For our study, we highlight their findings about female ability to better perceive low-intensity environmental stimuli, to be aware of environmental changes, to regulate or allocate their attention and impulses. In other hand, boys are greater to externalize behaviour problems, to be more active, less shy, and derive more pleasure than girls facing high-intensity stimuli.

In fact, in our society, there are two different stereotypes – one for women and another for men – that have been changing during the last years in order to promote gender equality and improve the gender balance in recruitments and career promotion and we believe that can start on the education process. Thus, being the PhD the final level of the academic formation, students should have the opportunity to experience other roles that take them out of their comfort zone and provoke them to discover new approaches to work that can open their research horizons.

So, we believe that sharing ideas and working in groups where specific roles are attributed to women and others to men will be a good way to exercise some missing skills and awake both for other interests, because in the working daily life it is important to develop collaborative skills in all people, especially in those who work with design.

For further research it should be interesting to explore the two PhD thesis that didn't use literature review as a research process, understanding the principles and fundamentals that they used to contextualize the scope of the work and also explore the courses taught on the doctoral program in order to reflect if they influence or contribute to the gender tendencies.

## REFERENCES


APGICO (2013, October). *Liderança feminina, bem-estar e inovação nas empresas*. Retrieved from http://www.apgico.pt/pagefiles/projectos/projecto1/RelatorioLIDERANAFEMININABEMESTAREINOVAONASEMPRESAS.pdf.

Else-Quest, N. M., Hyde, J. S., Goldsmith, H. H., & Van Hulle, C. A. (2006). Gender differences in temperament: A meta-analysis. Psychological Bulletin, 132(1), pp. 33–72.

European Union. (2019). *2019 Report on equality between women and men in the EU*. Luxembourg: Publications Office of the European Union. Retrieved from https://ec.europa.eu/info/sites/info/files/aid_development_cooperation_fundamental_rights/annual_report_ge_2019_en.pdf.

European Union. (2016, July). Gender balance on corporate boards. Retrieved from https://ec.europa.eu/newsroom/docum.

Pordata. (2017, August 02). *Doutoramentos realizados em Portugal ou no estrangeiro e reconhecidos por universidades portuguesas: total e por sexo*. Retrieved from https://www.pordata.pt/Portugal/Doutoramentos+realizados+em+Portugal+ou+no+estrangeiro+e+reconhecidos+por+universidades+portuguesas+total+e+por+sexo-237-4385.

Pordata. (2019, February 06). *População activa: total e por nível de escolaridade completo*. Retrieved from https://www.pordata.pt/Portugal/Popula%C3%A7%C3%A3o+activa+total+e+por+n%C3%ADvel+de+escolaridade+completo-1008.

Pordata. (2018, June 15). *População residente, estimativas a 31 de Dezembro: total e por sexo*. Retrieved from https://www.pordata.pt/Portugal/Popula%C3%A7%C3%A3o+residente++estimativas+a+31+de+Dezembro+total+e+por+sexo-5.

Pordata. (2019, February 07). *População residente com 15 e mais anos de idade por nível de escolaridade completo mais elevado: total e por sexo*. Retrieved from https://www.pordata.pt/Portugal/Popula%C3%A7%C3%A3o+residente+com+15+e+mais+anos+de+idade+por+n%C3%ADvel+de+escolaridade+completo+mais+elevado+total+e+por+sexo-333.

PwC. (2015, March). *Mulheres em Portugal: Onde estamos e para onde queremos ir*. Retrieved from https://www.pwc.pt/pt/publicacoes/imagens/2015/pwc-mulheres-em-portugal.pdf.

Rodrigues, M. L. (2015). Análise cronológica das políticas públicas: ruturas e continuidades. In Rodrigues, M. L. & Heitor, M., *40 anos de Políticas de Ciência e de Ensino Superior*. Coimbra: Almedina, pp. 25–50.

*Community-based Research*

*Research & Education in Design: People & Processes & Products & Philosophy – Almendra & Ferreira (eds)*
*© 2020 Taylor & Francis Group, London, ISBN 978-0-367-49296-0*

# Care for veterans and their healthcare

E. Morshedzadeh
*Department of Industrial Design, Virginia Tech, USA*

C.B. Arena, J.L. Robertson, A.A. Muelenaer & P. VandeVord
*Department of Biomedical Engineering and Mechanics, Virginia Tech, USA*

B.D. Hendershot
*DoD-VA Extremity Trauma and Amputation Center of "Excellence," Bethesda, USA*

E.A. Lianos
*Research and Development Service, Salem VA Medical Center, Salem, USA*

ABSTRACT: In design research needs identification should be investigated based on different communities. For this matter, proper and sufficient engagement of these communities is an undeniable core for design researchers. Healthcare in design education hasn't been fully explored because of its sensitive nature. This paper is a study in progress that creates a unique opportunity for Industrial Design and Biomedical students through an immersive course to experience the professional environment in healthcare. Considering the important role of the communities in user's need identification we chose veterans as target group because of their special conditions. This exceptional target group and their healthcare providers will bring another layer of emphasis on definition of needs in relation to the communities involved. This program will give any design educator an extraordinary experience in Community-Based Research and Design in Healthcare. This proposal has been granted by the National Institutes of Health starting from Spring 2020.

## 1 INTRODUCTION

Industrial Designers, by nature, collaborate and communicate with other professionals on different projects. Recently, with design and design thinking being a part of almost every project, professionals and educators have realized that having a team member with the design skill set can be effective on many levels. Studies show that under certain circumstances, design teams that use commercial groupware products or systems perform far better than teams that don't (Briggs, 2006). This realization in academia has resulted in the revolutionizing of design education in connection with other disciplines and created a wide range of opportunities for design faculty and students to collaborate alongside other fields of study. However, the role of these collaborative projects in most current design-program curricula are most often for graduate students and based on the faculty's research criteria.

In design for healthcare, an understandable dialogue between engineers and designers is especially important, and involving the community (in our case veterans and their healthcare providers) is crucial to developing medical devices that are both safe, effective and easy to operate. According to JR Goldberg (Goldberg 2007), the three main aspects of medical device design are (1) technical aspects and requirements, (2) human factors aspects for user interaction, and (3) aesthetic form for psychological influence. To achieve this level of comprehensiveness in professional industry, the culture of collaboration needs to be embedded into program curriculum and become a basic practice.

In this paper, we will discuss a new, collaborative, community-based course at Virginia Tech, involving undergraduate students from Industrial Design and Biomedical Engineering working directly with veterans, wounded warriors and their community of healthcare providers. We hypothesize that by joining the skillsets of biomedical engineers and industrial designers from the onset of product development and providing the access to clinical professionals and environments they will develop a unique problem-solving perspective that will improve outcomes in Senior Design courses and subsequent professional endeavours.

Since the U.S. Department of Defense (DoD) and the Department of Veterans Affairs (VA) have the combined mission to maintain the strength and readiness of the active forces while preserving the long-term health for those who served, this creates an opportunity for both the Industrial Design (ID) and Biomedical Engineering (BME) disciplines. While active-duty service members and veterans suffer from the same maladies as comparably-aged civilian population (e.g., heart disease, cancer, diabetes), they also have unique healthcare needs as a result of battlefield exposures that must be recognized and resolved (Spelman, Hunt, Seal, Burgo-Black, 2012). Creating this connection plays an important role in this course which creates an environment that includes the presence of healthcare professionals in the community as well as primary users. Also, through this practice and because of the specific criteria and needs for this target group the emphasis on the role of community in design products can be highlighted, and Industrial Design students can better understand and experience the importance of community-based research and design.

Additionally, the Veterans Administration (VA), the largest integrated healthcare system in the country (9.3 million enrolees), fosters the development of veteran-specific educational content that is now being emphasized in our healthcare-professional development curricula (Olenick, Flowers, Diaz, 2015) and this provides a great access and opportunity for exploration in this community.

## 2 DESIGN AND COLLABORATION IN ACADEMIA AND PROFESSIONALISM

Despite the fact that industrial designers have been playing an important role in professional practice, still their image in academia is not well-defined or well-established. Since many design programs are housed within Colleges of Architecture or Art, the profound and rigorous research methods used by Industrial Designers are not well-known to other disciplines, especially engineering (Beucker, 2004). It is safe to say that the discipline of design has two directions. The traditional definition of design is more focused on the artistic aspects while the broader definition of design encompasses any role that a designer can have during the design process. In the latter definition, designers are involved from the beginning of the project and they play an important role in defining the needs and proposing the solution based on the user's experience and interactions.

As it was mentioned, design faculty are creating similar environments by offering collaborative projects to students. The formation of faculty-student design groups from ID and BME in academia is an existing, but underutilized concept. Select examples include collaborations between Marquette University and the Milwaukee Institute of Arts and Design, Brown University and the Rhode Island School of Design, and at the University of Illinois Champaign-Urbana, and New Jersey Institute of Technology. Additionally, the Medical Device Innovation and Entrepreneurship Program at the University of Cincinnati includes a course with participants from business, industrial design, and biomedical engineering (Goldberg 2012). All programs recognize similar benefits of the approach, including an appreciation for complementary skills and acknowledgment that there are multiple ways to solve a problem.

In recent years, the ID program at Virginia Tech has been making great progress toward creating this experience for the students. For example, faculty have been actively pursuing collaborations with multiple disciplines on multiple projects, such as a group of students from Industrial Design, Mechanical Engineering, Biomedical Engineering and Public Health on "Designing a Neonatal Harness for Vital Sign Monitoring in Malawi." The resulting prototype received institutional approval to be tested on location, on actual users, and the goal is to

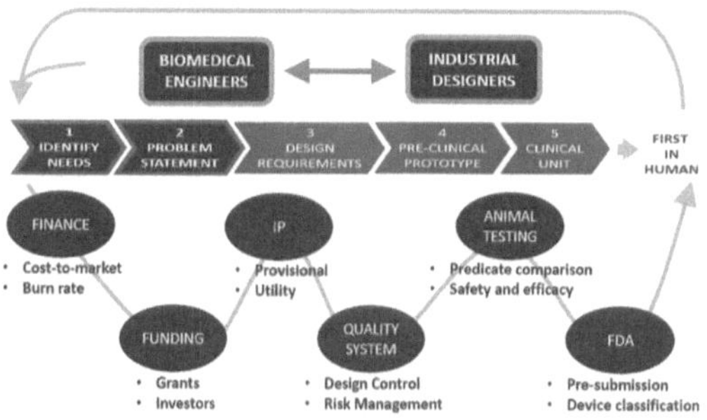


Figure 1. Proposed Path-to-Patient. Flow chart of product development milestones and accompanying parallel path of focus areas with collaboration between biomedical engineers and industrial designers.

eventually locally mass-produced on a large scale for Malawians. However, projects like this do not happen often and an established infrastructure within the curriculum might help provide a more well-rounded experience for undergraduate students.

Part of this infrastructure requires a well-established and well-defined connection and collaborations with professionals. Medical device design firms employ professionals with a broad range of expertise required to manage and innovate at each stage of the product development lifecycle. This includes Industrial Designers and Biomedical Engineers, both of whom are taught to analyze environments for design opportunities. However, in academia, interactions between these two groups are limited, since each group emphasizes different aspects of design.

The successful design of medical devices and software requires broad expertise in areas ranging from finance to a comprehensive understanding of the requirements of the Food and Drug Administration (FDA). Often, an overview of the intermediary concepts is presented in a Senior Design course. For example, students may be exposed to diverse topics including mechanisms of fundraising, intellectual property (IP), quality management systems, and animal/human testing. These subjects occur at various stages on the "path-to- patient" framework of product development (Figure 1) and are often supplementary to engineering and industrial design work.

We aim to test this hypothesis and address the goal of improving veteran healthcare by organizing an interdisciplinary training experience in the classroom. However, conducting this course is not possible unless there is an established connection and infrastructure with professional medical centers and their facilities. For this matter, this course has been designed not only between the two disciplines (BME and ID) but also in close collaboration with the Salem Veterans Affairs Medical Center (SVAMC), and the Walter Reed National Military Medical Center (WRNMMC) in Bethesda, Maryland.

## 2.1 *Background in veteran and their healthcare providers*

As mentioned previously, the U.S. Department of Defense (DoD) and Veterans Affairs (VA) have the combined mission to maintain the strength and readiness of the active force while preserving the long-term health for those who served. While active-duty service members and veterans suffer from the same maladies as comparably-aged civilian population (e.g., heart disease, cancer, diabetes), they also have unique healthcare needs as a result of battlefield exposures that must be recognized and resolved (Rábago, etal. 2016). These needs extend to military healthcare providers, scientists, and engineers who are actively treating and learning from our wounded veterans to optimize their quality-of-life (QoL) measures. For example, the DoD-VA Extremity Trauma and Amputation Center of Excellence (EACE) is the leading advocate for research and treatment of veterans with extremity trauma. EACE advances medical and scientific knowledge by identifying critical gaps and translating findings into clinical practice via multidisciplinary teams embedded within the Military Health System (Olenick, Flowers, Diaz, 2015). At Walter Reed National Military Medical Center, EACE manages a diverse portfolio of research, primarily focused on identifying unique needs associated with long-term care of traumatic injuries, understanding and mitigating secondary effects of extremity trauma, and optimizing QoL. At community-based Veterans Affairs hospitals, clinicians and therapists form in collaborative teams to understand the unmet needs of aging veterans and provide them with innovative healthcare solutions. The Salem Veterans Affairs Medical Center serves this role for veterans residing in Southwest Virginia.

As we mentioned, collaboration between Industrial Design and Biomedical Engineering is academia has not been fully utilized nor reached its full potential. Also, to the best of our knowledge, our program will be the first to align biomedical engineers and industrial designers at the start of the "path-to-patient" framework with a target goal of improving healthcare for wounded veterans.

## 3 STATEMENT OF AIMS FOR COLLABORATIVE INNOVATION

Professional healthcare design firms are always looking for employees with a variety of skills and knowledge in the product development lifecycle. The ultimate goal for this course is to create a foundation in design curriculum that trains Industrial Design professionals that can speak and understand a common language with engineers, specifically in the healthcare industry. This case study provides the chance to interact with users with special needs and unique circumstances. To this goal they have the opportunity to collaborate with DoD and VA clinics and laboratories to discover unmet needs and develop innovative solutions related to wounded veterans' healthcare.

Two major aims have been designed for this course:

Aim 1 – Complementary education in biomedical needs identification from different point of views: This course is being co-taught between two disciplines to junior students. The outcome of this course can lead them to their thesis project in the senior year. (See section 4. for more details)

Aim 2 – Clinical immersion in "wounded warrior" and veteran healthcare: In parallel with Aim 1, students will be placed in a variety of structured clinical and research settings and provided opportunities to apply their evolving skills in biomedical needs identification. Through partnerships with the SVAMC and WRNMMC, students will (1) participate in Grand Rounds presentations by clinicians/engineers/scientists and conduct de-briefing interviews to highlight challenges in their respective practices and pinpoint collaborators, (2) observe procedures and patient interactions and document the clinical workflow for offline analysis of design opportunities, and (3) explore state-of-the-art research facilities in rehabilitation, biomechanics, regenerative medicine, traumatic brain injury, oncology, podiatry, and hemodialysis to identify opportunities where new laboratory tools or interventions may assist with active duty and veteran healthcare.

After completing the Aims, participants will prepare a proposal on an engineering design need that is well-suited to enhance healthcare for wounded veterans. BME and ID students will continue as collaborative teams for the duration of their Senior Design projects. Taken together, this design program will offer a deep-dive into the best practices associated with identifying medical needs and will aid students in developing a professional rapport with patients, designers, and healthcare providers. In fact, students by engagement in this course can develop a unique skill in healthcare design that not only covers the traditional research aspects in design but also an introduction and exploration into other aspects such as regulatory and hazard analysis. (Figure 2).

This multidisciplinary approach also will help foster interprofessionalism (i.e., the practice of roles in various healthcare and biomedical settings), such that students will learn to work together and respect one another's roles in preparation for their future careers.

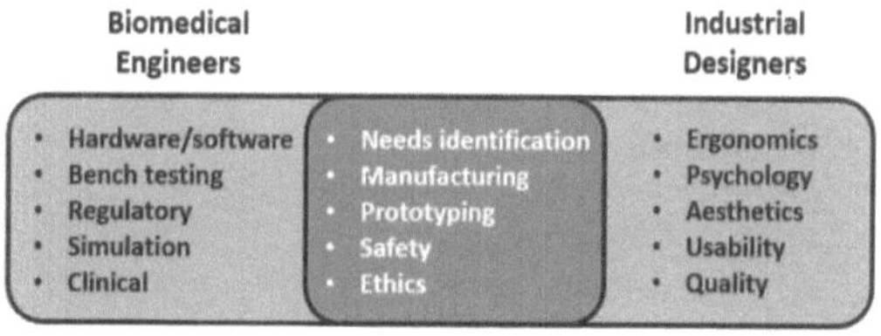


Figure 2. Complementary skills diagram. Adapted from, "Senior Design – biomedical engineering/industrial design collaboration in Senior Design projects," by JR Goldberg.

## 4 COURSE OUTCOME AND DELIVERABLES

As mentioned before, we are seeking two aims through this collaborative course: complementary education in biomedical needs identification and clinical immersion in healthcare for wounded veterans.

This course will be open to both BME and ID students during the Spring semester of their Junior year and the uncovered result of this course can lead them to their capstone project in senior year. Students from both programs, with junior-standing, will apply to enrol in the course during the fall. Applications will be reviewed and students will be selected for participation in the Spring. Recruiting emphasis will placed on students from underrepresented backgrounds. There will be teams consisting of students from both disciplines.

By the end of the first section of the course, which includes several lectures on a variety of related topics, each team member will be expected to function as an "expert" in a specific clinical focus area (rehabilitation, kidney disease, podiatry, or oncology) and perform background research before participating in a site visit. Therefore, there will be at least one team member that has gained a comprehensive knowledge of the associated focus area. This team member will be expected to develop a plan for their design research and lead the interactions and investigations with the clinical collaborators, potentially highlighting needs ahead of time from a literature review for further on-site validation. Anticipating the fact that the students will not have the knowledge or experience to fully understand the complexity of the medical problem, there will be lectures by the healthcare providers on disease state fundamentals. Since the veteran community is unique, this targeted group can bring a unique and valuable perspective for students.

Besides the traditional materials and lectures on design research, this course consists of a series of lecture and practice-based components provided for students such as:

Supplementary knowledge in unfamiliar but important subject in healthcare design (such as risk management and hazard analysis, device manufacturing, device regulation, intellectual property, market analysis)

Tour of the America Building, part of Walter Reed National Military Medical Center in Bethesda, Maryland. The America Building is the largest outpatient services building in the Department of Defense and houses both the Orthopedic and Rehabilitation Departments.

Learning forums with healthcare professionals (physical/occupational therapists, prosthetists/orthotists, and other clinicians/physicians).

Learning forums with biomedical engineers and other researchers (BS/MS/PhD). There are currently 7 biomedical engineers of varying backgrounds and training

Job shadowing/observation with biomedical engineers and other researchers (BS/MS/PhD).

*Some expected deliverables for this course:*

Create a database (minimum of 25 needs of stakeholders, existing solutions, and market)

Create a proposal for a capstone senior design projects by using the support information.

Students will function effectively on multidisciplinary teams of engineers and designers

Students will receive mentored training that will allow them to direct a conversation with a range of audiences, including veterans and healthcare providers in the workplace.

Patent landscape search to assess patentability
Regulatory report
User research sketchbook/notebook
Midterm presentations including an initial "needs database"
Grant proposal including needs specification
Final presentation (video format)
Report book
Self- and peer-evaluation surveys

## 5 CONCLUSION

To simulate a more professional environment for students in design education, creating a collaborative experience with other disciplines as well as professional industry is necessary. In many cases, designers need to work in teams with engineers especially in healthcare and at the same time understand the real needs of the community (veterans and their health care providers). As such, the right type of infrastructure with professionals and organizations involved needs to be in place. The history of collaborative projects at Virginia Tech has provided a good foundation for industrial design students to experience and learn different aspects of design research with regard to veterans and their healthcare providers. This semester-long, clinical immersion program has been framed to train students in recognizing and resolving the unmet needs of our "wounded warriors" and veterans. These students will continue as collaborative teams in their senior capstone projects that will result in a functional prototype to a veteran client or healthcare provider. Continual feedback will be obtained from these customers/partners to ensure the product effectively addresses their needs.

## ACKNOWLEDGEMENT

Information reported in this paper was supported by the Eunice Kennedy Shriver National Institute of Child Health and Human Development of the National Institutes of Health under award number R25EB027617.

## REFERENCES

Beucker, N. (2004). Research skills as basis for industrial collaboration in design education. In DS 33: Proceedings of E&PDE 2004, the 7th International Conference on Engineering and Product Design Education, Delft, the Netherlands, 02.-03.09. 2004 (pp. 185–192).

Briggs, R. O. (2006). On theory-driven design and deployment of collaboration systems. International Journal of Human-Computer Studies, 64(7), 573–582.

Goldberg, J. (2007). Senior design-biomedical engineering/industrial design collaboration in senior design projects. IEEE Engineering in Medicine and Biology Magazine, 26(3), 75–76.

Olenick, M., Flowers, M., & Diaz, V. J. (2015). US veterans and their unique issues: enhancing health care professional awareness. Advances in medical education and practice, 6, 635.

Rábago, C. A., Clouser, M., Dearth, C. L., Farrokhi, S., Galarneau, M. R., Highsmith, M. J., ... & Hill, O. T. (2016). The Extremity Trauma and Amputation Center of Excellence: Overview of the research and surveillance division. Military medicine, 181 (suppl_4), 3–12.

Spelman, J. F., Hunt, S. C., Seal, K. H., & Burgo-Black, A. L. (2012). Post deployment care for returning combat veterans. Journal of general internal medicine, 27(9), 1200–1209.

*Research & Education in Design: People & Processes & Products & Philosophy – Almendra & Ferreira (eds)*
 *ISBN 978-0-367-49296-0*

# Designing integrated solutions for resource-limited societies

S. Jagtap & T. Larsson
*Blekinge Institute of Technology*

ABSTRACT: Poverty is characterized by many different constraints at individual, institutional, economic, and technological levels. To satisfy unmet or underserved needs of people living in poverty, it is essential to address all the relevant constraints in the target context through the design of integrated solutions. Although previous studies in this field have indicated that designing such solutions requires involvement of many stakeholders, there is a lack of related research in this field. To address this, using a case study methodology, we analysed two design projects that have successfully responded to all the pertinent constraints in the resource-limited societies, supporting social and human development of the marginalised people. Based on the findings of these design case studies, we developed a framework of stakeholder inputs, representing a diverse range of stakeholders, embedded within or outside the context of poverty and contributing towards the design of integrated solutions.

## 1 INTRODUCTION

Poverty, a multidimensional issue, is inherently characterised by many closely intertwined constraints (UNDP, 2008; Jagtap et al., 2013). While some constraints are related to poor or non-existent facilities such as healthcare, sanitation, and physical infrastructure (e.g. Lenau and Hesselberg, 2015; Matlack et al., 2011), others are related to the economic and institutional aspects of the context (Banerjee and Duflo, 2007). Designers, aiming to improve life conditions of poor people, need to broaden the scope of their design efforts to consider the wider context of poverty, in order to address a multitude of relevant constraints in the target context (Aranda-Jan et al., 2016). In addition to satisfying core functions, solutions designed for marginalised communities must address several other requirements resulting from a wide range of deprivations in the target context (e.g. Jagtap and Larsson, 2018; Li et al. 2016). As such, these design solutions manifest in the form of integrated solutions (Devisscher and Mont, 2008; Jagtap et al., 2013). In the present paper, the term 'integrated solutions' is employed to refer to solutions that address all the pertinent constraints and deprivations in resource-limited societies in order to satisfy some needs of marginalised people.

Whilst the subject of designing integrated solutions has been extensively researched in the context of developed countries using many different names such as integrated products and services, total care products, and product service systems (e.g. Mont, 2002; Morelli, 2003; McAloone and Andreasen, 2004; Beuren et al., 2013), it has been given little attention in the context of marginalised societies in developing countries (Jagtap and Larsson, 2018). Most of the design research on integrated solutions has been undertaken in the context of developed countries. Design researchers have given little attention to this subject in deprived contexts such as marginalised societies in developing countries (Jagtap and Larsson, 2018; Aranda-Jan et al., 2016). Cases of integrated design solutions have not been widely explored and developed in these impoverished contexts. Although previous studies have indicated the involvement of many stakeholders in designing integrated solutions for resource-limited societies (e.g. Aranda-Jan et al., 2016; Jagtap et al., 2013), there is a lack of related research in this field. As such, this research aims at exploring different stakeholders involved in designing integrated solutions for resource-limited societies and at examining their design contribution in addressing relevant constraints in the target context of these societies. Because this field is relatively new and unexplored, we employed an exploratory approach using qualitative case studies (Yin, 1994; Eisenhardt, 1989). Based on the case study findings, we develop a framework of stakeholder inputs for resource-limited societies. The paper contributes towards better understanding of the ways in which design inputs of a wide range of stakeholders shape integrated solutions for resource-limited societies in developing countries.

## 2 RESEARCH METHODOLOGY

The aims of this research are to explore different stakeholders involved in designing integrated solutions for resource-limited societies as well as to examine their design contribution. Because this field is relatively new and unexplored, we employed an exploratory approach using qualitative case studies (Yin, 1994; Eisenhardt, 1989). This is in line with the previous studies that have aimed at exploring a wide range of design phenomena, e.g. some aspects of designing product service systems in developed countries (e.g. Morelli, 2003), how industrial designers contribute towards developing complex products (Kurvinen, 2005), and many other aspects of design processes in a variety of industries such as healthcare, aerospace, etc. (e.g. Lehoux et al., 2011; O'Hare et al., 2010; Jagtap and Johnson, 2010).

Two design cases, aimed at satisfying unmet or underserved needs of marginalised societies, were selected. This selection was purposive, driven by our research aims. The cases were selected as they offered rich diversity and learning potential, which are essential qualities in undertaking exploratory research (Eisenhardt, 1989). The cases ensured cultural and geographical variety; for example, the cases are from Asia and Latin America. The cases are from different design sectors. In addition, the cases yielded sufficient information to address the research aims. This sampling approach has some limitations, which are discussed further in the paper.

Secondary data collection on the two cases involved several steps (Yin, 1994), including an exhaustive search for academic publications, company reports, websites, newspaper and magazine articles, media interviews, and databases. Many studies in a wide range of fields, including, among others, design and business management have successfully used secondary data to address research questions (e.g. Jagtap et al., 2015; Shah et al., 2003; Zhang et al., 2015). The data collected in this research included, for example, background and performance of each design case, motivation behind designing the integrated solutions, how these solutions were designed, different stakeholders involved in their design, and their design inputs. The collection of data from many sources helped to triangulate claims made in these sources.

The data analysis was an iterative process, with the goal of exploring contribution of different stakeholders towards designing the solutions. Data displays were used to facilitate the process of data analysis (Miles and Huberman, 1984). These displays helped to place different constraints and functions that were addressed on one dimension and inputs of different stakeholders on another dimension. Aranda-Jan et al.'s holistic design framework (see Figure 1), consisting of a wide range of constraints and deprivations in resource-limited societies, was used to facilitate the data analysis because this framework is developed from a design perspective, based on a large-scale empirical study and review of the literature in this field. The framework consists of four major categories and related sub-categories of constraints that the authors acknowledge as open to refinements. It includes a wide range of constraints at individual, institutional, technological, and economic levels (see Figure 1). As the analysis progressed by iterating back and forth, we updated and refined the displays and conclusions. The displays and conclusions were cross-checked by two researchers to ensure reliability. After the first researcher analysed the data on a case, the second researcher reviewed the information and conclusions to ensure consistency.

Based on the case study findings, we develop a framework of stakeholder inputs for resource-limited societies, contributing towards better understanding of the ways in which

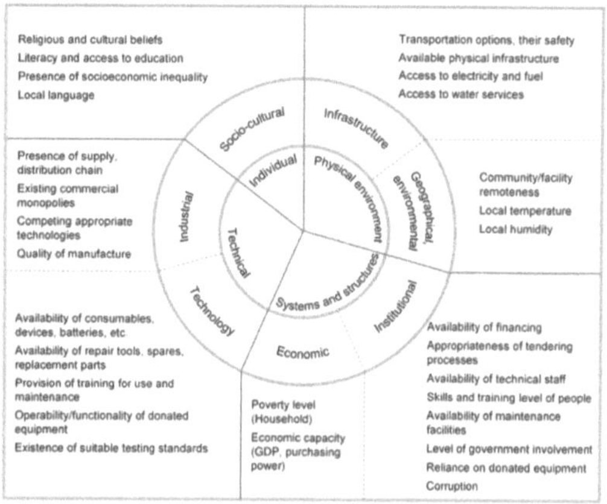


Figure 1. The holistic design framework for marginalised communities (adapted from Aranda-Jan et al., 2016).

design inputs of a variety of stakeholders shape integrated solutions for marginalised societies in developing countries.

## 3 CASE STUDY 1: HANDLOOM ARTISANS

Fabindia, a for-profit company founded by John Bissell and now managed by his son William, designed an integrated solution to support rural handloom weavers and artisans in India, with the aim of enhancing their productivity by alleviating constraints in producing handloom garments and fabric, while providing them access to non-local markets, mainly domestic markets in Indian cities (Fabindia, 2016a). These rural artisans faced many challenges, some were related to their productivity constraints and others were related to capturing value of their products by accessing markets. They lacked access to requisite technical facilities, hindering their ability to enhance productivity and quality of products. In addition, their limited educational and literacy levels, lack of access to market information, and inability to obtain information on aesthetic tastes and preferences of consumers contributed to weak demand for their products. They also faced competition from power looms (Sharma, 2007). The government of India attempted to support rural artisans and weavers by establishing the 'Khadi and Village Industries Commission' (KVIC) aimed at increasing rural employment and capturing value of artisans' products by selling them in *'khadi'* outlets – *khadi* refers to coarse, handloom spun cloth (KVIC, 2017). However, these khadi outlets could not capture attention of consumers, resulting into insignificant impact on the lives of rural artisans.

The integrated solution, designed by Fabindia, leverages the traditional skills of artisans. The solution skilfully balances artisans' upstream production skills and Fabindia's downstream retail and promotion activities. Fabindia's supply chain has a network of more than 40000 weavers and artisans spread across rural India, and has over 97 stores in Indian cities (Fabindia, 2016b). The design of artisan-shareholder plan and co-ownership-based business model helped the artisans in sharing a greater proportion of the value. In addition, Fabindia' collaboratively designed Craftmark label, aimed at giving identity to craft products, helped artisans to gain higher market prices (Kripalani, 2009). The solution revived the disappearing tradition of handloom weaving, while offering benefits not only to rural artisans but also to those employed in the distribution and retail activities.

Inputs from many stakeholders contributed to the integrated design solution. Fabindia works closely with the rural artisans, and its staff and managers working at stores and in the supply chain routinely visit these artisans in rural India. Fabindia has collaborated with professional designers. These designers, with deep understanding of fabrics and crafts, work closely with the weavers, offering them designs with clear explanation of the fabric colour, weave, and print. Consumption trends in urban areas and preferences of the growing prosperous class inform these designs. Beyond this consumer class, the designers also target a wider range of consumers. The designs attracted a diverse range of customers, including those whose purchases are not driven by ideological meaning of khadi garments (Bajaj, 2008). The designs given to weavers, informed by consumer tastes for aesthetic and symbolic qualities of garments, tackled constraints related to weavers' limited knowledge of market information. Besides working with professional designers, Fabindia included specialists in technical and manufacturing aspects of translating designs into feasible products. These specialists work with both rural artisans and designers to ensure feasibility of designs, tackling technological constraints.

Just as design inputs were critical in designing garments and offering those designs to weavers, they were crucial in selling products at Fabindia's stores in many cities across India. An inherent lack of consistency in texture and colour of handloom products posed a challenge in producing designs in bulk quantities, without variations in quality. To overcome this challenge, Fabindia developed a programme to educate customers at the retail end of the supply chain. Posters were designed to inform customers that intrinsic uncertainty in colour and finish of handloom garments elegantly shapes their ethnic beauty. Fabindia also organises events inviting artisans to speak about their craft (Majumdar, 2010). In addition to these design inputs at the downstream retail activities, Fabindia also developed educational programmes for rural artisans to enhance their knowledge about consumer expectations. To improve consistency and quality of products, Fabindia also worked with washing units in areas where weavers lacked access to adequate washing facilities in the dying phase of fabric production, making these facilities accessible to the weavers. These educational and infrastructural support-schemes targeted factors about artisans' limited knowledge and infrastructure available to them.

Another important design input came from Fabindia's collaborative work with many partners in handloom sector, resulting into the Craftmark Initiative. By working with several retailers and NGOs in craft sector, Fabindia established the 'All India Artisans and Craft Workers Welfare Association' (AIACA) (AIACA, 2017). AIACA developed the 'Craftmark Initiative', with an aim of giving handloom products a meaning of high value and expert human skill, permitting retailers of such products to use the 'Craftmark' logo to signify their products as authentic Indian handcraft (Craftmark, 2017). The Craftmark initiative raised confidence and provided independence to artisans. Design inputs, coming from Fabindia and other stakeholders targeted many constraints, including, those associated with institutional programmes, such as KVIC. Moreover, by working with India's large banks, Fabindia developed schemes to provide loans to weavers. In recent years, Fabindia has expanded its product portfolio by working with farmers and woodwork artisans (SPAN, 2010).

## 4 CASE STUDY 2: IRRIGATION SYSTEMS

Amanco, a for-profit company (Amanco, 2017), in collaboration with several partners, designed an irrigation system for marginalised farmers from the 'La Testaruda' community in Mexico (UNDP, 2008). The farmers in this community faced numerous problems. They were using traditional and outdated irrigation methods and relying on groundwater through boreholes and wells. These traditional methods were inefficient, and only farmers with sufficient water availability could grow crops enough for their livelihood, while others without access to such water deposits relied on rainy seasons. Furthermore, the productivity of these farmers was constrained by knowledge deprivations. For example, they lacked knowledge about productivity gains resulting from renewing their old lemon plants. They also faced problems due to institutional gaps, e.g. they believed that agriculture was not government's priority and they were not receiving public subsidies and resources.

Contribution of many stakeholders shaped the integrated irrigations system in the La Testaruda community. Ashoka, an organisation promoting social entrepreneurship (Ashoka, 2017), supported Amanco to collaborate with a locally embedded NGO, called 'Sustainable Farmers

Network' (RASA). RASA had 25 years of experience in rural projects. RASA's trust and legitimacy in the community helped Amanco to collect initial information on the community. Amanco and RASA, together with the marginalised farmers, developed and implemented the irrigation systems, while addressing many constraints in the region. These efficient irrigation systems contributed not only towards the growth of agricultural output and the farmers' income, but also helped reduce land erosion, water consumption, and irrigation time, permitting famers to give more time to other farm activities.

Many deprivations and limitations in the community required design inputs from several partners, ranging from Amanco – a company, to RASA – an NGO (Kapyepye, 2013). As a result of many value chain gaps and unfamiliarity about the community, Amanco engaged in bottom-up learning and in-depth contextual understanding early in the design process, contributing significant amount of its effort in these activities. The irrigation unit of Amanco organised social gatherings with farmers and their families. This not only allowed to gain insights into several aspects such as daily lives of marginalised farmers and their cultural beliefs, but also helped in leading activities such as making roads for excavation. In addition, by engaging in the activities of hydraulic design, topographic mapping and gathering related data from the region, the irrigation unit designed three types of irrigation systems, namely, drip irrigation, portable irrigation and micro-sprinkling. These systems were customised to the specific needs of marginalised farmers, addressing many diverse technological and socio-cultural factors.

The financial model was designed by Amanco and RASA. This involved 20% down payment by farmers in three instalments, 30% microcredit, and 50% public subsidy. RASA facilitated financial and commercialisation aspects, and engaged in related administrative and bureaucratic procedures for receiving public subsidies (UNDP, 2008). RASA's deep understanding of the region, its social familiarity, and its trusting position in the network were key contributing factors to its various roles in the design and implementation of many aspects of the solution.

Design inputs for contextualised awareness and promotion programmes came from Amanco. While Amanco designed and financed these programmes, they were implemented by RASA. Several methods such as word-of-mouth promotions, meetings, and exhibitions were designed in these programmes. Exhibitions were designed to demonstrate how irrigation systems operate and their benefits for low-income farmers. These exhibitions were set on plots lent by a few farmers. The promotion programmes were designed to assist farmers in recognising inefficiencies in their farm activities and benefits of irrigation systems. In these promotion programmes, RASA gained from existing farming cooperatives, and supported their development when they did not exist. In addition to the awareness and promotion programmes, which alleviated constraints about weak literacy and limited education of marginalised farmers, Amanco trained its own technicians and those from RASA to enhance and contextualise their technical skills for the community's specific needs.

The systematic design management assigned clear roles to various partners, including low-income farmers, who upon receiving necessary training and support from Amanco and RASA, replaced old lemon plants, paid 50 percent of the entire irrigation system cost (20% down payment and 30% microcredit), cleared the terrain to install the system, and actively engaged in the installation activities, tackling not only socio-cultural but also technological and economic constraints. The design of distribution and installation systems was a joint effort of Amanco and RASA, addressing technological aspects as well as those related to maintenance, repair, and associated skills and knowledge. RASA acted as a distributor of the irrigation systems, and equipped with training received from Amanco, supervised their installation by working with marginalised farmers. The role of design, with inputs from many partners, has been crucial in shaping this integrated design solution, addressing a range of economic, technical, institutional, and socio-cultural constraints and contributing to humanitarian, social, and economic outcomes for the farmers.

## 5 DISCUSSION AND CONCLUSIONS

In this section, we look across the two cases. To support discussion of our findings, the resulting framework of stakeholder inputs is presented here first (see Figure 2).

Bottom-up learning and direct engagement with marginalised communities is observed in the examined cases, e.g. Fabindia's collaboration with the marginalised weavers in rural India. Any context of poverty is characterised by many deprivations. Poor people have valuable skill in surviving in such deprived contexts (Jagtap et al., 2013). The case studies show that organisations aiming to design solutions to improve their life circumstances need to engage with them to learn about their context, different problems they face, etc. Engaging with poor people is not only useful in gaining insights into their life circumstances and in understanding their context, it also enables poor people to offer inputs in designing, implementing and operating the solution. An illustration of this is the contribution of the rural artisans in making fabrics and garments to sell through Fabindia's stores.

Actors such as NGOs, who are socially embedded in marginalised communities also have deep understanding of the local context. Such actors are knowledgeable about the needs of poor individuals and their social networks. In addition, they are trusted by the communities as development practitioners. As such, leveraging knowledge and trusted position of these actors offers many benefits in addressing various constraints. For example, Amanco benefited by working with the NGO RASA, which addressed several constraints by collecting information on the community, implementing promotion programmes, and distributing irrigation systems.

Whilst marginalised people and socially embedded organisations make important contribution to the design of integrated solutions, companies and professional designers, originating primarily outside the context of poverty, offer crucial inputs. These 'outsiders' bring in knowledge and resources in many areas, e.g. design and technology. These organisations can connect marginalised communities with their network, including non-local markets. This kind contribution of outside organisations is observed in both the cases, e.g. Amanco's technological knowledge in developing irrigation systems and Fabindia's network in non-local markets supporting weavers to sell their products.

In addition to for-profit and not-for-profit organisations, public organisations (e.g. governments) also offer important inputs towards designing solutions. An example is the role of

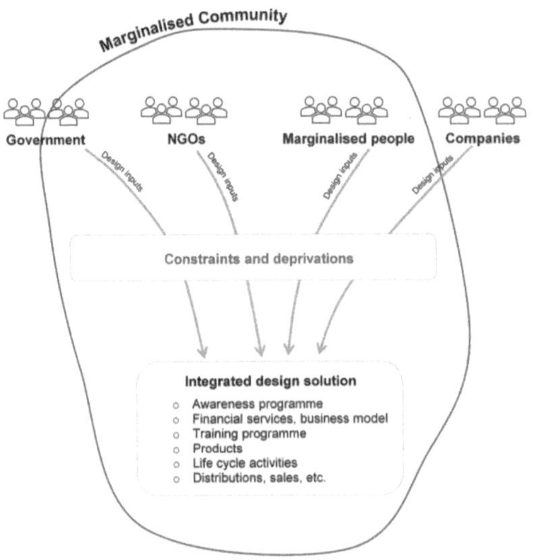


Figure 2. Framework of stakeholder inputs in designing integrated solutions for marginalised communities.

public subsidies in implementing the financial model designed by Amanco and RASA. Whilst not-for-profit organisations are generally embedded in the marginalised communities and for-profit organisations are not, local governments typically operate both within and outside of these communities. The government has a major responsibility of providing basic public services in all regions of the country, including marginalised communities, and as such they can support efforts aimed at the development of poor people.

There are some limitations to this work. The sample is based on the two successful design cases. Future studies can compare successful and unsuccessful design cases to explore differences between the involved stakeholders and their inputs. Whilst the collected secondary data was sufficient to address the research aims, future studies can use methods such as in-depth interviews and ethnographies (e.g. Ozanne and Saatcioglu, 2007).

In sum, we have analysed two case studies to explore stakeholder inputs in designing integrated solutions for socio-economic segments characterised by many constraints. Just as problems of marginalised societies are multidimensional in nature, satisfying their unmet needs may require multidimensional, integrated solutions, requiring inputs from a wide range of stakeholders.

## REFERENCES


Aranda-Jan, C. B., Jagtap, S., & Moultrie, J. (2016). Towards a framework for holistic contextual design for low-resource settings. International Journal of Design, 10(3), 43–63.

AIACA, All India Artisans and Craft Workers Welfare Association, http://www.aiacaonline.org/

Amanco, http://amanco.com.br/

Ashoka, https://www.ashoka.org/en

Bajaj, K., 2008. Fabindia—Style, substance and purpose. Business Today, November 16.

Banerjee, A. V., & Duflo, E. (2007). The economic lives of the poor. Journal of Economic Perspectives, 21, 141–167.

Beuren, F. H., Ferreira, M. G. G., & Miguel, P. A. C. (2013) Product-service systems: a literature review on integrated products and services. Journal of Cleaner Production, 47, 222–231.

Craftmark, http://www.craftmark.org/

Devisscher, T., & Mont, O. (2008). An analysis of a product service system in Bolivia: coffee in Yungas. International Journal of Innovation and Sustainable Development, 3(3-4), 262–284.

Eisenhardt, K.M. (1989) 'Building theories from case study research', Academy of Management Review 14(4): 532–550.

Fabindia – about us https://www.fabindia.com/pages/about-us/pgid-1124038.aspx (accessed January 2017).

Fabindia, https://www.fabindia.com/(accessed October 2016a).

Fabindia, Wikipedia, https://en.wikipedia.org/wiki/Fabindia (accessed December 2016b).

Jagtap, S. and A. Johnson (2010). Requirements and use of in-service information in an engineering redesign task: Case studies from the aerospace industry. Journal of the Association for Information Science and Technology, 61(12): p. 2442–2460.

Jagtap, S. and Larsson, T. (2018). Design and Frugal Innovations: Three Roles of Resource-Poor People. International Design Conference - Design 2018. Dubrovnik – Croatia.

Jagtap, S., A. Larsson, and Kandachar, P. (2013). Design and development of products and services at the Base of the Pyramid: a review of issues and solutions. International Journal of Sustainable Society, 5(3): 207–231.

Jagtap, S., Larsson, A., Hiort, V., Olander, E. & Warell, A., (2015) Interdependency between average novelty, individual average novelty, and variety, International Journal of Design Creativity and Innovation, 3(1), 43–60.

Kapyepye, M. (2013). Resource Mobilisation for NGOs in the Developing World. Adonis & Abbey Publishers Ltd.

Kripalani, M., 2009. Fabindia weaves in artisan shareholders. Business Week, March 12.

Kurvinen, E. (2005). How industrial design interacts with technology: A case study on design of a stone crusher. Journal of Engineering Design, 16(4), 373–383.

KVIC, Khadi and Village Industries Commission, http://www.kvic.org.in/kvicres/index.html

Lehoux, P., Hivon, M., Williams-Jones, B., & Urbach, D. (2011). The worlds and modalities of engagement of design participants: A qualitative case study of three medical innovations. Design Studies, 32(4), 313–332.

Lenau, T., and Hesselberg, T. (2015). Dry sanitation concepts with inspiration from nature. Journal of Water Sanitation and Hygiene for Development, 5(2), 330–335.

Li, X., Ahmed-Kristensen, S., & Daalhuizen, J. (2016). Identifying design requirements for emerging markets. In 11th International Symposium on Tools and Methods of Competitive Engineering (TMCE 2016).

Majumdar, R. (2010). Consumer behaviour: Insights from Indian market. PHI Learning Pvt. Ltd.

Matlack, C., Chizeck, H., Davis, T., and Linnes, J., 2011, "A Low-Cost Solar Disinfection Indicator for Safe Water," Global Humanitarian Technology Conference.

McAloone, T. C., & Andreasen, M. M. (2004). Design for utility, sustainability and societal virtues: developing product service systems. In DS 32: Proceedings of DESIGN 2004, the 8th International Design Conference, Dubrovnik, Croatia (pp. 1545–1552).

Miles, M.B. and Huberman, A.M. (1984) Qualitative Data Analysis, Sage Publications: Beverly Hills, CA.

Mont, O. K. (2002). Clarifying the concept of product–service system. Journal of cleaner production, 10(3), 237–245.

Morelli, N. (2003). Product-service systems, a perspective shift for designers: A case study: the design of a telecentre. Design Studies, 24(1), 73–99.

O'Hare, J., Dekoninck, E., McMahon, C., & Turnbull, A. (2010). Adapting innovation tools to the eco-innovation requirements of industry: case study results. International Journal of Design Engineering, 3(2), 172–194.

Ozanne, J. L., and B. Saatcioglu. 2007. Understanding subsistence marketplaces: Exploring the contributions of participatory action research. In Advances in international management series 20, ed. J. A. Rosa and M. Viswanathan, 111–32. Oxford: Elsevier.

Shah, J. J., Smith, S. M., & Vargas-Hernandez, N. (2003). Metrics for measuring ideation effectiveness. Design studies, 24(2), 111–134.

Sharma, E. K., 2007. Weaving woes. Business Today, November 4.

SPAN magazine, The Fabindia Story. July–August 2010.

UNDP (United Nations Development Programme) (2008) Creating Value for All: Strategies for Doing Business with the Poor. United Nations Development Programme.

Yin, R.K., 1994, Case study research, 2nd edn. Thousand Oaks, CA Sage Publications.

Zhang, L., Zhang, J., Duan, Z. Y., & Bryde, D. (2015). Sustainable bike-sharing systems: characteristics and commonalities across cases in urban China. Journal of Cleaner Production, 97, 124–133.

*Research & Education in Design: People & Processes & Products & Philosophy – Almendra & Ferreira (eds)*
*© 2020 Taylor & Francis Group, London, ISBN 978-0-367-49296-0*

# Fostering empathy through design thinking among children in rural Trinidad

L. Noel
*Tulane University*

T.L. Liu & T.R. Rider
*North Carolina State University*

ABSTRACT: Children from a fourth-grade class at a primary school in Trinidad and Tobago participated in a three-week summer camp with a design thinking- based curriculum. The study aimed to examine how children developed and practiced empathy during the design class. Qualitative data were collected from the children and the instructors, providing thick descriptions of the children's experiences during the camp. As anticipated, the findings showed that the children practiced empathy at several stages during the design process, where they had to learn to take the perspective of the target user of their designs. However, the less anticipated result was that the findings also demonstrated that the children had more opportunities to practice empathy during groupwork and critiques, where they empathized with the angst of their colleagues. These findings demonstrate how design education can be used to enhance the social development of children by providing varied opportunities for them to empathize with others.

## 1 INTRODUCTION

Empathy is the detailed awareness and perception of the experience of another person (Levine,2012). It is a condition that historically was necessary for the biological fitness and reproductive success of the human race (Preston & De Waal, 2001). The condition was required for humans to survive through empathic responses that would 'fast track' the acquisition of knowledge to facilitate survival. Empathy is also a skill that is needed in many professions, such as the medical profession, (Chen, Kiersma & Plake, 2015; Stepien & Baernstein, 2006; Penprase, Oakley Ternes & Driscoll, 2013), sales (Delpechitre, 2013), education (Warren & Hotchkins, 2014), and management (Somogyi, Buchko & Buchko, 2013). Designers also rely on empathy to identify problems and develop solutions for target users. Design thinking is an empathy-driven, user-centered, process-based approach to solving problems rooted in the work of Dewey and Montessori (Roth, 2017; Elmansy, 2016; Goldman & Kabayadondo, 2017). It can be applied in daily life situations in order to solve everyday problems and can lead to new objects, ideas, narratives or systems. Psychologists have provided evidence on the importance of developing the skill of empathy among children (Levine, 2012; Borba, 2017; Jones, Greenberg & Crowley, 2015) yet the explosion of bullying in elementary school suggests that society has not been successful in promoting empathy among children (Chen, Kiersma & Plake, 2015; Levine, 2012). The aim of this study was to understand how children could become more empathetic through exposure to design thinking in a design class.

## 2 METHODOLOGY

The research setting for this qualitative case study was a small public primary school located in a village in southern Trinidad and Tobago. Eighteen children took part in a 3-week design thinking camp during the 2017 July-August vacation in a rural town in southern Trinidad and Tobago in the Caribbean. The group was comprised of eight girls and ten boys. The children were primarily recruited from students of the incoming fourth grade class a public primary school. The anticipated age of the participants was 9 – 10 years old. This age-group was selected for the study since most upper elementary children master language and are proficient communicators, which would facilitate the feedback process (Finnan, 2009). The actual age range of the children was wider than anticipated, and ranged from 8 – 13. This was because one of the girls had skipped a grade, another had been held back, and a third child, the oldest, was recruited from the neighbourhood and not the school. Two boys and 1 girl did not complete the camp.

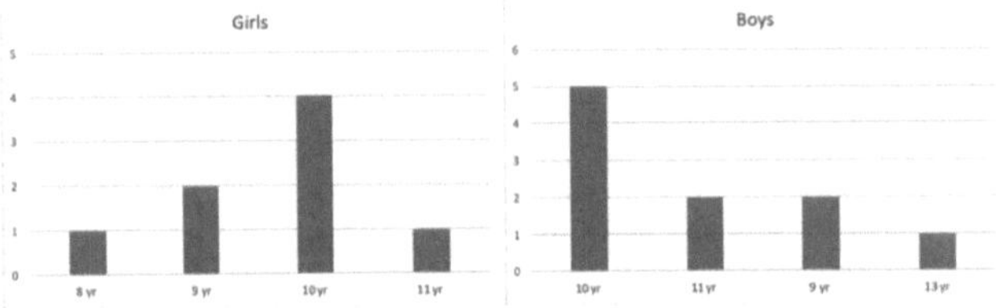


Figure 1. Number and ages of girls and boys who participated in the study.

The children participated daily in the Design Camp on afternoons from 12 - 3 p.m., while they did Math and English vacation lessons in the morning. The class size of the group ranged daily from 10 – 18 participants. There were two camp instructors. The first was an Industrial Designer and a lecturer in Design at a local university. She was not from the community. The second instructor was a primary school teacher with a bachelor's degree in Special Needs education and a master's degree in Curriculum Development. She did not have any experience in Art or Design as a teacher or a student. She was also a resident of a neighbouring community.

### 2.1 *The design curriculum*

The first week of the camp began with discussions around children's rights. The second week began with the needs of people in the community. The discussion in the third week was rooted in a critique of the local newspapers, where the students had to see is and how their part of the country was represented in the news. These discussions were used to help lead the children to critique the worlds around them. After the critique, the children were encouraged to dream about the ideal utopian scenario for the problem they were addressing. Finally, they designed solutions to achieve the change that they wanted to see for themselves and their village. This curricular approach was designed based on a framework of Critical Utopian Action Research, which is an approach that combines critical analysis with the vision of a sustainable, democratic lifestyle, and examples of lessons that promote critical discussions at elementary level found in the literature (Nutti, 2016). Age-appropriate themes that would engage the students' interests and lead to critical discussions were selected for each of the three activities of the design thinking workshop: the rights of the child, improving the school and improving the community.

During the workshop, the children discussed their design process and create their own model for the process post-discussion. They created persona profiles for the people that they were designing for after discussing the needs of these individuals, completing journey maps for them and conducting empathy interviews. After which they brainstormed about solutions using words and drawings, and they created preliminary prototypes using modelling clay.

Finally, they created concept boards with details of their concepts, and made a short presentation to their colleagues.

The schedule was as follows:
Monday: Discussion of the problem
Tuesday: Identification of the needs of the persona and development of the general concept
Wednesday: Development of preliminary 2D solutions
Thursday: Development of 3D prototype/Preparation for the presentation and critique.
Friday: Critique and feedback.

### 2.2 *Data collection and analysis*

In designing data collection, emphasis was placed on understanding the perspective of the children who participated in the study and on their experience while learning how to problem solve using design-thinking techniques. In the research design, the research and data collection methods were selected to ensure that the children would feel comfortable about sharing their perspectives throughout the process.

The data were collected through four main strategies: the children's daily journals, weekly focus groups with the children, observation and written reflections of the instructors, and finally analysis of the children's work.

### 2.3 *Selection of codes for data analysis*

For this study, a priori coding was used. A priori codes are predetermined codes and in this study were derived from the work of Borba (2012) and Davis (1983). Borba's headings for sections of her book on promoting empathy among children: a) developing empathy, b) practicing empathy and c) living empathy; provided a useful framework to analyse the evidence of empathy in the design class, and therefore became three broad categories for a priori codes. According to Borba, developing empathy, requires that children gain emotional literacy, develop a moral identity, learn to perspective take and develop a moral imagination. Practicing empathy, requires that children learn to control their emotions, practice kindness with others and learn to work with others. In living empathy, children learn to act on their empathy by learning to become changemakers and practicing empathetic acts regardless of the consequences, in the form of 'moral courage'.

Davis identified four main aspects of empathy, drawing upon research, which he used to develop the Interpersonal Reactivity Index (IRI) scale to measure empathy. The four aspects are perspective taking, fantasy, empathic concern, and personal distress. Perspective taking helps people to predict the behaviours and reactions of others by understanding their perspective and feelings (Borba,2016). Fantasy is the ability of people to place themselves into the context of a fictitious character, e.g., from a movie or a book (Borba,2016). People with high IRI fantasy scale scores are more verbally intelligent and emotionally reactive (Davis, 1983) and are more attuned to the feelings of others. The third criterion of the IRI scale is empathic concern, which assesses the individual's feelings for the good fortune or misfortune of others. (Davis, 1983; Tellez, 2017) Individuals with high scores of empathic concerns are less lonely, are shyer and more anxious, but are not selfish and boastful (Davis, 1983). The final criteria in measuring empathy, on the IRI scale is personal distress. Individuals with high levels of personal distress are often fearful and vulnerable, more introverted, shy and socially anxious and have low self-esteem (Davis, 1983).

Borba's model provided cues on how to build empathy among children, while Davis's model provided information on measuring cognitive and affective empathy. Based on the work of Borba and Davis, the categories for the a priori codes for this study are presented in the table below.

Table 1. Broad headings of empathy related a priori codes derived from Borba and Davis.

| Seeking Change | Concern for others | Managing Emotions | Working with others |
|---|---|---|---|
| Developing a moral identity | Perspective taking, Fantasy | Developing emotional literacy | Practicing kindness, |
| Developing moral courage | Empathic concern for the user | Managing emotions, | Collaborating with others |
| Giving children a voice | | Personal distress | Empathic concern for colleagues Perspective taking |

## 3 FINDINGS

The findings are discussed under the four major themes.

### 3.1 *Seeking change*

Children need opportunities to give feedback, listen to others, and to share converging and diverging views. This contributes to building their moral courage and moral identity, which play a role in building empathy. In the focus group, one boy stated that the design class was different to a regular day at school because they could share their opinions, while another stated that he felt that in a regular class they did not have that freedom.

The design presentations during the critique also gave the children a platform and a voice to promote change. Week one encouraged the children to reflect on their rights and the rights of other children around the world. The discussions around this problem drew them out of their worlds and made them more aware of life in other places. In weeks two and three, they focused on designing for the school and the communities. These activities made them examine the world around them from a point of view other than their own, and encouraged empathy, by making the children focus on the needs of others.

### 3.2 *Concern for others*

To develop the skill of empathic concern for the user, the children needed to take perspective. The specific design challenges helped the children build empathy by making them focus on helping others through their designs. Another element of 'concern for others' is the act of making others feel good about themselves and generally practicing kindness. In this design class, the children were observed helping and supporting each other more and more as the weeks progressed. They seemed to try to make each other feel good about themselves after the first few days of the study, where they tried to make each other feel better about their critiques, by complimenting each other on their posters and trying to give supportive feedback on how to get through the critique.

### 3.3 *Fantasy*

Fantasy helped the children to take perspective. During the first project on their rights as children, picture books on the 'United Nations Rights of the Child' were used to guide small group discussions. Three of the books depicted the rights using vivid artwork in different styles, while one book was illustrated with photographs of real people instead of illustrations. The photographs provoked a wider range of emotions from the children, than the illustration. The empathetic responses included emotions such as shock at the conditions of life of some of the characters, surprise at the absence of some of the rights that they had taken for granted, and sympathy for the people whom they felt were mistreated in the book.

Fantasy was again a theme when the children created personas for their design problem. In design and marketing, personas can be used to represent the target user of a product. In the first week, the children created a 'persona' based on themselves or a fantasy version of themselves.

Figure 2. Tia created this 'persona' for the first design challenge.

In the second week, as a group, they created a persona for a design problem related to their school. The groups worked together to develop four personas. The children discussed the personas in great detail when they were developing them. They referred to the personas by name throughout the week, to check what the needs of the personas were, and this interest in the needs of the personas was reflected in the final design.

In the third week, the children interviewed real people from the community. The challenge focused on developing designs to improve the community based on the problems that people spoke about. In their groups, they created interview protocols to interview the adults. After the interviews. They created the personas from the information that they had collected.

### 3.4 *Perspective-taking and empathic concern for the user*

The children's designs reflected a better understanding of the needs of the users as the weeks progressed. In the first week, most of the children seemed unable to focus on designing a toy for someone else, and even though they had developed personas, their solutions were focused on themselves and their own needs, rather than that of the persona. In the second week, since the problem area was their school, they were able to build on their knowledge of their needs as users of the school, to develop solutions that were relevant to the needs of the personas that they had created. Both boys and girls showed empathic concern for others during the class, empathizing with their colleagues' angst during the presentations and showing care for their well-being. Both genders also recognized the concern of others for them.

### 3.5 *Managing one's emotions ... and those of others*

One phase of developing empathy is developing emotional literacy (Borba, 2012). Being in tune with their feelings motivates children to care for others. The children expressed their feelings during the discussions facilitated by the instructors, during group work with their colleagues during the day, at the end of the day in their journal reflections and at the end of the week during the focus group discussions. Sometimes the children responded to prompt questions in their journals, and sometimes they expressed their feelings without prompting. During the first project on their rights as children, the children expressed many emotions during this activity as they reflected on the right to play, such as surprise, happiness, and sadness.

Figure 3. One child expressed in her journal that she intended to use her design to get better education.

Self-regulation of strong emotions is a factor in personal development and the development of empathy (Levine, 2012; Borba, 2017). During the design classes, children were required to cope with negative emotions at various phases of the process, such as when they received feedback during critiques, or when faced with the challenges of group work. The discussion led by the children during the focus groups showed that they learned over time to regulate their emotions better. When asked how the feedback made her feel in Week 2, one nine-year-old said it made her feel 'bad, sad, angry, embarrassed and ashamed'. However, her twelve-year-old colleague, noted that 'people who are not accustomed to design might feel disappointed when getting the negative feedback', demonstrating that he felt exposure to the design class had made them better understand the process of giving and receiving feedback from colleagues.

By Week 3, the children seemed to agree that they should use the feedback to improve their work and were grateful when others told them that their work 'did not look good'. They learned over the three weeks to control their emotions and move beyond the initial anger of receiving negative comments to being able to find the critical points in the negative comment to improve their work. One ten-year-old boy noted that in the regular (traditional) classroom, other students would laugh at them instead of helping them to improve, but the design class was different, showing how pedagogical features of the design class promoted a forum where students could share opinions, make suggestions and collaborate on solutions.

Acknowledging and managing personal distress is another element of managing one's emotions. The weekly critique placed the children in situations where they were confronted with their distress, and the personal distress of others. The children shared their fears of presenting their work with their colleagues. One girl shared with her small group how afraid she was of having to talk in front of everyone. Another girl suggested a strategy of pretending that she was in a more comfortable setting like a room in her own home instead of in the classroom. Over the three weeks, both boys and girls suggested to their colleagues many strategies on how to feel better during the critique, such as looking at the wall behind the audience instead of directly at the audience.

### 3.6 *Working with others*

The children who participated in the study were not accustomed to having to work together so frequently. In the first week, while creating a mindmap, the group with younger girls fought over everything, such as the colour of the markers that each group member used, where each group member wrote the idea that they wanted to add to the mindmap, the neatness of the mindmap, and what they were writing on the mindmap. The instructors tried to reassure the children that the neatness of the mindmap did not matter. In their reflection at the end of the day, the instructors surmised that the conflicts on the first day were related to the influence of the traditional work attitude at school where there is less group work, and where teachers also place a high value on order and rules, and less value on spontaneity.

Over time, the work environment improved as the children learned to collaborate more, and they realized that the instructors placed value on their group collaboration. During the camp the children collaborated on many activities such as discussions on the design and research process they would use each week and during problem framing and brainstorming. They also worked together but less collaboratively while making their prototypes and posters for their presentations. For these latter activities, they shared materials such as modelling clay and markers and coloured pencils, even though they worked on individual projects. Finally, the children were observed helping each other prepare for the presentation and critique on by giving each other feedback on the design and legibility of the posters and offering unsolicited help to improve the designs.

During the three-week design camp, the participants were forced to learn to regulate their own emotions and cope with negative emotions and the challenge of working with others, key skills needed in practicing empathy (Borba, 2017). Later on, they also had to manage their emotions as they received feedback from their peers on their ideas. Some of the feedback was positive and made them feel good, some feedback was also negative, and the students now had to understand how to 'sift through' the negative comments to figure out how to improve their designs.

### 3.7 *Collaborating*

During the design challenges, there was also a mix of group and individual work. The group work was challenging for the young children, and they complained about how difficult it was many times during their focus group sessions, and in their journals. Early in the design process, the children brainstormed collaboratively in an exercise called brainwriting where they worked as a group and passed their drawings to the person on the right who would then have to continue working on the same drawing. This collaborative drawing exercise was an extremely challenging activity for many of the children. They had to give up ownership of their initial drawings, as the drawings passed from one student to the next. Some defaced the drawings of their colleagues, instead of building on them. Many expressed their frustration at how they felt their classmates had treated their work both in their journals and in the focus group meetings at the end of the week.

Despite the difficulty of the group work, it also took the focus away from the individual and created a 'co-dependence' as everybody had a stake in everybody's success. One boy made a very insightful comment at the end of the workshop about the benefit of collaboration. "Miss, in the design class nobody can make fun of your work!" After some probing, he said that it was impossible to make fun of the work that was produced in the design class, since everybody had a stake in it, as everybody had contributed to it. It was, therefore, everybody's work, and not the work of one individual. He said that he appreciated that and indicated that he preferred the collaborative process.

### 3.8 *Perspective taking and empathic concern for colleagues*

The children were accustomed to laughing at each other and making fun of each other in the classroom. When they were encouraged to reflect on how others would feel at being laughed at, and when more positive behaviours were modelled and reinforced the classroom environment changed, and the group dynamics improved. There was an increased collaboration among the children as the week went on.

## 4 DISCUSSION

The design class with fourth graders created an environment where empathy could be fostered by the types of activities that participants went through to create user-cantered solutions. These guided activities included collaborating with other children, doing research and focusing on the problems of the users, and finally presenting their work and giving and getting feedback on their proposed designs. The four broad themes related to developing empathy that were used as codes in this study were: a) seeking change; b) empathic concern for others, including the user and colleagues; c) managing one's emotions and developing emotional literacy; and d) working with others. Elements of all four themes were seen in this specific intervention, likely related to the specific content of the design challenges, which had an activist undertone. The activist themes would account to some extent for some of the examples of empathy-building related to seeking change and activism, which affect some empathy building activities proposed by Borba (2017) such as developing a moral identity, developing moral courage and giving children a voice.

In this study, the design brief was developed in a way that was conducive to greater empathy since the products focused on social issues such as human rights and promoting change in school and the community. The work attitudes that were fostered in the class also would have played a role in facilitating the practice of empathy since the children had many opportunities for group work, and to give feedback on the work of others. The participants demonstrated their concern for others in both the products they developed and the way they interacted with their colleagues. Throughout the whole process, they had to learn to manage their emotions and to take care of the emotions of others.

One benefit of these types of design classes at primary school is that there are many opportunities to build empathy in children through diverse paths, such as promoting civic engagement in the participants (seeking change), promoting empathic concern for others, providing deep and emotional experiences as well as opportunities for collaboration. The design studio also provides opportunities to model empathic behaviour for the students to emulate by encouraging them to take perspective and imagine how their users feel or to imagine how their colleagues feel during a critique. Participants in this study responded positively to the modelling of empathic behaviour and became more empathetic with their colleagues over the period. Empathic concern for the user might be an expected outcome in a design challenge given the focus that the designer must have on the user. Empathic concern for colleagues, however, may be a less anticipated outcome.

These findings suggest that the combination of the design brief and the studio pedagogy create an environment that can facilitate the practice of empathy among children, rather than the isolated act of designing. Therefore, educators wishing to use design projects as a way of promoting empathy must also ensure that they are familiar with and employ design pedagogies such as the collaborative work, group discussion, presentations, and critiques.

At the start of the study there were also several examples of lack of empathy such as teasing, which suggested that at this age, the children already have developed non-empathetic behaviours. Non-empathetic behaviours, such as telling on others, and teasing and putting down others, are often rewarded in the traditional classroom by both teachers and students by the attention that the non-empathetic person receives from the teacher and students.

## 5 LIMITATIONS OF THE STUDY

This study took place within the setting of a school but took place as an extracurricular activity in a more informal environment over a period of three weeks. The shortness of the workshop was due to the difficulty in gaining access to a class for a more extended period during the academic year. The researchers recognize that three weeks is a short period for this type of study, and therefore these findings should be considered preliminary findings that can be used to suggest a framework for a more extended study. The design studio pedagogical style is a more relaxed style than a traditional teacher-centered classroom. Students who did not want to participate in this study could opt out, and as a result of this, it was not possible to fully engage some of the children in the study since they left before the end. Their feedback and attitudes could have provided significant insights, on why they were less engaged and whether a less engaged student would also have become empathetic over time.

## 6 CONCLUSIONS

This paper presented detailed accounts on the practice of empathy among children who participated in the design camp.

The findings indicated that participating children had practiced empathy in several stages in the design process such as in defining the design problem, in researching with target stakeholders and in developing appropriate solutions. Empathic concern for the user was an expected outcome in this study given the focus that designers have on the user in human cantered design. Empathic concern for colleagues, however was a less anticipated outcome of the study. In the design studio the instructors also modelled empathic behaviour for the students to emulate by encouraging them to take perspective and imagine how their users feel, or to imagine how their colleagues feel during a critique. The children revealed in their own words that they were also able to practice empathy through the act of collaborating with other students, which they would not have typically done in their regular classroom. They empathized with their colleagues who were afraid of making a presentation to the group during the critiques. They learned to find ways of providing gentle feedback on the designs

Fourth grade was selected as the period for the intervention because of the developmental and cognitive phase of development in which the children should have been. However, the evidence of lack of empathy from the children at the beginning of the study seems to suggest that empathy building activities at school (design-based or not) need to take place at an even younger age. Fourth grade might be late for this type of intervention. In a world where lack of empathy has increased social problems such bullying, the authors propose that design education and an introduction to design thinking at an early age can create many opportunities for children to develop this much-needed skill.

## REFERENCES


Borba, M. (2017). Unselfie: why empathetic kids succeed in our all-about-me world. New York, NY: Touchstone.

Chen, A., Kiersma, M., Yehle, K., & Plake, K. (2015). Impact of an Aging Simulation Game on Pharmacy Students' Empathy for Older Adults. *Am J Pharm Educ American Journal of Pharmaceutical Education, 79*(5). Retrieved November 14, 2015, from http://www.ajpe.org/

Davis, M. H. (1983). Measuring individual differences in empathy: Evidence for a multidimensional approach. Journal Of Personality And Social Psychology, 44(1), 113–126. doi:10.1037/0022-3514.44.1.113

Delpechitre, D. (2013). Review and assessment of past empathy scales to measure salesperson's empathy. *Journal of Management and Marketing Research, 13*.

Elmansy, R. (2016, March 10). Design Thinking in Schools: Building a Generation of Innovators. Retrieved March 06, 2017, from http://www.designorate.com/design-thinking-in-schools/

Finnan, C. R. (2009). The upper elementary years: Ensuring success in grades 3-6. Thousand Oaks, CA: Corwin Press.

Goldman, S., & Kabayadondo, Z. (2017). *Taking design thinking to school: how the technology of design can transform teachers, learners, and classrooms*. New York: Routledge.

Jones, D. E., Greenberg, M., & Crowley, M. (2015). Early Social-Emotional Functioning and Public Health: The Relationship Between Kindergarten Social Competence and Future Wellness. American Journal of Public Health, 105(11), 2283–2290. doi:10.2105/ajph.2015.302630

Levine, M. (2012). Teach your children well: Parenting for authentic success. New York: Harper.

Nutti, Y. J. (2016). Decolonizing Indigenous teaching: Renewing actions through a Critical Utopian Action Research framework. *Action Research*, 1–23. doi:10.1177/1476750316668240

Penprase, B., Oakley, B., Ternes, R., & Driscoll, D. (2013). Empathy as a Determining Factor for Nursing Career Selection. *Journal of Nursing Education J Nurs Educ*, *52*(4), 192–197. doi:10.3928/01484834-20130314-02

Preston, S. D., and Frans B. M. De Waal. “Empathy: Its ultimate and proximate bases.” *Behavioral and Brain Sciences*, vol. 25, no. 01, 2002, pp. 1–72., doi:10.1017/s0140525x02000018.

Roth, B. (2017). Foreword In S. Goldman & Z. Kabayadondo (Ed.), Taking design thinking to school: how the technology of design can transform teachers, learners, and classrooms. (pp. xvi–xviii). New York, NY: Routledge.

Somogyi, R., Buchko, A., & Buchko, K. (2013). Managing With Empathy: Can You Feel What I Feel? *Journal of Organizational Psychology*, *13*(1), 32–42.

Stepien, K., & Baernstein, A. (2006). Educating for empathy. *Journal of General Internal Medicine*, *21*(5), 524–530.

Tellez, F. A. (2017) Empathy Expression and Development in the Context of Industrial Design Education.” (Doctoral dissertation, North Carolina State University).

Ip, C., & Hotchkins, B. (2014). Teacher Education and the Enduring Significance of “False Empathy”. *Urban Rev The Urban Review*, *47*, 266–292. doi:DOI 10.1007/s11256-014-0292-7

*Research & Education in Design: People & Processes & Products & Philosophy – Almendra & Ferreira (eds)*
*© 2020 Taylor & Francis Group, London, ISBN 978-0-367-49296-0*

# Empowering design innovation in Albania: Curriculum design methodologies and approaches

J. Dhiamandi, V. Perna, S. Jojic & E. Curraj
*POLIS University*

ABSTRACT: The paper has an objective to define what is intended as 'Innovation in design' in the context of Albania. Throughout this work, the authors are going to examine the tools of how universities and academia, in this case, POLIS University of Tirana, can foster innovation through research and design curriculum. The concept of innovation in this context of crisis is going to be examined as a matter of research methodology, education vs. industry product development, and customization of social needs. This challenging process of the continuous collaboration between universities and business can lead to a new definition of innovation using research methodologies like design thinking, user-centered design drawing a bottom-up process toward product design.

## 1 INTRODUCTION: WHAT IS DESIGN IN THE CONTEXT OF CRISIS

In the field of design, as in many other areas of studies, the research process builds up a strong foundation for future educational programs. This paper aims to showcase the connection between design research and design education in Albania, where "Design" only has recently become a discipline of study and a field of research, thus establishing itself as a novel discipline in the national realm, and emerging - since the 1990 - as a result of the liberalization of the market economy. The latter steadily increased in the import of products from abroad and, as a consequence, has enhanced sensitivity and awareness of the role of design, especially in the aesthetic, functional, and communicative improvement of goods and services.

To economically sentence, there is a need to explain why the field of design suffered a severe waste of great potential to be developed due to wrong reasons. Indeed, such a rapid and pervasive diffusion of imported products and lifestyles has limited the ability of the local industry to make a transition from the regulated model ruled by the state to the competitive market, leading to the phenomenon that fields related to design - such graphic, product, advertising, etc. - are mostly inducted from abroad, but not stimulated by the current domestic research concerning the discipline. One of the most exciting challenges for design culture in Albania today is to promote the recovery of awareness about the value of the public good that, nowadays, is remarkably reduced in terms of public perception because of the memory of the government's overwhelming and pervasive presence of before 1990. Another aim is to affirm the importance of design as a tool for promotion, innovation, and social development.

For this reason, and since we are involved in the design from different points of view - either academic or market-related ones - we want to use our personal experience at POLIS University to foster further discussions regarding this topic from and within the context of Albania. POLIS is a small university oriented in the discipline of design research, which, through its Tirana Design Week[1] (TDW), tries to create an open platform for research,

[1] Tirana Design Week - organized by POLIS University since 2013 - is the very first event entirely dedicated to design in Albania and is created with the aim of tackling the manifestations of creativity, with a specific focus

education, and industries. The University's mission is to position itself in the regional debate developing a deeper understanding of what do innovative design processes mean to activate a virtuous cycle that could recompose the current fracture in the national Triple Helix Approach[2] framework (University-Industry-Government). For this purpose, Innovation is seen as a catalyst that could define a holistic process (Schumpeter et al.,1980) composed of three different phases: invention (when ideas are generated), innovation itself (putting ideas into practice), and diffusion (the widespread application of the innovation, in relationship with the markets' dynamics and mechanics). Following this premise, the objectives of this paper are threefold: on the one hand, to show - through direct experiences - how it is possible to develop a synergic approach that can put together theoretical speculation and applied research in a young local academic institution; on the other hand, fostering the concept of innovation not only in design research and methods but also for creating new businesses model to engage private stakeholders and production systems. As we do so, we also want to develop a shared nuanced vocabulary that can be useful to empower different actors to create a solid common ground to propose new business research-oriented models.

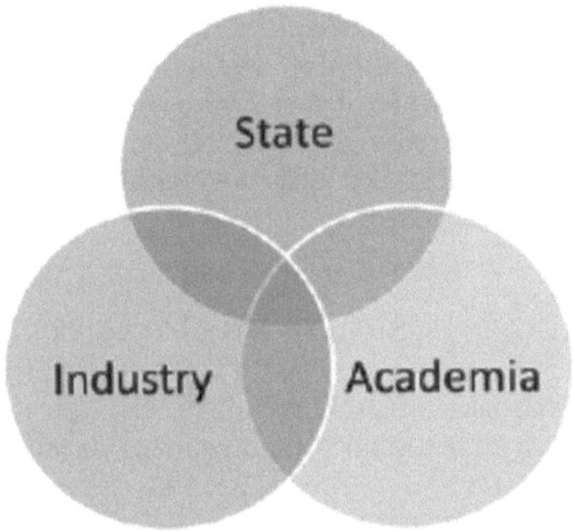


Figure 1. The triple helix model in its 'balanced configuration' where Academia works in synergy with the industry and the state (Source: Etkowitz 2002, 2004; Etzkowitz, Leydersdorff 1998, 2000).

## 2 DESIGN INDUSTRY AND INNOVATION IN ALBANIA

As mentioned in the introduction, since the 1990s, the liberalization of the market economy has led to a steady increase in the import of products from abroad. On the one side, this fact has enhanced sensitivity and awareness of the role of design, especially in the aesthetic, functional, and communicative improvement of goods and services, while, on the other, has

on the contemporary context. Further information about the event can be found at the following link: http://www.tiranadesignweek.com/.

[2] The Triple Helix Model moves its steps from the New England University-government-industry efforts - since 1920 - to renew a declining industrial economy, convened by the region's political leadership. At the core of the model there is a need to identify the generative source of knowledge which is necessary to enhance innovation, entrepreneurship, and regional development (Etzkowitz, Zhou, 2008). The theorization of the model happened during the 1990s (Etzkowitz, Leydersdorff), when the several interactions between universities, businesses, and governments, fostered the rise of new intermediary institutions such as technology transfers and science parks. The Triple Helix starts from the interactions within every actor in its 'initial role' (university - research; industry - commercial goods; governments - regulating markets); the more the three entities connects one another, the more each of them develops some characteristics of the other institution, originating new hybrid actors involved in the market.

limited the ability of the local industry to make a transition from the regulated model ruled by the state to the competitive market. After the fall the Communist regime, it has been challenging for Albanian designers to specialize and gain experience in the field of product design in the local industry, while it was easier for them to participate in the development of the economy through the practice of communication design and graphic design, disciplines for which there is a growth in social awareness. The shift from a socialist-inspired economy, regulated by the state, to the current, distinctly liberal one, has been more than a phenomenon of a historical transition and has assumed the value of a cultural dialectic that is still strongly perceptible in social behaviors. One of the most interesting challenges for the national design culture in Albania is to promote the recovery of awareness concerning the value of the public good, which, nowadays, is very reduced in terms of public perception because of the memory of the government's overwhelming and pervasive presence of before 1990. Another aim is to affirm the importance of design as a tool for promotion, innovation, and social development.

Currently, professions and professionals' skills are losing their solid boundaries and application; many graduates in Art, Scenography, Sculpture, Management, Marketing, etc. reorient their careers by undertaking professional design courses, master professional, master of science or research doctorates in design. Meanwhile, social media or application programs, and different websites related to design, are helping to increase the knowledge about the design, but as a negative consequence, they are also 'producing' internet users that deliberately pretend themselves to "designer" or building barriers to allow the designers to make the important decisions. Indeed, if we live in an Era that design is an additional value of production, then why is the designer's position in Albania is not parallel going as such? If design instills in the client requirements, why is the designer's role isn't firmly established?

To understand this gap between industries and academia is essential to understand the contemporary production processes in Albania. Retailers with the sizes of hyper and supermarkets and the B2B[3] knowledge market placed the role of the designers as one of the important tools to be competitive in the market, any moment oriented towards the market. The Albanian design processes in itself don't have this structural role in contemporary enterprises. If the companies can't afford to have the designer in the company's structure, they should outsource it from the designer companies, because the designer is no longer distanced from the consumer, both can give life to processes of reciprocal co-action based not on "consumption" but on relationships and conviviality (Illich, 1981).

| | Research system | Entrepreneurship and SMES (non-research based innovation | | Research based innovations | |
|---|---|---|---|---|---|
| | | Programmes | Institutions | Programmes | Institutions |
| Croatia | Complex | Complex | Complex | Complex | Complex |
| Serbia | Complex | Complex | Complex | Moderate | Moderate |
| FYR Macedonia | Familiar | Moderate | Familiar | Beginner | Moderate |
| B&H | Moderate | Familiar | Familiar | Beginner | Moderate |
| Montenegro | Familiar | Beginner | Moderate | Beginner | Beginner |
| Albania | Beginner | Beginner | Beginner | Beginner | Beginner |
| Kosovo UN Res.1244 | Infancy | Infancy | Infancy | Infancy | Infancy |

Figure 2. Tentative categorization of WBC by the maturity of NISs (innovation infrastructure and programmes) - Source: Švarc 2012.

On the other hand, scholars agreed that the evolution of design and its education is equidistant. There are countless reasons supporting this growth but some phenomena, in particular, are forging the world of design education: from the development of design schools (private or government eLearning) to laboratories and studios that allow students to pursue a fully experimental education; from the regulation of public education

[3] Business to business also called B to B or B2B, is a form of transaction between businesses, such as one involving a manufacturer and wholesaler, or a wholesaler and a retailer. Business to business refers to business that is conducted between companies, rather than between a company and individual consumers.

pushed by the respective countries' education ministries to the increasing relationship between the business sector and academia.

Both academia and enterprises must design and develop collaborative mutual links with the final aim to be innovative and competitive by ensuring students' work placements and enriching product development with students' ideas. They must draw a common agenda to be respectively respected. The case study University of "Polis" 'in Tirana is attempting through different practices to minimize this lack of knowledge through different courses, workshops, Labs on site.

## 3 TOWARDS INNOVATION IN ALBANIA – HISTORY AND DATA

Nixon and Helms (2002) suggest that partnering between corporations and Universities may offer the best of both worlds but still there is a quantitative evidence showing that this system isn't proper working into the curriculum development, research and development, lifelong learning, and mobility. Over the last decade policies of Albanian, HE institutions to involve or/and approach to market and its demands, have slightly changed. Although there are from both, academia and business sector, consistently emphasizing the importance of collaboration, knowledge exchange, and potential of common research and development, still their actions to minimize the gap, are too general.

Existing industries in Albania are still rigid organizations based on centralized management "heritage" from the centralized economic system are unable or unwilling to embrace the importance of designer's role on production or market and to give design culture the space it needs to reform production and the processes that are part of it. According to INSTAT 2018, businesses registered at the end of 2018 counted 162,835 active enterprises, and only 25% of the total number is farmers (registered to tax offices). The "active enterprises" by economic activity and size is showing that; the economic activity 34.9% is the total producer of good (agriculture, forestry and fishing, industry, construction) and 65.1% is the producer of service (trade, accommodation and food service activities, information and communication etc).

The potential of Albania market is yearly increase, globalization and investment in technology as Meister (2000) state had a swift and substantial impact on the way businesses need to have a high skilled and flexible workforce. "Manufactured products ", in the period 2013-2017 occupy the bulk of exported and imported products. In 2017, the share of exports of these products is 83 %, increasing by 21.9 % compared to 2016 (INSTAT 2018).

The main purpose of the universities should be to prepare students for the labour market, and as Kitson (2009) state, university plays in stimulating innovation and economic growth, but instead of familiar obstacles there should be a cooperative model that enables business sector to stay continuously connected to early stage research and to accelerate the translation of that research into new products that drive economic growth. (Lutchen,2018) In order to understand better the background of innovation in Albania a survey has been undertaken.

Survey with the designer furtherance the emerging need of reforming, Question "how much do they (designers) impact in the innovation of the company:" from 1 to 5, 80% of asked individuals replied 2. About the question "are the participants in the important decision making of the company?" from 1 to 5 only 27% answered 4, 42% answer 3 and the rest 1 and 2. But from the other part, the same questions were asked to the management of the case study enterprises if the designer working in their enterprise is impacting the design-making on the company, 64% answered "Yes". The rest of those, answered "No", the reasons according to this group are; i) the designers still are utopic to meet design with technology, ii) limit knowledge on technology, iii) cost, is not taken in consideration or a good designer is expensive iv) lack of market survey knowledge. For the group of those who have responded positively, addressed the source of the designs from their global competitors, then local ones and less than 5% from their deep research on innovation. The question if "the designer is a participant on the important decision making of the company" only 29 % answered "No". Conclusions, why no, are: i) the management of the company thinks that designers are just to "draw" not to know more, ii) another reason is that the designer is not a division in the company that they could be part of decision making, iii) because the management doesn't know that a designer has competences on market research, communication skills, branding and marketing. There is low penetration of industrial designers in a manufacturing firm settled Abdel-Malak & Brassard (2008) but the Ministry of Economy, Integration and Export in Canada didn't delay to publish two studies on design and designer role in the economy. It is to be highlighted that the synchronization programme of governments, universities and media can bring healthier and promptly closer the designer with the enterprises

Production on the other hand is undertaken with the help of resources which can be categorized into natural resources (land), human resources (labour and entrepreneur) and manufactured resources (capital) (Economics discussion, 2018). An additional value on the human resource is the design process. For sure the customers, the production techniques, the management etc, are important on the Human resource factors, but the challenge starts by educating and training the end users. It is challenging in Albanian context where the design process before the free economy was driven from up-down, it was controlled from the centralized economy and the end users were not participants. But after the free economy, the designs started slowly bottom-up. More competitive prices, freedom in choosing quality and design, and increasing the commodity in purchasing process allowed the end-user to bring dynamic to the production process. Albeit, in Albania context slowly and sometimes not in the right direction. One reason is that new designs, called often in the daily use "modern" referring to current context designs, were not in accordance with the end-users recognizing different value capacity, the slowly updating production technology and raw material. But for the Western Balkan countries, this was called innovation, it has to do more with technology upgrading, innovation and R&D as factors for economic growth and social progress have emerged as new policy issues in the WBC (Radosevic, 2014). After the fall of communism, due to the economic background, Albanian customers were driving from the cheapest prices, but even after two decades still the lower prices lead the priority of choices. In this light, the innovation in design and the role of designer evanesce.

Until years 2005's consumer shopping behaviour recalled. The number of global retail store arrival, shopping malls, and retailers increased and they're focused on the products design and store design increased. The end-user is increasingly involved in the process of co-production of industrial products and services. Therefore, Albania as the new industrial development country the ministry of education is requesting the universities to enlarge the design courses curricula and the number of students applying in these courses is growing steadily from year to year. Young people in Albania are the first target group aware to undertake or extend professional design training. The best art design universities in Albania are taking in their particular focus on design thinking. Subjects like: Branding, Marketing, Product design, Product development, architecture design etc. aimed to approach closer the curricula with the market demands. There is great interest in design's ability to offer methodologies and tools that can give shape to ideas and concepts in the pre-industrial phase of productive processes. While many important multinational companies have arrived in Albania during the last decade, what do they apply in their home countries are yet too far away the methodologies applied in Albania market as; in-house training of their intellectual workers, and being organized in "creative workshops" that aimed at stimulating Albanian HE institutions are not ambitious, their demand towards business sectors is to ensure a high success rate of employment but how it could be without knowing the market demands?

Until now almost non-existent was the attempt of HEIs to invite the business sector before designing the curricula. Methodologies to understand and build the business sector priorities could be different approaches; meeting, workshops seminars and after a deep analysis could be shown the impact in the curricula. These initiatives will lead to the development of a common research and development impacting the knowledge exchange activities while shifting the traditional teaching and reach activity.

In our journey focused on the relationship within the domestic design culture and the role of our University in its enrichment and dissemination it is important to underline that, in Albania, the word Innovation stands for more than just a fancy definition and occupies a fundamental role to position the country in a worldwide healthy and multi-actors design market. The future economic growth of this small nation of the Western Balkans, passes through the capacity of fostering innovative design processes that could activate a virtuous cycle to recompose the current fracture in the national Triple Helix Approach framework (University-Industry-Government), and engage different actors in a long-term research agenda dealing with this purpose (Leydesdorff, Lawton Smith, 1990).

Recent studies from international academics (Nientied, Karafili, 2016) highlighted how the need to reflect upon what 'innovation' means in the WB is stressed by multiple levels and stakeholders: researchers, policy-makers, businesses, etc. More precisely, Albania is in many aspects behind every kind of innovation-driven processes. In 2015-2016 WEF Competitiveness Ranking of 140 countries, Albania occupied the $118^{th}$ position in the pillar of 'Innovation', while in other studies has been ranked as 'underperformer' (WEF, 2014), or as a beginner in the field of innovation-based R&D (Švarc, 2012) The currently active technological industries are not able to absorb the flux of new ideas coming from Western Europe, and the amount of highly-educated emigrants - together with the low number of yearly patents produced by the market - does not predict an increase of the already small R&D investments (WEF, 2015; World Bank, 2011; World Bank, 2013; Despotović, Cvetanović, Nedić 2014). Another data that is important to point out is that, according to the World Bank (2013), Albania has only 245 researchers per million of population, a number which represents less than 10% of the EU average of 3,166 researchers per million of population. Furthermore, despite the WB government's actions to foster R&D and science and technology (World Bank, 2011; Marinkovic & Dall, 2014), only 1.3% of Albania exports are connected to high-tech manufactures, a number sensibly lower than developing countries in the Sub-Saharan areas (5.3%).

As a consequence, the connection between academia and private/public stakeholders is becoming weaker every day, causing a gap where exciting research has no access to proper systems of national/regional funding–and infrastructure–to be disseminated. In addition to this, the involvement of the government has been far limited in the field of innovation for two main reasons (Nientied, Karafili 2016): on the one hand, the necessity - after the fall of the Communist regime in 1991 - to set up the state apparatus and institutions, that left out the country from the quick development phase that invested the Western Europe during the 2000s - and, on the other hand, as pointed out by Bahiti and Shahini «[...] *business environment suffers from a high level of corruption, serious shortcomings in the judiciary, and very weak institutional and law enforcement capacity*» (2002, 2010).

Within this critical condition - beyond just dealing with the concept of innovation in Albania from a European perspective - there is a need for a complete paradigm shift regarding this topic, developing a country-specific model that could take into account the inner problems of the national economy and create an innovative (this might be a playful statement, but useful in our case) framework that could engage and empower multiple actors. What we aim for in this paper, is the diffusion of a 'everyday innovation' approach - a term coined by NESTA (Patterson, Kerrin, Gatto-Roissard 2009) - rather than other policy models already in phase of implementation that are revealing themselves only a copy of European ones and scarcely efficient (Niented 2016, cfr. Švarc, 2012).

The model we are currently implementing in our everyday work - either as professors at POLIS University or as professionals - focuses on the connection between theoretical speculation and practical approach, and on fostering innovation not just as a result but as a daily practice to orientate regular activities rather than a specialty just for the few.

What we argue for is to propose innovation as a catalyst to define a holistic process able to play a dual role: empower the national market in dealing with the specific necessities of low/mid income country; lead to a curricula development in the field of education and academia to grow a future generations of experts where innovation is not just for a specialized niche, but an everyday practice. To clarify our point of view, we will present the work of the INNOVATION_Factory, an anti-disciplinary think-thank at POLIS, and a series of case studies engaging our students to better understand our role *vis-a-vis* this topic.

Following the social economical context and market's needs, POLIS University defines itself as "Space for Thinking", and focuses on spatial development policies, applied design and technology, as well as leadership and entrepreneurship programmes. The University's mission is to provide "knowledge, technology and leadership for people". Although in its traditional aspects the design curriculum intends to foster specific educational aspects of the discipline, it is also forged by the sum of values gathered from learning activities over time. Eisner (2004) states that the concept of curriculum stands in the selection and organization of learning activities, which give direction to the cognitive development of students. Moreover, the scope of policymakers dealing with curriculum design also lies in new approaches and means to promote the university. Thus *"The curriculum is the description of a set of activities that brings relevance and vitality to the classroom. The outcome of applying a curriculum is to improve the quality of the programs defined by that curriculum."* (Eisner, 2004)

Still a very young University, POLIS has taken into account many other well-known university cases into developing its overall state of implication towards curriculum development, such as 'Bauhaus". While exploring the "Bauhaus" curriculum design approach, its main objective was in combining aesthetics with usefulness, attempting in unifying principles of mass production in combination with individual artistic approach, influencing upon subsequent developments in the field of art, architecture, graphic design, interior design, industrial design, and typography. As far as the schematic teaching approach, "Bauhaus" methods of teaching remained in replacing the traditional schematic approach of pupil-teacher relationship and form social and community bond.

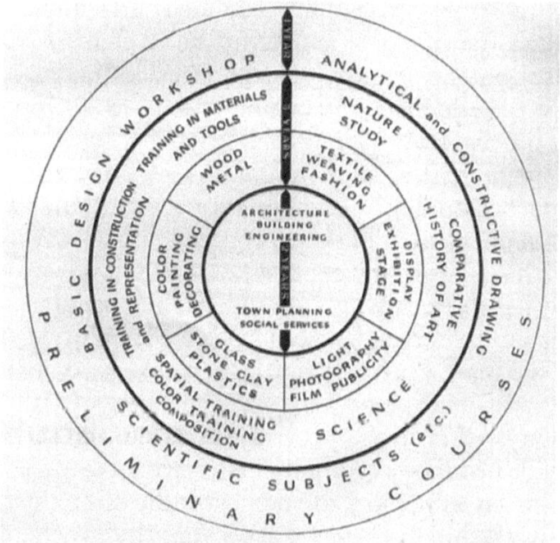


Figure 3. Bauhaus ideal course structure and student pathways (Gropius, 1919).
Source: Bauhaus-Archiv / Museum für Gestaltung.

Nevertheless, "Ulm School of Design", gained international recognition by emphasizing holistic, multidisciplinary context of design beyond the Bauhaus approach of integrating art, craft and technology. The subjects of sociology, psychology, politics, economics, philosophy and systems-thinking were integrated with aesthetics and technology. Known as the "Ulm Model", it focused on linking basic core courses with theoretical disciplines, influencing today's design education, especially industrial design, by pioneering the integration of science and art, creating a teaching of design based on a structured problem-solving approach.

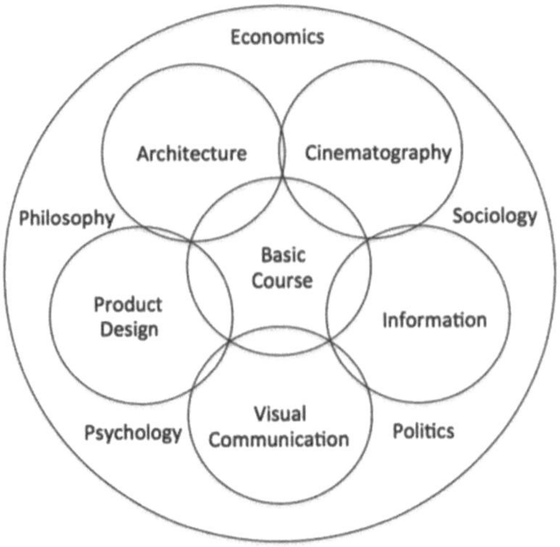


Figure 4. Schematic teaching of HFG Ulm. The school design was characterized by formulating a scheme-based education in art and science. Source: Ulm school of design.

Although the definition of the curriculum can be very hard to circumscribe, at POLIS University its content development is seen in three dimensions where the overall concept is linked with the different approaches of learning and teaching, although curriculum and pedagogy tend not to be the same thing.

The above-mentioned dimensions are the following:

1.formal curriculum; 2. informal curriculum.; 3. hidden curriculum.

In the formal curriculum the outcome stands in the content of what is taught while the informal approach focuses on the support or other services based on student learning and results. The last category - the hidden curriculum - refers to a deeper understanding of the organizational culture of the place.

Thus, for the development of the study programs of "Art and Design" and "Applied Design" the above approaches are taken strongly into account and, when thinking about the content and the progress of curriculum in these programs, three aspects are considered: disciplinary learning; interdisciplinary learning; capabilities/attributes development.

This approach aims to reinforce three peculiarities of the programs which do focus on the pure field or art and design or either applied design such Flexibility, Quality, Relevance, Inclusion, which are some of the main aspects where policymakers dealing with curriculum development focus on.

The concept of flexibility refers to the process of decision making regarding and its content, which are linked with pedagogy and administrative structures that enable change. On the other hand, quality is seen as the key to emphasize the disciplinary expertise and quality of excellence in relation to the institution. We are aware that, with a fast-changing economy in the design field and with the actual situation of the market, students which graduate nowadays are faced with more complex challenges that require them to have a proactive attitude and several new characteristics. For this purpose, the development of the curriculum needs to provide a common interest in learning outcomes between the student skills and attributes derived by future employers. Then, in order to design a successful educative structure for the designated programs, the curricula and their content need to reinforce students' backgrounds and experiences.

For these aims, the ones at POLIS University focus on creating a strong continuation between the areas of subjects and topics during the course year, divided between core courses in Fine Art for the Bachelor Programme and Design one's for the master Program, and rely on the aspect of the formal curriculum approach.

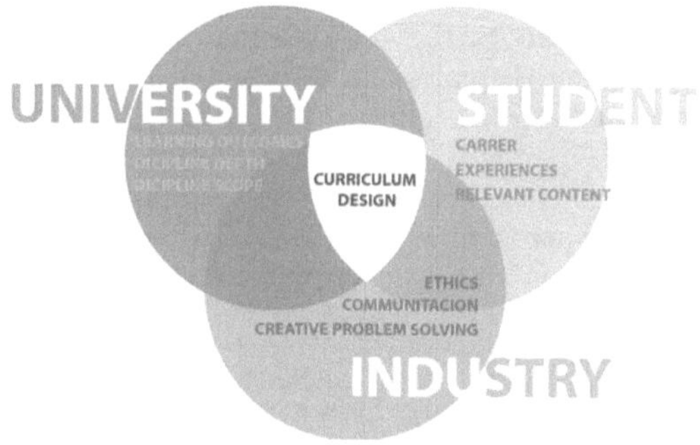


Figure 5. POLIS University - curriculum design in the study programmes of art & design and applied design. Source: POLIS University.

Meanwhile, the aspect of informal curricula approach has been implemented with new innovative approaches in the academic environment, such as the implication of IF Innovation Factory as a tool to stimulate furthermore the research and services of students focused on learning outcomes within the given courses. Finally, hidden curriculum is guaranteed through the strong relationship that the University has established with the Design Industry in Albania. The cooperation between the industry and POLIS stimulates also small hints of the current culture of the field related to the market, thus enhancing the possibility of having indirect feedback for the program curriculum. In conclusion to the aspect of Designing Curriculum for Art and Design Programs and Applied Design, the role of the University stands in highlighting the main scope of fields and depth within the areas focused of the curricula by defining its outcomes. On the other aspect, the student is stimulated and educated by the relevant content of it while, at the same time, enriching their experiences within the program and establishing a career by internal and external stimulator, hence activating the industry in terms of ethics and communications between University-Student-Industry, and creating problem-solving solutions for both parties and the University towards curriculum development.

## 5 INNOVATION FACTORY: INTERDISCIPLINARITY AS A RESEARCH APPROACH IN DESIGN

What is fundamental to point out, is the precise definition of 'innovation' that moves the research of the unit, and of POLIS itself. Even though many explanations for this word currently exist, and it is seen either as «*the successful exploitation of new ideas, a process of turning ideas into reality and capturing value from them*» (Tidd & Bessant, 2013; Epstein, Shelton, Davila, 2012), or as the differentiation between two phases: *exploitative* (refining, leveraging and extending existing knowledge) and *explorative* (searching for new alternatives or unfamiliar, distant and remote expertise). Indeed, these researchers described innovation as a process made of three phases: invention (when ideas are generated), innovation itself (putting ideas into practice), and diffusion (the widespread application of the innovation). Therefore, according to their work, we do not see innovation as a 'plug-in' that can be uncritically attached as a sort of label, but as a process that must lead both theory and practice. More precisely, the pillars of IF - intended as acceleration forces able to structure its contents and to lead innovative projects' developments - are:

**a systemic and holistic conception** of contemporary knowledge, in the pursuit of omni-comprehensive design notion;
**a deductive what-if method**: a non-linear working methodology, that roots in a dynamic and lateral (De Bono 1980) way of thinking composed by a series of questions;

**a pluralist research methodology**: we see the world as many interconnected, multidisciplinary, complex, self-adaptive systems across scales and dimensions that are unknowable and largely inseparable from the observer and the designer;
**proactive behavior (proactivity) approach**: an attitude that pushes to operate with a view to a future event rather than linearly reacting to a present situation;
**information technology-driven approach**: not disconnected punctual task-oriented interventions, but synergic ones able to trigger new structural, formal, and relational, inquiries;

Together with these strong concepts (Höök, 2012), IF developed a chart of keywords and actions that will lead the unit research agenda in the years to come. The idea is to set a series of opposing pairs that generate a paradigm shift from the past to foster a forward-looking worldview dealing with a holistic and multilevel concept of innovative processes:

Table 1. Innovation principles for the INNOVATION_Factory.

| | |
|---|---|
| Emergence over Authority | Complex, bottom-up systems to substitute top-down authority |
| Localism over Globalism | Express and test large scale phenomena in localized contexts to foster reflection and solutions |
| Disobedience over Compliance | Needs for question authority issues and think independently |
| Complexity over Linearity | Dynamic and lateral way of thinking instead of a linear cause-effect one |
| Diversity over Expertise | Non-traditional synergic teamwork more productive than the simple sum of an individual |
| Risk over Safety | Foreseeing innovation is not riskless, but the latter should be balanced and transformed into new value |
| Ecological over Productive | Connecting humans and nature at different scales instead of separation of (eco)systems |
| Compasses over Maps | Strict fixed maps become less valuable than vision, cultures, ideas |
| Non-reductionist over Reductionist | Holistic point of view where every phenomenon is interpreted as a complex set of relationships |
| Systems over Objects | The necessity to use systems as a key to interpret the complexity of the world |

The highlighted principles and the above chart, were useful to show how it is possible to deal - in a country still far from the European standards like Albania - with the topic of innovation in an original and peculiar way. Our main aim is not to rely on pre-made work packages uncritically inherited from other countries but to develop a national framework for innovation in accordance with the *Albanian National Strategy for Science, Technology, and Innovation* (STI) 2017-2022, and the *Action Plan 2016 - 2020 "Support for the Development of Innovation Policies Based on the Triple Helix Approach*[4]". Moreover, in conclusion of this paper, a series of case studies - the first one dealing with a design-n-built shelter, the second one with a mobile app for an Albanian company - conducted by us are presented to the reader to understand deeper how for us innovation is not a matter of outcomes but as a process, that can influence research in a 360 degree way and be the linking points between academia, businesses, and investors.

### 5.1 *The solar capsule experience - Merging together university and businesses*

After describing the IF methodology and pillars, it is beneficial at this point to highlight one of the several projects that have been developed through the years following the philosophy that characterizes POLIS University and its several departments. The Solar Capsule is an example of an environmental design-n-built prototype that was developed during Tirana

[4] Developed by the Ministry for Economic Development, Trade, Tourism and Entrepreneurship (MEDTTE), with the support of GIZ - ProSME Project

Architecture Week 2016, and that involved in its process both professors and students from the University (Di Raimo, 2016). The objectives of the project were two-folded: on the one hand, it wanted to involve the students in the meaningful experience to design a small capsule by using an environmental approach and through the re-use of recycled and waste materials; on the other hand, it aimed to highlight the importance of a stronger connection between universities and private companies in the construction field, that were dealing in selling technological components. Many specific examples are chosen in order to be useful in understanding the methodology of both IF and POLIS in the development of innovation-based processes that involve multiple stakeholders in a complex prototyping and design process. Indeed, not only it aimed to empower the students towards contemporary fields of research such ecology and environmental issues, but also took the concrete chance of a construction experience to fill the gap between universities and businesses towards a mutual and beneficial exchange process for both of them, starting from the awareness that creative sectors can help businesses under several aspects (Müller, Rummer, Trüby, 2008).

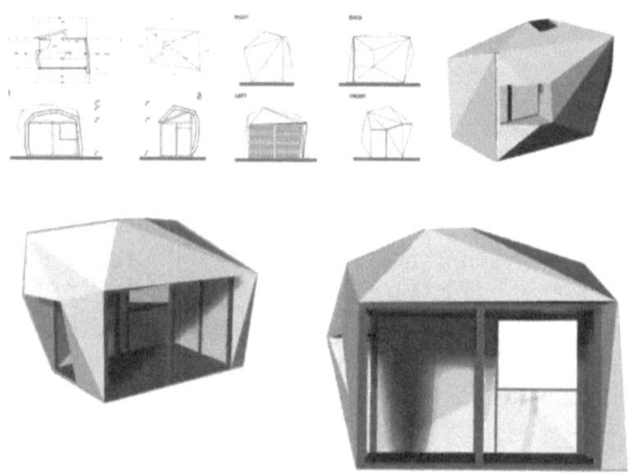

Figure 6. Solar panel presentation board
(Source: Di Raimo 2016).

## 5.2 *Between teaching and practice*

As part of the Scientific Master in Applied Design, students are engaged through several disciplines to guide them in positioning themselves as young creatives. IF is involved in it with the Interactive Design Class in the 2 nd year of the MA. With the rise of technology and innovation, there is the need to possibly discover many other fields of studies, such as Interactive Design (IxD), Human-Computer Interaction (HCI), and Digital Game Design.

These new arising subjects, especially in Albania, tend to empower students assumptions towards of the so-called 'rise of the platform society (de Waal, 2014), which, based on the 'hacker ethic,' can empower the citizens to organize themselves around issues, bringing about a sharing economy, a participation society, or civic economy. Innovation deals with many different problems and can be intended as a way to activate new social and economic processes together with business partners and private stakeholders; to solve the contemporary crisis with the implementation of new media; to speculate and find new research insight and disruptive solutions that can empower the surrounding environment. The methodology applied is RtD (Research Through Design) methodology (Zimmerman, Stolterman, Forlizzi 2010) intended one of the many methods and epistemologies that are leveraged in the broader field of design research, and has to do with the use of design practice as a form of scientific inquiry, and latter is rooted in Dewey's definition of learning as relevant and practical, not just passive and theoretical.

Specific case studies show how, within an academic-oriented activity, small but synergic actions can be activated to bridge the gap between stakeholders and students of our curricula. From the get-go, the frame of this class as a case study, as a mere speculative process but also to ground into a real-world experience that on the hand could empower the students on innovative thinking and, on the other, could enrich the Albanian market with strong and meaningful collaborations.

Figure 7. Interactive design class (II year, master in applied design) - Brainstorming phase for NOVA, the topic of the semester is the creation of a smartphone mobile application and, since we are dealing with almost graduated students, involving in the process of some real clients that could deliver a real design brief to the class and help them with their specific insights in the development of the product. That step was fundamental, because, in order to empower the curricula towards innovation and a business-oriented model, there is the need to involve real stakeholders that could benefit from our approach and, at the same time, from which the students can benefit too.

## 6 CONCLUSIONS

In conclusion to this paper, design-based research is methodology evolved near the beginning of the 21st century and was heralded as a practical research methodology that could effectively bridge the chasm between research and practice in formal education. The first decade of this century has seen the emergence of a new research methodology for education research—design-based research to empower innovation. A number of respected education researchers and special issues of well-known journals have celebrated the potential of this approach to make a significant difference in the quality and utilization of education research, in order to increase the impact, transfer, and translation of education research into improved practice. In addition, it stresses the need for theory building and the development of design principles that guide, inform, and improve both practice and research in educational contexts.

At the same time, the literature and the available data on the design and role of designers in Albania are almost non-existent; this article offers the first trace to enrich the knowledge. The industry is living a historical shift of its role in society and production through the admittance of the new technologies. Along the crisis of industry, knowledge, and creativity come to be a primary force capable of generating value and innovation. The latest technological revolution is questioning the way we produce, manufacture, distribute, and fund everything, from small to big objects we deal with in our everyday lives. The process of digitalization is opening new challenges but also opportunities for rethinking what and the way we produce towards a society that is more environmentally sustainable, socially equitable, and democratic (Imbesi, 2014). In this case, design research is challenging the disciplinary fields to understand hybrid knowledge, which is growing "in-between." To create a better understanding of the user's needs, the design should integrate multidisciplinary competence, in the innovation of products and services field in order to analyse and to map the change of the contemporary societies. By mapping the needs, the emotions, the desires of the contemporary user, different research approaches like design thinking, ethnographic design and user-centred design helps to understand that the cognitive products are more immaterial, informational, virtual, and in that there is something more to understand what lies beyond the prototypes.

For this purpose, the curricula of POLIS University bases its education process in a Design-based research approach. An approach methodology that effectively bridges research and practice in academic education. Education should be connecting knowledge with creativity, this leads to surprising and valuable results for individuals, for organizations, and for society as a whole. The study allows us to concur with Dede, Ketelhut, Whitehouse, Breit, and McCloskey's (2009) claim that "Design-based Research" offers a 'best practice' stance that has proved useful in complex learning environments, where formative evaluation plays a significant role, and this methodology incorporates both evaluation and empirical analyses and provides multiple entry points for various scholarly endeavors" (p. 16). However, as promising as the methodology is, much more effort in this and other areas of education research is needed to propel the type of education innovation that many of us feel is required. The work presented and discussed in this paper is for us a way to position ourselves and to communicate and share ideas concerning the education philosophy, and highlight how to empower and foster the rise of a synergic ecosystem in Albania.

## REFERENCES


Eisner, E. W. & D.Day, M., (2004). *Handbook of Research and Policy in Art Education.* London: Lawrence Erlbaum Associates.

Imbesi, L. (2014) *DESIGNING FOR A POST-INDUSTRIAL SOCIETY, Rome.*

Dede, Ketelhut, Whitehouse, Breit, and McCloskey's 2009.

Abdel-Malak, S. & Brassard, L. (2008). *Synthesis of the results of the survey on the perception and use of industrial design by Québec manufacturing companies.* Electronic document. Montréal: Développement économique, innovation, exportation.

De Bono, E. (1970) *Lateral Thinking: Creativity Step by Step*, New York: Harper and Row.

Despotović, D., Cvetanović, S., Nedić, V. (2014) Innovativeness and Competitiveness of the Western Balkan Countries and Selected EU Member States, in *Industrija, vol. 42, br. 1, 5*, Economic Institute, Belgrade, pp: 27-4.

de Wall, M. (2014) *The City as Interface: How New Media Are Changing the City*, Rotterdam: nai010 publishers.

Di Raimo, A. (2016) Solar Capsule, in *Forum A+P. [EN]visioning Future Cities, vol. 16*, pp. 50–54, Tirana: POLIS_Press.

(Economicsdiscussion, n2018) www.economicsdiscussion.net/production/factors-of-production-land-labour-capital-and-entrepreneur-national-income/541

Epstein, M. J., Shelton, R., Davila, T. (2012) *Making Innovation Work: How to Manage It, Measure It, and Profit from it*, New Jersey: Pearson FT Press.

Etzkowitz, H. (2002) Incubation of Incubators: Innovation as a Triple Helix of University-Industry-Government Networks, in *Science and Public Policy, 29(2)*, 115–128.

Etzkowitz, H. (2004) The evolution of the entrepreneurial university, in *International Journal of Technology and Globalization, 1(1)*, 64–77.

Etzkowitz, H., Zhou, C. (2008) *University–Industry–Government Innovation and Entrepreneurship*, London: Routledge.

Etzkowitz, H., Leydesdorff, L. (1995) The Triple Helix – University-Industry-Government Relations: A Laboratory for Knowledge Based Economic Development, in *EASST Review, Vol. 14, No.* 1, 14–19, Rochester, NY.

Etzkowitz, H and Leydesdorff, L. (1998) *A Triple Helix of University-Industry-Government Relations, Industry and Higher Education*, August.

Etzkowitz, H and Leydesdorff, L. (2000) The Dynamics of Innovation: From National Systems and 'Mode 2' to a Triple Helix of University-Industry-Government Relations, in *Research Policy, 29*, 109–123.

Goldschmidt, G. (1991) The dialectics of sketching. *Creativity Research Journal*, *4*(2), 123–143.

Heskett, J., (2005) *Design: a Very Short Introduction*, New York: Oxford University Press Inc.

Höök, K., Löwgren, J. (2012) Strong concepts: Intermediate-level knowledge in interaction design research, in *ACM Transactions on Computer-Human Interaction* (TOCHI), 19(3):23.

Illich, I. (1981). *Per una storia dei bisogni*. Milano: Mondatori.

International Trade in Goods and Enterprises 2018, INSTAT 2018.

Kitson M., Howells, J., Braham, R., and Westlake. S. (2009), *The Connected University: Driving Recovery and Growth in the UK Economy*, NESTA, London.

Leydesdorff, L., Lawton Smith, A. (1990) The Triple Helix in the Context of Global Change: Dynamics and Challenges, in *Prometheus 32(4)*, Milton Park, UK: Taylor & Francis Online.

Marinkovic, I., Dall, E. (2014) *R&D and innovations in Western Balkans, Moving towards 2020. Vienna*. Retrieved from http://wbc-inco.net/object/document/13962/attach/PUBLIKATION_WBCINCO_web.pdf

Massey, A., (2008), *Interior Design Since 1900*, London: Themes &Hudson Ltd (UK),. p: 65.

Meister, J. (2000) *Corporate universities: Market-driveneducation*. Journal of Business Disciplines,1, 1527-151X.

Müller, K., Rammer, C. Trüby, J. (2008). The role of creative industries in industrial Innovation. Centre for European Economic Research, in *Discussion Paper No. 08-109*. Retrieved from http://ftp.zew.de/pub/zew-docs/dp/dp08109.pdf;

Munari, B., (1966) *Design as Art*, London (UK): Penguin Books Ltd;

Nientied, P., Karafili, E. (2016) Towards a Pragmatic Perspective on Business Innovation in Western Balkan Countries: The Case of Albania, in *International Journal of Business and Management Vol. 11, No.* 3, Richmond Hill: Canadian Center of Science and Education.

Nixon, J., & Helms, M. (2002). Corporate universities vs higher education institutions. Industrial and Commercial Training, 34(4),144–150.

Patterson, F., Kerrin, M., Gatto-Roissard, G. (2009) *Characteristics & Behaviours of Innovative People in Organisations*, NESTA Policy and Research Unit;

Radosevic, S., (2014) Technology Upgrading and RTD Challenges in Western Balkan Region: Issues and Policy Options, UCL, School of Slavonic and East European Studies, in *Marinkovic, I., & Dall, E., (Eds.) (2014) R&D and Innovation in Western Balkans: Moving Towards 2020, Centre for Social Innovation, ZSI Austria*, pp: 59-66.

Schön, D. (1983) The Reflective Practitioner: How professionals think in action. New York: Basic books;

Švarc, J. (2012) Comparative analysis of the innovation capacities in the WBC with emphasis on joint cooperation needs in the field of innovation, in *Conference: Proceedings of the 5th International Conference for Entrepreneurship Innovation and Regional Development, ICEIRD-2012*, Sofia. Retrieved from http://wbc-inco.net/object/event/5694/attach/06_svarc_WBC_Innovation_2011-12-01.pdf

Tidd, J., & Bessant, J. (2013) *Managing innovation, integrating technological, market and organizational change*. Chichester: Wiley.

World Bank (2011) *Regional R&D strategy for innovation in the Western Balkan countries-key issues and implications for technical assistance*. Washington, D.C..

World Bank (2013) *Western Balkans, regional R&D strategy for innovation–country paper series Albania*. Washington, D.C.

World Economic Forum (WEF). (2014). *Enhancing Europe's competitiveness, fostering innovation-driven entrepreneurship in Europe*. Geneva. Retrieved fromhttp://www.weforum.org/reports/enhancing-europes-competitiveness-fostering-innovation-driven-entrepreneurship-europe.

World Economic Forum (WEF). (2015). *The global competitiveness report 2015/2016*. Geneva. Retrieved from http://reports.weforum.org/global-competitiveness-report-2015-2016/economies/#economy=ALB

Zimmerman, J., Stolterman, E., Forlizzi, J. (2010). An analysis and critique of Research through Design: towards a formalization of a research approach. In *Proceedings of the 8th ACM Conference on Designing Interactive Systems (pp.* 310-319*)*. ACM.

*Research & Education in Design: People & Processes & Products & Philosophy – Almendra & Ferreira (eds)*
*© 2020 Taylor & Francis Group, London, ISBN 978-0-367-49296-0*

# Author Index